INSCRIBED BODIES

Inscribed Bodies

Health Impact of Childhood Sexual Abuse

by

Anna Luise Kirkengen

Specialist in Family Medicine, M.D., Ph.D., General Practioner in the city of Oslo, Norway
Former Research fellow at the Institute of General Practice and
Community Medicine at The University of Oslo, Norway

KLUWER ACADEMIC PUBLISHERS
DORDRECHT / BOSTON / LONDON

Library of Congress Cataloging-in-Publication Data

Kirkengen, Anna Luise.
Inscribed bodies : health impact of childhood sexual abuse / by Anna Luise Kirkengen.
p. cm.
Includes bibliographical references and index.
ISBN 0-7923-7019-8 (hbk. : alk. paper)
1. Child sexual abuse. 2. Sexually abused children--Health and hygiene. 3. Sexually abused children--Mental health. I. Title.

RJ506.C48 K57 2001
616.85'8369--dc21

2001033722

ISBN 0-7923-7019-8

Published by Kluwer Academic Publishers,
P.O. Box 17, 3300 AA Dordrecht, The Netherlands.

Sold and distributed in North, Central and South America
by Kluwer Academic Publishers,
101 Philip Drive, Norwell, MA 02061, U.S.A.

In all other countries, sold and distributed
by Kluwer Academic Publishers,
P.O. Box 322, 3300 AH Dordrecht, The Netherlands.

Printed on acid-free paper

Printed in the Netherlands.

To Karin Lisa
and Martin

Acknowledgments

I feel indebted to the following persons:
Christian F. Borchgrevink, professor emeritus, University of Oslo, for encouraging me to systematize my curiosity;
Hanne Haavind, professor of psychology, University of Oslo, for actively and explicitly supporting my study project;
Tordis Borchgrevink, anthropologist, Institute of Sociological Research, Oslo, for carefully and wisely guiding my research;
Trond Berg Eriksen, professor of philosophy, University of Oslo, for skillfully criticizing my thinking and writing;
Eline Thornquist, co-fellow and physiotherapist, University of Oslo, for sharing her knowledge and her visions with me;
Åsa Rytter Evensen, assistant professor, University of Oslo, for advocating my demands and needs on my way;
Karin Dolven, institute secretary, University of Oslo, for patiently editing my list of references;
Susan Schwartz Senstad, writer and family therapist, Nesodden, for giving my reflections an appropriate language;
Wera Sæther, poet and psychologist, Oslo for continuously thinking-along with me during the research process;
Anders Seim, physician and friend, Fagerstand, for his critical and supportive reading of my thesis;
Jaqueline M. Golding, professor of psychology, University of San Francisco, for encouraging my approach and sharing her research experiences;
Jóhann A. Sigurdsson, professor of family medicine, University of Reykjavik, for advocating the publication of my thesis;
Linn Getz, physician and researcher, University of Reykjavik, for reviewing my thesis in the Scandinavian Journal of Primary Health Care;
Knut A. Holtedahl, professor of family medicine, University of Troms, for his engagement, support and encouragement;

Martin Kirkengen, research scientist, Norwegian Computing Center, Oslo, for his patience, skills and time during the process of formatting the text; *Elizabeth A. Behnke*, phenomenologist and writer, Bellingham, WA, USA, for her invaluable final review and advice.

I want to thank the following institutions:
The "Northern" and "Western" incest centres for cooperation and open-mindedness;
The Institute of General Practice, University of Oslo, for hospitality and fellowship;
The Norwegian Research Concil for economical support of my research.

Contents

Chapter 1

APPROACHING LIFE-WORLD EXPERIENCES

INTRODUCTION

The present study represents an investigation into three aspects of the human life-world and their inter-relationship.[1] The first of these is integrity violation, explored within the framework specific to sexual abuse in childhood. The second aspect is the impact of sexual violation on health, explored within an epistemology of embodiment. The third aspect is biomedical intervention into illness stemming from sexual violation, explored within the frameworks of critical medical anthropology, feminist theory and a phenomenology of the lived body.

1. ETHICAL CONSIDERATIONS

The study's tripartite exploration problematizes *central ethical issues.* These issues concern: the theory of current western biomedicine, the methodology and practice of biomedical knowledge production, and the normative impact of valid biomedical knowledge on both research focus and clinical practice. My considerations are based on the empirical material and the results of this study. Biomedical knowledge production is grounded, almost exclusively, in a naturalist cosmology. In such a philosophical framework, ethics is deemed separable from that which is regarded as nature. Since nature and natural matter is what biomedicine seeks to investigate, ethical dimensions of human life are defined as separate from the issues of interest. Thus definitionally subordinated, ethical considerations are not seen as influencing biomedical theory. They are not part of the groundwork of biomedical research. They do not inform the criteria for the choices of research issues. Issues regarding ethics are raised only to address the etiquette of research practice. Applied to research as rules of form, ethics are intended to secure correct conduct on the part of the scientists and their proper performance of research activities. They are addressed explicitly whenever research funds are allocated, study results published, or participants' consent requested. Ethics thus are seen to concern social or legal questions only. They form an addendum to the research design, *ergo* are of only decorative, or at best marginal, relevance to research. They are, of course, acknowledged as being of central relevance to issues of values. However, because issues of human values are seen as a principal source of contamination, and subjectivity is deemed a general source of mistake, the theoretical framework excludes them and sophisticated methodologies are developed in an attempt to avoid them.

The present study provides evidence that it is precisely these measures, those intending to secure a value-neutral production of knowledge in biomedicine, which obstruct the disclosure of social pathology and its impact. Consequently and implicitly, valid biomedical knowledge actually contributes to maintaining a culturally constituted silence concerning certain practices, certain abuses of power. When measures taken in order to help or cure do not result in healing the people who ask for help, there are two possible outcomes: Either the measures are re-evaluated so that what was considered valid knowledge but which has proven itself inadequate is no longer considered valid. Or, no such re-evaluation takes place and, consequently, no revision of medical theory results. This outcome might be taken as evidence that the field of medicine is more interested in maintaining its epistemology than in adapting to real results. Were that the case, it would severely discredit medicine as an institution of power.

The study addresses *a medical wrongdoing* which is implicit in the presumably correct practice of presumably valid medical knowledge. This medical wrongdoing only becomes evident through an analysis of patterns and constellations which are subjectively guided and situated, applying a logic which continually adapts to a societally silenced violation experience. Such an adaptation, however, cannot be acknowledged by a discipline which disregards the human subject, and which, in both theory and method, values only formal logic and scientific rationality. Within that framework, reliable and valid knowledge about the nature of the human body, and of human health, ought to be arrived at only through methods for objectification and fragmentation. Through these processes of objectification and naturalization the biomedical body is constituted.

Those human conditions which are embedded in interpersonal relations, societal values, and culturally constituted meanings, are, through the very logic of biomedical theory, made invisible. The logic of its dominant methodology also renders them incomprehensible. Finally, they are deemed ignorable or irrelevant since values and meaning are non-issues according to objectivist science. The result is that the power implicit in social rank and the humiliation of human beings due to abuses of power are turned into non-medical topics, making medicine, inevitably, blind to the adverse effects which abuse has on human health. This becomes even more the case whenever the practice of such abuse is either societally legitimized or culturally taboo.

In other words: *By degrading ethical reflections to the status of mere etiquette, "scientific" prejudices acquire normative force. This will continue to be implicitly the case within medical research and practice until and unless questions of ethics are given primacy.*

Any exploration of the impact of sexual boundary violation on health, within societies whose members are socialized into a predominantly Western culture, must also cast light on biomedical epistemology and biomedical theories of the human body. Since these operate to constitute the way the human body is to be understood, and how the idea of health is to be conceptualized in society, they also inform the norms concerning health and social functioning. As a consequence, they help determine the criteria for what the society will define as deviant. Thus, while presuming to be a scientific discipline, *biomedicine is, in fact, normative.*

As medicine is a respected societal institution, and in its guise as a science, the normative character of biomedical epistemology accrues crucial influence. It affects central decisions with regard to what is, and what is not, to be considered relevant in drawing medical conclusions. Purporting to apply objective scientific knowledge while actually applying societal norms, medicine as a practice maintains the mandate to define the categories of ill health and malfunction. By defining these categories, medicine has the right to include any conditions which meet the categorical criteria. Thus, according to the rules of formal logic, medicine also has the power to exclude those conditions which fail to meet those criteria. This distinction between "proper" and "improper" states or conditions plays a role in every medical decision. The norms of biomedicine are embedded in the practice of any medical examination and treatment, and affect every living person who addresses a medical institution in the role of a sick patient. Through the application of these norms, distinguishing the "proper" from the "improper" within a formalized societal context, medicine has the power to stigmatize people who ask for help for "improper" conditions. While acting in the name of giving help, medicine may, in fact, violate a person's dignity. But even those who present apparently "proper" conditions may risk stigmatization if presumably appropriate medical interventions prove ineffective. According to objectifying medical theory, such measures ought to result in a predictable outcome. If they consistently do not, the most probable question is not, "what is wrong with medical judgment and medical theory?" but rather "what is wrong with this patient?" Failures stemming from the foundations of professional judgment, namely medical knowledge acquired by applying rules requiring objectivity, are more likely to be attributed to those whose condition fails to improve. In other words: *Medical norms exclude, marginalize and then stigmatize.*

Basic prerequisites in objectivist methodology are defined, discrete, and thus calculatable entities. These so-called variables result from measures which objectify and fragmentate both the body and of the patient's social reality, which, in turn, necessitates relations being broken and their continuity interrupted. Of course, these ruptures and interruptions do not occur within the actual body or in the realities of social life. However, due to biomedical theory and methodology, they do occur at each step in the production of medical knowledge: in the modes of approach; in the tools for data collection; in the choice of statistical calculations; and, in the final interpretation of the study results. Subjectivity, both that of the researcher and of the persons being researched, is considered to be the major source of contamination in objectivist methodology, and so, by definition and method, is avoided. However, in this final step of the process, the interpretation of results, subjectivity is unavoidable if one is to make the outcome meaningful; it can not help but reflect either the position of a single researcher or the culture of a team. This particular kind of contamination, however, is seldom problematized in a research tradition which purports objectivity. Such a research practice has a triple impact on its studied objects, be they aspects of the human body or of social life: the objects become alienated; they will be misinterpreted; and, through the legitimized generalization of the study results, they will be misrepresented.

Meeting suffering people with medical measures valid only for other categories of people, is not likely to diminish the patients' suffering. Any misinterpretation of the very nature of a health problem that the medical profession is mandated to solve, or of a malfunction it is empowered to treat, must add to the likelihood that inadequate measures be taken. In other words: a wrong interpretation results, with a high degree of probability, in medical wrongdoing. No matter how valid the knowledge is presumed to be or how appropriate the treatment measures are deemed to be, such a practice is still unethical. Its failure to cure or to comfort will reveal its inadequacy. However, what is wrong cannot be read from knowledge or from interventions taken. In other words: *The wrongness is embedded in the practice, which means in a theoretically conditioned blindness which does harm in the name of science and of helping.*

The present study represents an exploration of the long-term impact of societally silenced violation, as exemplified by one specific modality, that of childhood sexual abuse. Within a framework of phenomenology of the lived body, and with hermeneutics as methodological tools, the dynamics between violation and suffering, and between the suffering person and the representatives of the health care system, have been made visible. The trajectories from violation to adaptation, from adaptation to the body, and from the body to sickness, have been made comprehensible. They have emerged, complete with their convincing internal logic and meaningfulness. Violation embodiment has been shown in its concreteness. This means that the nature of human suffering in the wake of sexual assault is non-symbolic. Violation inscriptions are not random but rather follow patterns which can be interpreted as Gestalt. Violation embodiment is shaped through subjective, situational logic and rationality. Therefore it holds within it its own validating power. My exploration of violation embodiment has opened an avenue to an understanding of the experiential body as contrasted to the biomedical body. The body into which violation has been inscribed, is not only a mindful body, as is valid for every lived body. The inscribed body is highly "problematic". It is an "obtrusive" body, demanding and absorbing attention due to unpleasant sensations. It is temporally very "insistent" as may be expressed in pain or dysfunction. And it is certainly a body "mutely testifying" the socially silenced abuse. [2]

The present analysis of life in the wake of childhood violation addresses the topic of boundaries. Violence, in every modality, also symbolic violence in the sense in which Pierre Bourdieu uses the term,[3] is the practice of disrespect of another person's integrity. To practice disrespect means to humiliate another person. Such a humiliation may take the form of neglect or abuse, in all the various ways these may occur. Thus, it is not of primary importance whether a humiliating act aims at violating the person's physical, psychic, sexual, intellectual, social, ethnic, cultural, moral or legal integrity. Such distinctions are secondary to the fact that, whenever humiliation occurs, it involves and affects the whole person in her or his being-in-the-world.

The results of my study imply that sexually violated human beings do not suffer primarily from the assault which has been perpetrated upon their bodies but rather from the violation done to their dignity. Humiliation and degradation are at the core of their suffering. Shame, guilt, and stigma become the frames for their perceived selfhood, powerlessness, horror and pain the components of their sickness. Continuous adaptation to threat, fear and defeat are their predominating efforts. Intermittent breakdowns or a permanent exhaustion of their ability to adapt to violation are the states which bring violated people into the sphere of medicine.

The impact of childhood sexual violation on life and health was explored in depth through dialogues between thirty-four adult persons, thirty women and four men, and myself. All the subjects, who ranged in age from seventeen to seventy years, were clients at two Norwegian incest centers. The interviews, which totaled more than eighty hours, were tape recorded and transcribed verbatim by myself. Relevant hospital records for sixteen of the thirty-four people interviewed were added. Extensive field notes from my frequent visits to the incest centers completed the material.

Through analysis of the texts and interpretation of the topics addressed, seven consistent patterns emerged in every interview, though in a variety of modalities. As in classical music, these seven "themes" presented in at least thirty-four "variations" each. These variations demonstrated how something is known *and* new, stable *and* changeable, general *and* particular. I could read from the notes how the same motif, the basic pattern, reemerged in each variation. The plentitude of variations attested to both the persistence and the variability of the motifs. In other words, these seven phenomena were characterized both by regularity *and* unpredictability. They illustrated something general about how it is to be a violated human being, *and* something particular about how one specific person is affected by a violation experience. In total, the themes and their variations made comprehensible the logic implicit in every violation experience, and the adaptability of each particular violated person.

Hidden and silenced violations persisting in both the person's body and life became visible whenever their lived meanings were reactivated. Violent, abject, uncanny, and painful sensations of the past were perceived as if in present tense whenever current sensory perceptions so closely resembled those of the past that they could not immediately be identified as different from the earlier ones. Consequently, what had been experienced there-and-then could apparently be perceived as here-and-now, thus unrecognizable as a Re-experience. The distinction of tenses would disappear. The horror, fear and pain of the past would be *as-if* present. Nobody else would be able to share this perception of reality. Anybody else, and definitely the objectifying medical gaze in the sense of Michel Foucault[4], would judge the horror, fear and pain as lacking reason, meaning, and causal substrate.

Biomedical theory is built on four basic criteria for correct biomedical judgment: categorical distinction, linear time, irreversible cause-effect relation, and individual sickness. The present study provides evidence that silenced violation embodiment invalidates these four criteria. The first of these criteria, categorical distinction, was not met. While all the seven patterns of the phenomena could be linked to violation, sickness, and treatment, the logic in the sequencing of those three categories did not prove to be valid. Their sequential order could be supplanted by a situated logic. Sickness could provoke violation, suffering could terminate abuse, treatment could be violent and result in chronicity, or therapy could represent abuse. Thus, "violation", "sickness" and "treatment" changed content due to changing contexts. The meaning they held was constituted situationally. Thus, they lost their function as categories.

Nor was the second condition met, that of linear time, since simultaneity and the breakdown of the distinction of tenses characterized the violated person's universe of experiences. The third criterion, causality, was also lacking as cause and effect could be experienced as identical, or interchangeable. The fourth precondition, individuality, was not met as violated persons would identify with other violated people or feel violated by proxy. Thus sickness and disability could result from assaults not even directed at that particular person.

Since violation embodiment did not meet these four preconditions, medical misjudgment became highly probable with improper treatment as a frequent result. This, however, was shown to be but part of the adverse effect of medical intervention with victims of violence. The other aspect emerged when juxtaposing the violation history with the documented medical history, rendering visible the medical transformation of violated persons into sick patients.

Consequently, the present study provides evidence that psychiatric and somatic medicine, by means of a naturalist theory and an objectivist methodology, contribute to the societal silencing of social and relational pathology. *The medical sickening and chronifying of violated people, in the name of helping and of science, is its most controversial result.* This fact accentuates the urgency for the scientific biomedical community to appreciate human beings as purposeful, as capable of constructing and communicating meaning. Medical research which disregards these central characteristics alienates human subjects forcibly. Such degradation is indefensible. *Its scientific legitimacy must be regarded as subordinate to ethical principles which prohibit integrity violation in the name of science.*

Actually, my exploration of socially silenced sexual boundary violation with regard to health impact has revealed both particular and general dynamics, mutually constitutive, between individuals and society, and between violated persons and health agencies. The violated human bodies have emerged as both bodies unto or into which violence is inscribed in a literal, graphical, etching-like sense, and bodies which are inscribed into the social politics which silence the main impact of patriarchy, namely the societal structures of domination and objectification. The etching impact of silenced violations, and the sickening impact of the medical aid offered to the violated persons are, indeed, metaphorically spoken: The double costs of patriarchy. Sexually violated bodies are, in this double sense, Inscribed Bodies.

2. CLINICAL BACKGROUND

I have been a General Practitioner in a western part of the city of Oslo since 1975. From the very beginning, the majority of my patients, almost 90%, has been women of all ages. [5]

One kind of professional experience has generated most frustration, the most intense feelings of incompetence, and the greatest amount of insecurity about decisions concerning my patients' life course. The core of it might be grasped most appropriately via the term *discrepancy*. The feeling of being in touch with different kinds of discrepancies had most often arisen in connection with those of my female patients who presented to me health problems, complaints or bodily sensations which were difficult to label and difficult to treat. Various discrepancies between phenomena could present as follows: the intensity of a woman's complaint - and her obvious high level of social function; her localization of a pain, sensation or discomfort - and my knowledge of the anatomy of the human body; the kind of her illness - and the results of my diagnostic measures; her description of the problem's development - and the medical models of pathogenesis; her suppositions of the cause or source of her bad health - and the medical concepts of etiology; my therapeutic approaches - and her lacking experience of any positive change or improvement; the total diagnostic and therapeutic investment - and the obviously failing cure, non-lasting release or unregained health.

Many of those patients whose health problems provoked in me this feeling of discrepancy, reported difficulties concerning the accomplishment of their daily tasks. In the course of our interactions for diagnosis and treatment, they needed short- or long-term sick-leave certification. Some remained disabled and asked for assistance to apply for a disability pension. During the process of preparing my medical advocacy to prove the patients' eligibility to receive pensions for disabilities due to chronic illnesses, I used to invite the women to an extended consultation about their lives. These consultations were meant to provide a frame for exploring in detail what had happened to them in different life phases and in relevant relationships.

Gudrun was one of those women. She told me the story of her life at a point when she felt unable to do her work, which included having full time responsibility for a group of ten people. She had been receiving treatment for years, with her level of functioning continuing to decline. Four chronic diseases had, one by one, gradually disabled her. Sixty years old, she applied for a disability pension. She had been sexually abused in childhood, and she herself formulated the core of her pain like this:

> As far as my chronic and disabling muscle pain, I recognize how, since childhood, I've been tense and on my guard. I've been constantly listening, holding my breath, expecting humiliations to hit me at any time from everywhere.

Gudrun had offered me a possibility to learn, to look with more awareness for certain signs, and to listen more attentively for particular remarks. Her reflections upon the impact of hidden life experiences made me more sensitive to perceptions of other important yet non-thematized "histories" in my encounters with several other patients.

Anna was one of these women who expressed a particular fearfulness and sense of avoidance. During gynecological examinations, I had regarded her exceptionally high level of tension. The paper cover on the chair used to be wet from her perspiration. Once, while having inserted an intra-uterine device, her entire body suddenly broke out in perspiration. An intense shivering spread from her legs upwards, culminating in cramps in her legs, arms and pelvis. She lost consciousness, and her body twisted in the chair. When she awoke, Anna told me that she had been raped by her husband when she was eight months pregnant. The day after, she had experienced an extremely problematic delivery - causing her to decide never again to become pregnant.

Ulla came for contraceptive pills although she had no partner. She wanted to be protected, as, "you can never know what might happen." I noticed how calmly she spoke but did not draw inferences from that ambiguity. Later, I could not recall why I had failed to see that which was a cue to something she found impossible to say. A few months later, however, she came complaining of abdominal pain. Our conversation gradually brought out a disturbed eating pattern which had started some years ago, when her stepfather had moved out. He had raped her several times. After that, Ulla felt extremely vulnerable to possible sexual approaches since her mother had recently entered into a new relationship with a new male partner.

Karin reported suffering frequently from intense urticaria, rashes and itching. It always began at her legs and thighs and would spread to her stomach and chest. She would feel desperate inside her burning, itching, and blistering skin. Gradually she had understood that nylon stockings provoked the skin reaction. While eliciting her associations to such stockings, she began to talk about frequent trips with her uncle from the time she was about ten years old. During their travels, he used to stop and then order Karin to put on a pair of lady's stockings. "And he then would use my legs," she said, almost inaudibly. Sitting in the car while he masturbated between her thighs, she would press her feet against the inside of the door, stiffening her legs completely. She added that a few years ago, an orthopedic surgeon had advised her to have her Achillean tendons operated since she had suffered for years from a painful stiffness and heaviness, and from nightly cramps in her legs. But she had not felt convinced that such an operation would be appropriate, and had been hesitant to agree.

Gina had been misshapen from childhood and walked uneasily. When reviewing the records of the first five years of our relationship, her complaints seemed to cluster around: respiratory tract infections, which she often explained by her lack of outdoor activities due to her deformation; pain in her limbs, which she attributed to the various tensions in her body, due to her deformation; and, acute attacks of nausea, which she explained as an imbalance in her body, due to her deformation. This resulted in her taking frequent sick leave and undergoing a variety of medical interventions. I did not grasp her metaphors of ugliness and unworthiness until, when again visiting for acute and disabling nausea, she gave me a cue about her husband which made me ask: "Does he batter you when he is drunk?" Hastily, she answered: "He mostly pushes and shoves me a bit. But my balance is so bad. So I fall. But he's never hit me when I'm lying down." All of a sudden, her undefined pain, nausea and shortness of breath were not inexplicable any more. If a body is being pushed and loses its balance and falls down, the result will be painful limbs. If escaping from being hit demands lying down, an incapacitating nausea is a rational reason to stay in bed until the danger has passed again. Gina never considered protesting against the man to whom she felt deeply indebted as he tolerated her misshapen body, and she would not demand that her integrity be inviolate.

Else considered it to be within her husband's rights for him to "punish" her if she was too sinful in thought and deed, according to her parish's principles of how men and women ought to live. The religious codex for male and female honor gave him the right to batter her; she deserved it. Sometimes she would phone and ask for sick leave certification without being able to come to my office. Gradually, she became disabled by pains in different parts of her body, by tiredness, by an eating disorder, by persistent diarrhea, nausea and vomiting. Due to chronic pelvic pain, endometriosis was diagnosed and an ovarectomy performed, causing the need for hormone replacement therapy. Next, a hypothyroidosis had to be treated. When one of the lumps in her breasts proved malignant, a mastectomy and lymphonodulectomy were performed, resulting in pain that kept her sleepless. Once, when visiting to be treated for this pain, she told me that her husband stopped beating her after she got cancer.

Ingween had asked twice within a few weeks for sick leave certification because of incapacitating nausea and continuous vomiting in connection with an approaching weekend job. I therefore asked her which problem she was "unable to swallow." I expected a report about conflicts at her department related to the weekend schedules. Ingween gave her answer hesitatingly: "My husband is usually all right, I mean normal. On Saturdays, however, when I'm at work, he borrows such violent porn videos. When I come home late, he's been watching these videos while drinking a lot. Then I know what he'll make me do. So I start vomiting the very moment I see my name listed for weekend charges. Do you understand what I'm saying?" I did. She received her sick leave certification, and then we talked about the necessity of informing the male head of the department in order to prevent similar situations. She agreed and managed to communicate how important it was for her to avoid Saturday night shifts.

Elisabeth made frequent appointments due to a fear of cancer. Her anxiety was triggered by pain. The expressions of her pain would differ, as would their localization and intensity. Elisabeth mocked herself, saying, "Here I am again with my pain. Now it's in my gall bladder. I can't touch my upper stomach. But it's not unlike that pain I felt in my pelvis when I thought there was something dangerous going on in my right ovary." Although she saw a pattern of pain, she would not calm down until she had been examined. After a long period of disabling lower back pain and incapacitating fatigue, I invited her to recall experiences of danger. Addressing the room, she said, "Well, the first was a relative. He raped me when I was fourteen during a motor-bike ride he'd invited me on. During my second pregnancy my husband began to beat me. I became psychotic after the delivery and had a long depression. Later he forced me to have intercourse, even while I was pregnant with my third child. We divorced after nine years. For many years I was single. Through friends, I met a man who later raped me. I have been attacked by several men and have escaped only because someone else intervened.

Please, tell me, what do you think? Why has this happened to me? Does it prove that I'm a woman every man can use as he pleases?" I had no answer.

Cecilie came to me when I had been in practice for about ten years. Though considering myself an experienced doctor, I could not cope with a patient like her. She presented problems and offered challenges I did not have the skills to meet. Despite our both being health professionals, during the first years of our relationship we spoke completely different languages. We held completely different views on what problems we faced, how they ought to be met, what was reasonable to do, and who was responsible for what. Obviously, our patient-doctor relationship was not according to the textbooks. In retrospect, it seems that we inhabited divergent realities without being aware of it. Despite many hours of consultation, we had never come to any agreement as to how things were interrelated, or what lines existed between current problems and earlier experiences. I was convinced I had sufficient knowledge about Cecilie's life and conditions. She was convinced that she had told me what was relevant to enable me to help her. Still we disappointed and frustrated each other again and again. As Cecilie's health worsened, we decided to have a talk in her home, in an attempt to change roles, behaviors, aims. When telling me her life history, it became obvious how both the course of her life and her illness history had been influenced by the experiences of incest, of repeated abuse by friends of her parents, and by a rape by a cousin during adolescence. She had never talked about the incest, but had told about the other events. Her revelations had been met with mistrust and blame, which silenced her totally. When Cecilie had to encounter her rapist as a witness in a family conflict, the meeting provoked the first of her suicide attempts. Until then, she had been unable to acknowledge the long-term impact of repeated integrity violation.

Based on an accumulating concern resulting from my increasing knowledge about assaulted women, I felt a strong need to improve my professional skills at comprehending the impact of boundary violations, particularly sexual boundary violation at a young age, on adult women's somatic and mental health, well-being, self-esteem and sexuality, as well as their personal relationships to significant others. As the information about sexual abuse which occasionally came to light had proven to be relevant and important to the process of understanding ill health, I hypothesized that my unawareness of a prior history of sexual trauma might have adverse effects.

When consulting the literature in 1989, I learned that during the last two decades, sexual trauma, and especially childhood sexual abuse, had become a topic of multi-professional interest. A considerable number of studies had been performed concerning its frequency. [6] Likewise, a growing body of knowledge concerning the short-term,[7] and also the long-term impact of such abuse could be reviewed. [8]

Obviously, until then, most studies explored conditions arising from the mental impact of abuse. Depression, anxiety disorders, self-destructive and suicidal behavior, dissociation disorders, drug abuse, eating-disorders and victimization by accident, crime or sexual violence had shown to be associated with a history of childhood sexual abuse. [9] Some studies had revealed a high occurrence of somatization in those who were abused, interpreted as an expression of post-traumatic stress disorder. [10] Recent research had dealt with the long-term impact of childhood abuse on somatic health, and on gynecological illness in patients in clinical settings. There had appeared significant associations between sexual abuse in childhood or adolescence and pelvic pain, pelvic inflammatory disease, urinary tract infections, dysmenorrhoea, dyspareunia, and vaginal discharge, and, apart from the gynecological diagnoses, asthma and headache. [11]

All the above-mentioned studies had been based on female populations in psychiatric or gynecological care, or in pain clinics. To my knowledge, however, as of that time no study of childhood sexual abuse experience among patients in a general practice setting had been performed. Consequently, I decided to perform such a study among female patients in my own practice. I presumed a gynecological consultation to be an appropriate frame for anamnestic questions about adverse sexual experiences in general, and sexual abuse in childhood or adolescence in particular. I assumed the introduction of these topics would evidence my professional concern and an awareness of their impact, and thereby might facilitate the reporting of such experiences among my patients.

During half a year from 1989 to 1990, all women in a defined age group attending for problems and complaints they themselves had defined as gynecological were invited to be interviewed by me directly after the examination. Among 117 women, 85 agreed to participate in a semi-structured interview linked to the gynecological examination. Among the interviewed women, nine reported being approached by men who exposed themselves indecently. Twenty-four women had experienced sexual abuse in terms of physical contact before the age of eighteen, many of them repeatedly, either by different men or by the same abuser for years. Of all these women, whom I had known from anywhere between several months to twelve years and whom I had been convinced I knew well, only one had told me about her abuse experience prior to the interview. (Kirkengen et al. 1993)

The first conclusion from the study resulted from the analysis of the structured answers to the questionnaire. By means of a regression analysis, a positive association could be shown between the experience of sexual assault in childhood or adolescence and a history of pelvic pain, as well as a history of gynecological surgery. Furthermore, the majority of women with a history of both pelvic pain and at least one gynecological operation, were abused women. Only four of them, which equals one out of six, had neither. This was in contrast to the absence of both in 43 of the 61 non-abused women, which equals almost three out of four. Consequently, these results matched the previously acquired findings from studies among gynecological patients in specialist care. This could allow the conclusion that information about sexual abuse is accessible in gynecological encounters in a general practice setting.

The second conclusion was based on the experience of the reactions to the questions concerning childhood sexual abuse. Concerning their sexual debut, all women responded by recalling their exact age at that time. Almost two out of three, however, expressed sadness, disappointment or even anger linked to this event. They distinguished between their first voluntary intercourse, although obviously rather an adverse event for a large number of them, from any unwanted sexual approach. A few women expressed doubts when reflecting upon the possibility of having experienced any sexual assaults. These women told of having actually avoided sexual contact, having been disinterested in sexual relations or having felt endangered by sexual advances; they suspected that this bore witness to the existence of past negative experiences which they were unable to recall.

My third conclusion was based on my continuous observation during the interviews. Although I saw the majority of the women struggle with ambivalence and insecurity, none refused to answer. And although some of them cried when telling their stories, none gave me the impression of being shaken by my question, the topic, or the reporting. There were no breakdowns or any other reactions so often mentioned as a probable and adverse result of an approach to the topic. None of the women expressed feeling that the question was an attempt to invade her. None got angry or irritated, or mentioned being trapped by being asked questions more intimate than she had expected, or more detailed than she could accept. The topic of adverse sexual experiences was perceived as relevant to a dialogue about gynecological health linked to a consultation they themselves had asked for. Some expressed that they had acknowledged more about themselves through reflecting while telling about the abuse. This, they said, felt difficult but not dangerous, since, as one of the women put it, "It was not new for me. I've been carrying it and have known it since it happened. It was only new for you."

My fourth conclusion concerned the very special way sexual boundary violations, and especially childhood sexual abuse, becomes an existential trauma embedded in and influencing aspects of a person's life. I was told of a great variety of abuse performed by trusted people and the way these abuse experiences were presented showed the array of aspects of their impact. The women did not just answer yes or no to the question about unwanted sexual experiences. They linked the stories they told to circumstances and events, thus showing how the abuse was woven into the context of life as it had been when the events described occurred. These descriptions of context rendered the reports consistent. They made comprehensible how the abusive relationships and the insulting acts had become the hidden background for the women's life course, their choices, decisions, problems and preferences - and not only those concerning sexuality and interpersonal relations. The women reflected upon the abuse experience as the hidden and unrevealed precondition for a web of consequences, influencing major aspects of their personal development in terms of self-esteem, confidence, trust and health. Until hearing these narratives, I had been ignorant not only as to the extent and variety of such experiences but also of just how traumatic they were, and of the extent to which they might shape not just the health but also the lives of those experiencing them.

The study was not designed to elicit narratives about abuse in biographical context beyond those framed in the pre-structured alternative answers regarding unwanted sexual experience. These narratives were introduced by the women themselves. During the first interviews it became obvious that some of the questions invited comments. Therefore, I immediately began to take notes. The questions concerning abuse elicited comments regardless of whether the final answer was no or yes. None of the women remained indifferent, and several expressed concern on behalf of their own children, or persons they knew had been abused, or children in their professional care. Those who had an abuse experience described in detail what had been done by whom, where and when. Most narratives were uninterrupted. A few of the women searched for reassurance that they were not mistaken in defining what had been done to them as abuse. Some explored the impact of repeated abusive episodes before stating which had felt most abusive. A variety of constellations emerged, going beyond the boundaries of the questionnaire.

The 24 abused women presented experiences which, due to their diversity, did not lend themselves to being grouped. Their narratives contained, in different combinations, all the categories of sexual abuse classification used in the international literature. All narratives included physical contact, either as a single or repeated event, continuous or intermittent, intra- or extra-familial, attempted or completed penetration. The women reflected extensively upon the details, circumstances and consequences of their experiences. Their reports heightened my awareness about the overall presence of abuse experience among my female patients, and about the relevance for me as their primary care physician of acknowledging these experiences.

These narratives of abuse, which I encountered more or less unintentionally and which the study's design prohibited me from elaborating, engendered in me the desire to learn more by focusing on that sort of source material in a more appropriate way.

In 1990, a study about the sickness histories of seven female patients in tertiary psychiatric care was published. These women between 22 and 39 years of age, all sexually molested as children, had a mean of eight operations and eighteen contacts with non-psychiatric consultant teams. More than two out of three of the operations had not shown pathological findings. The authors underlined the possible connection between a history of childhood sexual abuse, somatization as previously reported by other authors, and a history of excessive sickness including a high number of unnecessary diagnostic and surgical interventions. [12] The authors also referred to a study from 1953 about the frequency of operations among women diagnosed as hysterics presenting close to a fourfold number of operations as compared to other women, and a similarly unsatisfying outcome of the majority of the operations as described above. [13] These two studies, although differing in point of departure and approach to the issue of surgery among female psychiatric patients, stressed the importance of improving the interpretation of chronic or intermittent pain, as well as the criteria for invasive diagnostic and surgical interventions among female patients frequently presenting pain as a major symptom. This message was supported by three studies focusing on a possible link between childhood sexual and physical abuse and somatic complaints in adulthood. [14]

In 1991, the first book addressing the issue of encounters with sexually abused persons in primary care appeared. The author, an English female General Practitioner, specialist in family planning and community pediatrics, drew attention to the considerable difference between the available studies from specialist care and psychoanalytic perspectives on the one hand, and the primary care setting on the other. Her accumulated and systematized knowledge about the issue and its presentation was meant to heighten the awareness of abuse, and to lower the resistance to approaching it, among General Practitioners and health care professionals in primary care. [15]

In 1991, the issue of medical problems among sexually abused women was addressed in two studies focusing on gynecological health, and on overall health with an accent on gastrointestinal symptoms, which had already been addressed by Drossman and colleagues. [16] Studies concerning mental health, somatization, and their different combinations in relation to a history of sexual abuse in childhood, presented a view of dissociation as an expression of trauma experience and a mediator of trauma impact in its different forms. [17]

In 1992, on the background of the Los Angeles epidemiological catchment area study, the author concluded the impact of childhood sexual abuse (CSA) on the mental morbidity of a population to be as follows:

> ...data indicate that a history of CSA significantly increases an individual's odds of developing eight psychiatric disorders in adulthood. On the community level, however, it is estimated that 74% of the exposed psychiatric cases (i.e., those with a history of CSA), and 3.9% of all psychiatric cases within the population can be attributed to childhood sexual abuse. [18]

The same year, a study about sexual abuse among female patients in general practice was published, based on a randomly selected group of 1,443 women attending a family practice: 22.1% reported sexual abuse according to the definitions of the study. Focusing on a possible association between childhood sexual abuse, amount of health problems, somatization and health-risk behaviors, the group of abused women differed significantly from the non-abused for all three aspects mentioned. The severity of the abuse experience was positively correlated with the severity of health problems. Additionally, the extent of support the abused women had received in connection with disclosure of the abuse, correlated inversely with the number of gynecological problems. The authors concluded as follows:

> *Fewer than 2%* of the sexually abused women had discussed the abuse with a physician. To identify and assist victims of sexual abuse, physicians should become experienced with *non-threatening* methods of eliciting such information when the medical history is obtained. [19] (italics mine)

The same year, the fact of the *patients' reluctance* to talk with their physician about hidden sexual trauma, was confirmed by a study in primary health care: 405 consecutive women were included, 147 of these general practice patients, and 258 students attending a health care center for a health maintenance visit. Of the general practice patients, 47.6% reported sexual contact victimization during their lifetimes, of the students 57%. Thirty percent of the general practice patients and

44.9% of the students felt, "they would not be comfortable discussing the experience with medical personnel." The authors concluded:

> A history of past sexual victimization can have a significant impact on a patient's health status, yet it is not commonly disclosed to medical personnel. Primary care clinicians are in a key position to identify sexually traumatized patients and to facilitate appropriate care or referral necessary. Accurate assessment of sexually victimized patients may *minimize unnecessary medical treatment and enhance their overall health status.* Studies are needed in more representative primary care settings that include male patients, use more direct assessment methods, and compare the health status of sexually victimized patients and their rates of health care utilization. [20] (italics mine)

Not only patients were shown to be reluctant to mention their sexual abuse histories. Physicians were even more reluctant than patients to address the topic, and reluctant to report the abuse once it had come to their attention. The obtaining of a sexual trauma history in psychiatric institutions of secondary or tertiary care, although regarded as necessary, had been shown widely lacking in the studies mentioned above. The results of two studies among professionals in a child care institution showed several factors influencing professional reporting and evaluating of sexual abuse: having been abused as a child oneself; the gender of both child, abusive parent, and professional; the sexual activity involved; and, personal emotions. [21] Yet another study directly addressed the influence of the therapists' resentments and concluded:

> The professionals who helped adult, former incest victims were hampered by shortcomings in knowledge and skills as well as their *own emotional resistance.* [22] (italics mine)

In 1992, the issue of physicians' own emotional resistance against the disclosure of and the engagement in the health consequences of sexual violence, especially when displayed within family life, was dealt with in an ethnographically designed study. Physicians in primary care were in-depth interviewed by a female colleague about their attitudes to domestic violence with regard to recognition and intervention. The majority of the physicians admitted feeling a lack of comfort, fear of offending, powerlessness, loss of control, and time constraints to be the major barriers to a more aggressive approach. Among the male doctors, 14% had themselves experienced violence at some time during their lives, and among the female doctors, 31% had.[23] In a North American community practice study, among 476 consecutive women, 394 agreed to participate. Of these, 38.8% answered affirmatively to questions related to domestic physical violence. Only six women in this sample had ever been asked about physical violence experience by their physicians. The authors concluded:

> Although spouse abuse is common, physicians rarely ask about it. Physicians should be trained to detect and assess abuse among female patients. [24]

In 1992, two frequently cited studies, and a comprehensive proposal for health research for women[25] resulted in a US Council Report focusing on the role and responsibility of physicians in developing medically adequate responses to domestic violence as it affects women at all ages, in the form of therapeutic and, possibly more important, of supportive and preventive measures. The Council Report addressed in particular the effect of prior child sexual victimization as a risk factor for all types of future violation:

> Perhaps most troubling is the increased risk of women who have been sexually abused as children to be revictimized later in life by both strangers and intimates. [26]

Until 1992, the debate about problems in sexual abuse research had concentrated upon methods and definitions in studies concerning abuse prevalence. Now, theoretical issues became explicit, relating to estimation and understanding of abuse impact. Beitchman and colleagues reviewed and compared study results of short and long-term impact, while Briere revised the methodological approaches to the different aspects of sexual assault, stressing the necessity of a "second wave" of studies with a more sophisticated methodology than those he termed the "first wave," designed to register the amount of certain types of assaults rather than to understand their human impact or social dynamics.[27]

Still, Brier's quest for sophistication did not challenge the currently predominant reductionist methodology but rather left unaltered the classification of abusive events established in the sociologically inspired and informed studies. Nor did he address the limitations in psychiatric studies of a disease and pathology oriented focus. And he seemed by no means concerned with the emerging indications of an excess of sickness among the abused, which was formerly understood as somatization and thus regarded as contributing to psychiatric morbidity, despite its provoking many physical interventions.

When recalling the more or less unsolicited narratives of integrity violation which arose in my previously mentioned study, I was struck by how many were the result of acts which are defined as criminal. However, only one of those I spoke to had seen her offender sentenced. Despite such gross violations, my female patients would *make light of* their experiences by calling their reactions "hysterical" and inappropriate. This was the case even when they described how obstructing the experience had been to later development and their freedom of choice or movement. Even reports of rape attempts were concluded with remarks such as: "Fortunately, nothing really nasty happened."

Their narratives contained a variety of descriptions of *wrong touch*, revealing a high awareness of the varying meanings of touch depending on how, when, where, by whom and for what purpose. Few women used the word fondling (*beføle* in Norwegian), the most usual term for improper sexual touching in literature. The wrongness of the touch was embedded in the intention of the one doing the touching. The message in the touch was unmistakable. As was that of the *wrong gaze*. The women's awareness and situational feeling of improper touch and gaze in connection with abusive acts appeared as a relevant indicator. Issues of wrong touch and gaze had an alarming impact for many of the women.

The women mentioned also that in the presence of others, wrong touch or gaze often had been "covered up" by a *strange laughter*, a cue for reframing in the sense of Goffman,[28] indicating, "this is play, fun, nothing serious, just a joke." Thus, the situation was characterized as harmless. The obvious lack of respect shown the girl's bodily integrity was converted by the laughter of the one disrespecting it. Her righteous rage against a violation of her boundaries was *silenced*. Her perception of wrongness was tricked.

Suppressed rage in the wake of sexual trauma was discussed by several of the women, but in a variety of ways and often in combination with shame, guilt and self-blame, emotions linked, and in a variety of ways, to the politics of silencing. Thus, an effect of violating acts was to impose *shame, guilt and blame* on the one who was violated; this cultural inversion of the source of the forbidden emotions had, in turn, generated *adaptive strategies or searches for help "in disguise."* If these strategies evoked a response from people the girl trusted, they could potentially serve to protect her, even if telling about the abuse did not result in terminating the abuse. Most often, however, telling had been impossible, or the girl had been disbelieved and reproached. Distrust and blame seemed to hold powerful destructive potentials.

The women reflected upon individual or repeated exploitive, threatening, frightening or insulting situations linked to acts carrying sexual meaning for the person initiating them. While doing so, they introduced the context of sexual abuse. This elucidation of context yielded information as to the women's cognitive, emotional and intellectual stage of maturity when the abuse occurred, and a recall of spatial, temporal and relational circumstances. It also included reports of consequences as regards personal conduct, interpersonal relationships, emotional distress and social inhibitions. The array of constellations of abuse, and, most remarkably, the differentiation of its meaning in every single situation, went far beyond what could be registered in preformed and standardized categories. In addition, it would have been impossible to extrapolate from those categories a classification of severity which would correlate to the women's own reflections upon the influence of abuse on their lives.

After having performed the preliminary study and compared my impressions with the tenor and terminology in the research of the field, I was certain as to the unacceptable limitations of the predominant methodology with regard to abuse impact appreciation.

3. THEORETICAL FRAMEWORKS

3.1 EPISTEMOLOGIES

Until the end of 1992, studies had addressed the probable relationship between unwanted sexual experiences in childhood or adulthood and a history of frequent mental health problems, including somatization disorder.[29]. Only to a lesser degree, however, a history of somatic complaints, such as gynecological and gastrointestinal ailments and reports of chronic pain, had been explored.[30]

My preliminary study had been structured according to models derived from prevalence studies about sexual violence.[31] However, the unstructured information in the narratives connected to abuse recall rendered the women's own impact-evaluation inconsistent with the objective severity rankings in abuse research. Their reflections concerning their own responsibility, guilt, shame and self-blame, their suppressed rage and active contribution to denial or belittling of harm, differed from the ranking scale in current research. Their bodily and emotional expressions while recalling hidden memories indicated an impact different from that of the literature concerning abuse categories and grades of severity.

The framework of the study, however, offered no options for an analysis of patterns in the possible relationships between sickness and sexual abuse experiences. Furthermore, the design did not provide access to sources of information outside my practice. Despite, or possibly because of, those frustrating limitations, I became aware of four salient topics:

- Previously unrevealed sexual assault experience seemed relevant to an understanding of mental and somatic health problems.
- Hidden boundary violations seemed to be a probable source of impaired self-esteem and well-being, not yet properly acknowledged by society and the health care professions.
- Assaultive events could be conceptualized as a source of alterations of body images and interpersonal relationships.
- Reactions to silenced boundary violations, manifesting as adaptation strategies, could contribute to the appearance of sickness as these

strategies were not interpreted within the context of the implicit sexual and gender code of the society and its forces toward silencing.

These four topics were all related to the dynamics between sexual abuse and health, and between the assaulted person and the health care system. They generated a question of relevance for health care professionals: *is a person's medical history adequately comprehensible without insight into the person's life history?*

Since my study-experiences had confronted me with the limitations of the traditional objectivist framework, the next question was: *how can a human science epistemology be applied in a medical research setting seeking to understand life experiences?*

Transformed into practical measures, the third question was: *how best to address and cooperate with persons who themselves are aware of sexual assault experiences which they are willing to reflect upon along with a medical researcher?*

The available body of knowledge concerning mental or somatic health impact of sexual assault was mainly based on sociological and biomedical methodology defining the abusive acts as discrete events, classified according to presumed estimates of severity.[32] The predominant theory of biomedical research through the last two centuries had been based on the Cartesian legacy of the separated concepts of res extensa, the outer, physical, material, and as such corporal aspects of human existence, and res cogitans, the inner, psychic, immaterial, and as such the mental aspects. In the frame of research on human beings, the physical body had become the central object for scientific inquiry. The non-physical aspects of human life, defined as the arena of the human sciences, had been of less interest for physicians engaged in research.[33]

With regard to the actual presenting problems, it seemed as if the Cartesian framework might almost be viewed as the origin of a special kind of obstacle. The very concept of the human body in current medical theory and practice could itself be the source of it. The questions arising in my professional experience addressed the interfaces between social life, hidden violation, personal experience and individual interpretation, all of which interacted in the shaping of health problems. These health problems were probably of the kind physician and phenomenologist Richard J. Baron called *"diseases in search of anatomicopathologic facts"* (Baron 1985).

Likewise, it seemed obvious that a traditional biomedical approach would not provide an insight into the dynamics I wanted to explore. The core of these problems has been addressed by the philosopher Edmund Husserl as follows:

> The exclusiveness with which the total world view of modern man, in the second half of the nineteenth century, let itself be determined by the positive sciences and be blinded by the 'prosperity' they produced, meant an indifferent turning away from the questions which are decisive for a genuine humanity. *Merely fact-minded sciences make merely fact-minded people...* In our vital need - so are we told - this science has nothing to say to us. It excludes in principle precisely the questions which man, given over in our unhappy times to the most portentous upheavals, finds the most burning: questions of the meaning or meaninglessness of the whole of this human existence. (Husserl 1970:5-6) (italics mine)

Husserl founded the European phenomenological tradition, and later, it has been taken in distinctive directions by the philosophers Martin Heidegger, Maurice Merleau-Ponty, and Emmanuel Levinas. Merleau-Ponty has built his work on the basic acknowledgment of the world as perceived. He writes:

> The world is not an object such that I have in my possession the laws of its making; it is the natural setting of, and field for, all thoughts and all my explicit perceptions. Truth does not 'inhabit' only the 'inner man', or more accurately, there is no inner man, *man is in the world, and only in the world does he know himself.* [34](italics mine)

According to Heidegger, phenomenological research is guided by "mindful wondering about the project of life, of living and of what it means to live a life."[35] "What it means to live a life", especially a life with illness, has been extensively explored by philosopher S. Kay Toombs. In her phenomenologically grounded reflections on the different perspectives of illness she writes:

> The foregoing phenomenological analysis has demonstrated that illness and body mean something significantly and qualitatively different to the patient and to the physician. This difference in perspectives is not simply a matter of different levels of knowledge but, rather, it is a reflection of the fundamental and decisive distinction between the lived experience of illness and the naturalistic account of such experience. It has been further noted that this difference in understanding between physician and patient is an important factor in medical practice - a factor which has an impact not only on the extent to which doctors and patients can successfully communicate with one another on the basis of a shared understanding of the patient's illness but, additionally, on the extent to which the physician can address the patient's suffering in an optimal fashion.
>
> If the physician is to minimize the impact of this difference in perspectives and develop a shared understanding of the meaning of illness, it is vital that he or she gain some insight into the patient's lived experience. (Toombs, 1992:89)

Phenomenology is concerned with the understanding of human life experiences and asks: "What is this kind of experience like?" It attempts to formulate insightful descriptions of the way in which human beings experience the world pre-reflectively, which designates a state before experiences are formed into taxonomies, classifications or abstractions. Thus, phenomenology allows insight into the human life-world as it is experienced and might be described, and it allows the generating of hypotheses about the nature of human understanding.

As the meaning of human life-world experiences is the central interest of all efforts within phenomenological scientific endeavors, the researcher will encounter human values into which all human actions, choices, intentions and experiences are embedded. This also means that the researcher has to be aware of his or her own "presence" in the entire research process. [36] As such, he or she is inevitably in the field of ethics and morals. Doing research "carries a moral force." [37]

Humankind has always, and in different modes, tried to express the necessary unity of human existence and human experience. As this is so, phenomenologists, when 'writing their world,' have been very conscious about language as a means of mediating that which had to be described in order to be grasped.[38] Baron points to the core function of language:

> Phenomenologists are acutely aware that words, especially everyday words, carry with them whole universes of philosophic presuppositions. Thus, though it sometimes involves great violence to language, phenomenologists try to be careful about the way that they use common words. So, for example, *body#* in phenomenological terms becomes *embodiment#*.[39] This serves to de-emphasize the physical body with its assumed subject-object split and instead to create an understanding of our bodies as they are given to us: agents of our consciousness that are capable of action on the plane of our experience that we have come to call the 'physical' world. It gives us a way to think about illness that allows us to take suffering as seriously as we take anatomy. *It permits us to practice medicine in a way that remains faithful to the human needs that create medicine.** (#author's italics) (*italics mine) (Baron 1985)

"Remaining faithful to the human needs that create medicine", was the paramount ethical imperative guiding the present study. Its predominant aim was, in the words of Baron, "to reconcile scientific understanding with human understanding". (Baron 1985) My main purpose was to understand how human beings integrate hidden and forbidden experiences into their lives and into their lived bodies. Therefore, I sought to encounter the 'stuff' of which shame, humiliation, guilt and pain are made, to paraphrase Shakespeare. I searched for insight into the structuring elements of integration in cases of sexual assault experience. I intended to ask people about crimes or social taboo violations, threats, terror, anxiety and violence. I would encounter the world of social stigma, the hidden, ignored, denied and silenced, and I prepared to ask for, and try to understand, aspects of human life hitherto unmentioned by the other person and unknown to me. How should the setting be designed, which kind of "information" would suffice, and how should the procedure be described? What kind of "data" were desirable in an investigation wherein nothing would be given (lat. dare, datus), but everything would depend on conceptualization and interpretation?

How could I practice openness in a systematic way, and especially the kind of openness required for the topic? According to Gadamer,[40] the essence of the question is the opening up, and keeping open, of possibilities. A dialogue-based, unstructured interview seemed to provide an appropriate frame for the issues of interest. Interviewing is, according to Mishler, "an occasion of two persons speaking to each other, a form of discourse, a joint product, a representation of talk, and a revelation of our theoretical assumptions and presuppositions about relations between discourse and meaning." (Mishler 1986:vii) [41] His reflections about the "essence of interviewing" is part of a debate about interview research. He argues against interview concepts which suppress dialogues and which are based on the suggestion that similar stimuli (standardized questions) evoke comparable responses (calculable answers). He addresses the, "disregard of respondent's social and personal context of meaning," based on a presupposition that contexts are technical problems, which renders them unacknowledged as "essential components of meaning-expressing and meaning-understanding processes." (Mishler 1986:viii)

The methodological limitations of standardized questionnaire interviews has recently become a topic of concern in the field of public health. The topic is addressed in an editorial in the European Journal of Public Health (1995;5:71), as follows:

> I think that the application of qualitative methods or approaches, such as grounded theory and phenomenology, may contribute a lot to the further development of concepts and understanding. *The huge mass of survey and questionnaire type of research on for example the links between knowledge - attitude - behavior regarding alcohol drinking have not helped much to reduce hazardous drinking.* [42] (italics mine)

The true origins of certain problems implicit in the medical knowledge accumulated via traditionally accepted methodology, however, are not clearly localized. The predominant mode of response to adverse outcome of medically indicated interventions or actions is a quest for yet more research of the same kind, based on the trust that the way as such is good but that eventual failures are the result of inadequate methods the details of which must be even more refined and sophisticated. This is addressed in an editorial in Family Practice 1995:

> The way we ask questions, and the nature of the questions themselves, define the answers we get and delimit both our research agendas and our practice. Focus on the biological sciences and the study of causation without attention to the evaluative sciences leaves a gap between knowledge and clinical practice. [43]

The author stresses that to include, for example, phenomenology in the repertoire of medical research is desirable since one might otherwise encounter, "the danger *of testing an unacceptable intervention among the wrong people against reliable but inappropriate measures and supplying information that is not taken up in practice.*" (Kinmoth 1995)

In a standardized interview, questions and answers are stripped of context abstractions for both the asking researcher and the answering individual; the result may be that they lack all relevance to the universe of meaning to which the study was intended to give access. The appearance of exactness may veil an insufficiency of knowledge derived from correctly performed research. Even researchers who explicitly promote other than mathematically based approaches in human health research have left this special problem untouched, as becomes obvious in an editorial in BMJ on this topic:

> Traditional quantitative methods such as randomized controlled trials are the appropriate means of testing the effect of an intervention or treatment, but a qualitative exploration of beliefs and understanding is likely to be needed to find out why the results of research are often not implemented in clinical practice. [44]

The author does not address any eventual inappropriateness of accepted research methods. He is convinced that "discovery" is possible in one way only. No consideration is taken of the theoretical possibility that the mode of "discovering" and the tools applied might possibly, in themselves, be the main source for the successive failures of implementation. According to the principles of logic, the consequences drawn from neutrally and validly discovered facts ought to result in what "they should." If they do not, this "failure," apparently not logical and thus unexpected in a realm of linear causality, will be perceived of as a new entity, independent of that which came prior to it.

Making the "failure" the object of exploration leaves its basis untouched and its actual structure, informed by what were supposed to be facts, undebated. Consequently, the focus will not be directed by the question *what is wrong?* but rather *what went wrong?* The inadequacy of the latter question will also remain invisible. Therefore, the "failure of implementation" has become the failure of those doing it or those responding incorrectly. Applying a different mode of approach for analysis of *what went wrong and why, despite correct knowledge,* is only hopefully, but not necessarily, a solution. No method can fill a so-called gap as long as there is a clear absence of awareness of how the gap is constructed and what constitutes its characteristics.

The first to criticize the accepted technique of formal interviews in social science research, working in marketing and in researching consumer behavior, was Lazarsfeld in 1935. His central point was the gap between frames for the interchange of questions and answers in everyday life on the one hand, and in research settings on the other, a concern shared by Mishler:

> I am suggesting that the varied and complex procedures that constitute the core methodology of interview research are directed primarily to *the task of making sense of what respondents say when everyday sources of mutual understanding have been eliminated by the research situation itself.* (Mishler 1986:3)

Mishler problematizes the description of instruments for data collection, their validation prior to the study expressed in mathematical terms for reliability, the criteria for selection of participants, the statistical tools, the calculation of the necessary numbers of included subjects according to these tools, and the step-wise analysis of variables to assure their impact purity. He conceptualizes that these measures, part of every scientific study on human behavior and meant to demonstrate the *presence* of all criteria granting validity, serve to compensate the *absence of context.* According to him, all efforts of instrumental refinement and statistical sophistication address an implicit or silenced concern about a basic uncertainty regarding the meaning of questions and the meaning of answers.

Meaning is contextually grounded. To remove the context methodologically matters. The coding of isolated responses to isolated answers is an "as if" exercise, allowing answers to be treated as if they were independent of the contexts producing them. To remove context and compensate with "objectifying" tools may veil multiple sources for error. Any interpretation of coded answers to preformulated questions brings with it the necessity for reintroducing context into what has been decontextualized in the process of data collection. According to procedural logic, however, what is reintroduced is not, in fact, the context which has been "stripped away," from which different answers than those given might have emerged. The answers might represent a diversity that is made inaccessible in type and degree by the applied methods. The answers will be attributed meanings during the process of coding and during interpretation of the results of the analysis. Both steps will be intensely related to the competence of the researcher, the former step being dependent on and informed by language competence, and the latter by professional and social competence.

Consequently, it becomes more likely that the tacit presuppositions embedded in the researcher's professional culture will influence the outcome of the study to a degree which is at best undefined, and at worst exceeds that of the information arising from those whom the researchers purportedly were eager to hear and to study.

Interviews are speech events, informed by the realm of language and the rules of social interaction and human relations. The setting of an interview is more than a mutual exchange of clusters of words in which there alternate what is called a question and what is called an answer. What is going on is not verbal exchange alone, but interaction which is structured by, "cultural patterns for situationally relevant talk," as Mishler underlines:

> ...these rules guide how individuals enter into situations, define and frame their sense of what is appropriate or inappropriate to say, and provide the basis for their understanding of what is said. (Mishler 1986:11)

The "rules which guide, define and frame," have been analyzed by Harold Garfinkel and exemplified, among others, in a study on coder's rules in the construction of clinical records.[45] The setting is formalized in terms of a distinction of roles, one person doing the asking, and one person giving answers, creating an asymmetry between the persons involved, and an imbalance of power.

The elements of distinction, asymmetry and imbalance are most likely counterproductive to the explicit aims of interviews which are: the opinions, wishes, demands or feelings of the interviewees. The content of social interactions according to their defined frame, the consequences of underlying organizational principles, and the meaning developed within such frames is explored in depth and analyzed by Erving Goffman. He developed the concept of key, and of keying as an analytical tool for the study of interaction within social frameworks. He writes:

> Social frameworks provide background understanding for events that incorporate the will, aim, and controlling effort of an intelligence, a life agency, the chief one being the human being. Such an agency is anything but implacable; it can be coaxed, flattered, affronted, and threatened. What it does can be described as *'guided doings.' These doings subject the doer to 'standards,'* to social appraisal of his action based on his honesty, efficiency, economy, safety, elegance, tactfulness, good taste, and so forth...[46] (italics mine)

The "guided doing" in medical research according to the rules of scientific tradition, is known to the researchers. The frame of a research setting in medicine, however, often includes persons neither socialized into the tradition nor explicitly informed about the rules and their purpose. To become involved in a "doing" which is guided by "implicit" rules, present in the organization of the setting, creates a fundamental imbalance between the participants in this very particular social activity. The one who knows gains primacy in all aspects of the interaction. The position of the other must, by the logic of the setting, be subordinated in all the aspects of the activity. Such a research setting is quite similar to other frameworks where power is displayed and obedience demanded.

This problem has been addressed in the theoretical and analytical work of both Garfinkel, Goffman and Freire. According to Freire "this way of doing research takes away from respondents their right to 'name' their world."[47] Garfinkel argues that interview research, by excluding the biographical rooting and contextual grounding of respondents' personal and social webs of meaning, bears a resemblance to a "degradation ceremony," [48] and Goffman calls it "an identity stripping process. "[49]

The distribution of power in the research interview, reflecting the doubly ranked roles of those involved, is of importance. With her point of departure in the asymmetry of power, sociologist Ann Oakley has contributed to a radical critique of its impact on the production of knowledge about human social conditions and health, stating:

> ... the paradigm of the 'proper' interview appeals to such values as objectivity, detachment, hierarchy and 'science' as an important cultural activity which takes priority over people's more individualized concerns. Thus the errors of poor interviewing comprise subjectivity, involvement, the 'fiction' of equality and an undue concern with the ways in which people are not statistically comparable. (Oakley 1984:38)

Oakley points out how the core characteristics of the accepted sociological research interview, the "virtues" themselves, have an unacceptable impact on the interview situation. She outlines how the act of research becomes an act of pretense, as the "right" way is given more status than the "adequate" way. She is explicit about mutuality, reciprocity and attachment being key, challenging "... the interviewer to invest his or her own personal identity in the relationship" (Oakley 1981:41). According to her experience and conviction, there is probably "*no* intimacy without reciprocity." (Oakley 1981:49) Consequently, her conclusion is that personal involvement is not only not a dangerous bias - it is "the condition under which people come to know each other and to admit others into their lives" (Oakley 1981:58). In her understanding, involvement does not represent a source of error but a source of experience which is not otherwise available.

Due to the aforementioned methodological choices, individual's narratives about their health and illnesses, and their experiences from encounters with representatives of the medical professions are the material of the present study. Thus, two groups of phenomena - and their intersections - are of particular interest. These are illness, health and medicine on the one hand, and individuals, society and culture on the other. The frame of the encounters between individuals and institutions is that of medicine as it is practiced in the Western countries. Consequently, this framework has to be made explicit and be explored as regards its fundamentals and its impact on those who practice it and those who seek counsel.

Sociologist Deborah R. Gordon, in her reflections upon the "tenacious assumptions in Western medicine" writes:

> While biomedicine has successfully created and hoarded a body of technical knowledge to call its own, its knowledge and practices draw upon a background of tacit understandings that extend far beyond medical boundaries. The biological reductionism by which modern medicine frequently is characterized is more theoretical than actual; in its effects, biomedicine speaks beyond its explicit reductionist reference through the implicit ways it teaches us to interpret ourselves, our world, and the relationships between humans, nature, self, and society. (Gordon 1988:19)

Medicine is embedded in a web of human-made assumptions about how the world is ordered, how things may be known, and how basic values inform human order. As such, the field of knowledge and the practices within the field are, as Gordon states, "eminently and irreducibly social and cultural." The growing body of knowledge from fields other than medicine has, during the latest decades, made this embeddedness visible, despite the fact that medicine itself maintains its assumption that both the accumulation and the application of knowledge occur uninformed and uninfluenced by value-based choices.

The encounter between sick people and medical professionals is an encounter within a naturalist cosmology. Within such a cosmology, a number of distinctions are implicit, rendering what is conceived of as *natural* something else than and distinguished from the *supernatural, conscious, cultural, moral, social, and psychological.* Consequently, biomedical practitioners approach sickness predominantly as a natural phenomenon, they legitimize and develop their knowledge using a naturalist methodology, and they see themselves as practicing on nature's human representative - the human body.

A naturalist cosmology distinguishes natural from supernatural, opposing that which is matter to that which is mind. Reality is expressed through materiality which excludes spirit. As reality almost equals physicality, the physical body is distinct from the non-physical mind. *The logical consequence is a medical practice wherein the definition of "real" illness corresponds to degrees of physical traces, which leads to distinguishing between health and illness by means of physical materialization.*

From this distinction between matter and mind results the separation of nature itself as objective, and any experience of nature as subjective. This distinction makes of nature a given, endowed by its own meaning, which is distinct from the meaning it holds for the society or for individuals. Consequently, a naturalist cosmology makes a distinction between the things, the primary objects, and the subjects who encounter them.

In the practice of medicine, this distinction is expressed in the distinction of signs, perceived of as objective indications in the patient's body and, as such, as unambiguous, and of symptoms as the patient's subjective complaints, and, as such, as ambiguous.

By distinguishing object from subject, and since nature is neither emotional nor psychic, emotions become a dubious way of understanding nature. Consequently, although emotions are considered to be a possible origin of human sickness, the emotionally initiated sickness is now distinguished from disease as a psychosomatic or functional illness. Nature as physicality is approached with questions about "how it works," and not "what it means" for humans,[50] since human meaning is subjective, residing in the mental sphere and in individuals as some kind of private affair. *Thus, in medical research and research-based medical practice, objectivity and meaning belong to separate worlds.*

The consequences of this theoretically grounded separation of disease and illness-as-lived, has been described as follows by S. Kay Toombs:

> The phenomenological description of the manner in which illness is apprehended in the "natural attitude" has disclosed that illness means much more to the patient than simply a collection of physical signs and symptoms which define a particular disease state. Illness is fundamentally experienced as a global sense of disorder - a disorder which includes the disruption of the lived body (with the concurrent disturbance of self and world) and the changed relation between body and self (manifested through objectification and alienation from one's body).
>
> Further reflection upon this global sense of disorder which *IS* the lived experience of illness in its qualitative immediacy, discloses that the lived experience exhibits a typical way of being - a way of being which incorporates such characteristics as a loss of wholeness, a loss of certainty, a loss of control, a loss of freedom to act, and a loss of the familiar world. Such characteristics are intrinsic elements of the illness experience, regardless of its manifestations in terms of particular disease state. (Toombs 1992:90)

The next distinction is the assumption that being is manifest in objects, and not in relationships or in processes, which means that a thing's identity is given with or in the thing, and not through that of which it is a part nor into which it is embedded. This presumed independence from a larger context, the atomistic view, has found its expression in the scientific notion of the variable, the entity, the relatively separate item.[51] The atomistic view has three major consequences: First, when relationships are ignored, the whole may be perceived of as determined by the sum of its parts instead of the parts being determined by the aspects of the whole. Second, as self-determined, parts may be removed from their context without altering identity. Third, as they may be removed without changing, their relations are external, and not internal, which means they affect each other as causal and not as mutually constitutive or creative. *In medicine, this means that disease, as an entity, is not regarded as constituted by a particular context, but as an effect of a distinct cause. This creates certain paradoxes and inconsistencies in the classification system,*[52] as what is taken into consideration, although with differing degrees of impact, is physicality. Emotionality and mentality are rendered incomprehensible, although they are regarded as possible causes mediated by as yet unknown paths.[53]

Another separation of influence is that of nature from culture. Diseases, biological facts, represent a universal dimension to which culture is external. Consequently, and keeping in mind the previously explored fundamentals, *medical taxonomy is conceived of as purely descriptive of natural conditions, and not prescriptive according to cultural assumptions.*

This view has been challenged by medical anthropologists who have described how both sickness experience and sickness presentation are deeply embedded in cultural concepts.[54] There exists a huge body of knowledge concerning the influence of the different ontologies on body perception and health concepts. Allan Young describes how medical anthropology approaches this diversity by conceptualizing disease, illness and sickness as designating different positions or perspectives on bad health.[55] He refers to Arthur Kleinman who coined the term explanatory model (EM),[56] which I shall use in the next section, when reading Hanna's migraine headaches within the extended body of the family. Young extends Kleinman's concept of EM to include the fact that cultures provide people with ways of thinking which are simultaneously models of and models for reality.[57]

The influence of culture on sickness experience and presentation has been further explored within so called "culture-bound syndromes." This notion has been challenged recently by Robert A. Hahn. He claims that the term implicitly renders all other sicknesses as non-culture-bound, which has to be viewed as inconsistent with regard to the accumulated awareness of the taxonomical diversities in different medical traditions and concepts.[58] By attributing cultural causality only to "exotic" human malfunctions, those constituted health problems apparently without cultural causality, such as hypertension, lower back pain, cancer, coronary heart disease, anorexia and obesity, to name just a few, remain unexplored as regards the way in which they are culturally constituted or conditioned. In addition, the meaning of medical rituals and the symbolic function of medical tools is likely to remain unproblematized within Western medical practice as these kinds of phenomena belong, presumably, only to "culturally" determined medicine.

Yet another separation within a naturalist cosmology has a major impact on the theory and practice of biomedicine: the distinction between nature and morality. This distinction, and the consequent disentanglement of sickness from guilt, sin and amorality, was once perceived as one of the major liberations scientific medicine had brought about. However, increasingly, current knowledge about the interrelatedness of certain diseases and certain conduct has reintroduced morality into the scientific sphere. The concept has changed. Sin is no longer the paramount explanation. The contemporary notions depicting the individual as the origin of and responsible for disease are "lifestyle" and "risk." This shift from an explanation of diseases as caused by biological conditions has brought with it new paradoxes and inconsistencies. *Now, medicine views cultural influences as causes of diseases, either due to unrestricted abundance and consumption, mediated by inappropriate conduct, or as expressed in individual neglect of danger and risk, creating the new personality type, that of the "risky self."*[59] Morality is reintroduced, and again individuals are responsible for their bad health, provided their disease belongs to the group of those conditions which *are defined as preventable on the individual level and which are calculated as individual risk factors in an additive interaction.*

Closely interwoven with this is the impact of the naturalist separation of nature and society. In naturalist cosmology, natural order is distinct from social order. Thus, accumulated knowledge about "laws of nature" is perceived as autonomous, beyond the social sphere, and as such uninfluenced by society. Michel Foucault, in his analyses of social institutions of power, has shown how the separation of "natural" and "social" has become an instrument of control, lending legitimacy to institutionalized suppression and marginalization. [60] Groups of people or individuals may thus legally be disadvantaged with reference to "natural" difference, which, in turn, is presented as the cause for "social" difference. By defining the natural as prior and uninfluenced, the "social" becomes secondary and consequential. Thus, social control can be veiled by an appearance of representing the natural order. *In medicine, this has led to the view that diseases follow rules independent of social conditions. Consequently, the origins of the renowned excessive morbidity of certain groups can be attributed to group behavior or lifestyle.*

These premises hold three important consequences in medicine:

1. as a field of knowledge about natural states, medicine *can perceive itself as uninfluenced* by societal values in its knowledge production;

2. as professionals helping diseased people, physicians *can have confidence in themselves as being uninfluenced* by societal priorities in their practice;

3. as a social institution trusted to secure health for all members of society, medicine *can remain unaware* of its institutionalized social power.

The final important separation implicit in naturalist cosmology is that of the truth of nature as unrelated to time and space. The universal truth of nature is conceived of as being independent, "abstractable" from a particular time or place.[61] *In medicine, this manifests itself in a preoccupation with abstractions from observations - entity, cause, disease, diagnosis, averages - leading to abstractions of these into concepts, classes and models. By conceptualizing health and sickness in abstracts, the human body can be analyzed and treated as if independent of life context, instead of as embodiments of life-time experience.*

In naturalist epistemology, the process of acquiring knowledge about "how things really are" is a process of and with the mind. Things cannot be known in immediate bodily perception, because bodily experience is subjective and, as such, untrustworthy. True or valid knowledge is only secured through a detached observer and a decontextualized natural object. Research is based on the analysis of observations in laboratory experiments or randomized trials. The analysis is, in one way or another, a process of transforming invisibility, the truth behind or in the objects, into visibility. Michel Foucault has explored the central role of vision and of the observing gaze in the history of modern medicine.[62] Vision as the most detached of the senses, allowing physical distance and still presence, allows the properly distanced position to objects which, by being observed according to certain rules, will reveal their inner secrets, their truth.

The emphasis in medicine on vision as the optimal or superior of the senses for the acquisition of knowledge about medical objects is reflected in medical technology; its tools represent expanded eyes enhancing visibility. The trust in vision's revelatory capacity has resulted in the disregarding of other sensory information, within knowledge production, in medical education, and in medical practice. An almost unlimited confidence in vision is contrasted by a lack of confidence in that which is only perceptible through confluent sensory impressions. Expanded medical eyes are directed toward physicality made visible. The belief in visibility as a verifier renders dubious or unreal whatever cannot be made visible. *As only visibility equals reality, any validation of bodily knowledge in medicine is dependent on making the body's "matter" visible. If no matter has been made visible, the bodily experience is made invalid. This is the case for all kinds of subjective bodily sensations. They are not real unless they have been seen visually. Consequently, an array of phenomena, such as pain, have become a problem for medicine.*

Pain is an individual perception, provoked or nourished by various conditions, some apprehensible, some not. Even where there is a "cause" in terms of visible matter, the intensity of pain cannot be read from that matter, nor can the impact of a pain on the individual's life and functioning be deduced from it. Where there is pain without visible matter, however, a double problem emerges. As all bodily knowledge and all subjective phenomena, pain is untrustworthy, and furthermore, its impact cannot be calculated. Consequently, a person experiencing an incapacitating pain of an indeterminate nature represents a quintessence of naturalist epistemology in medicine, creating professional tension, activating conflicts, and provoking a particular exercise of social power. There are two options for a medical "solution" of this conflict. The first is an intensified effort to make the matter of the pain visible through extensive diagnostic interventions. The second is an intensified effort to make the patient invisible by marginalizing or through the denying of gratification. *However, neither medical "solution" represents a solution for the person in pain. In fact, either of them may both intensify the pain and increase the person's disability, thus contributing to the pain's chronicity and increasing the patient's dependency on others.* The constructed conflict, created by naturalist epistemology, will not be recognized as being constructed unless medicine acknowledges the limitations of this particular theoretical frame.

Laurence J. Kirmayer has explored the role of psychosomatic medicine as a unifying corrective to dualistic biomedical view.[63] Likewise, a biopsychosocial model has been developed by Engel in an attempt to grasp what "falls out of" the medical grid and the classification system of diseases.[64] However, there has not been a high degree of awareness about the incongruity existing between the worlds of thought whose models can more easily be formulated than enacted. The combination of two or three words, either forming a new compound word or linked by a hyphen, creates the illusion of ease of composition and overlooks the fact that this word alone is not the same thing as the actual epistemological change which the word suggests has been enacted.

Considering these theoretical inconsistencies, any "epidemiology" of psychosomatic diseases must, by necessity, be problematic. Improperly defined illness is not easy to register, count and calculate. Furthermore, there is an apparent gender difference in the amount and distribution of psychosomatic illness. Likewise, there seems a gendered split of psychological causation to be at work. This cannot be understood unless one considers the impact of social roles on people, and on physicians' judgments based on implicit social norms. The difference between what men and women are supposed to represent in terms of traits and conduct can be conceptualized as a formative force in the development of "psychosomatic illness," or rather the "undefined" and "unspecified" health problems. This socio-culturally constructed difference itself may contribute considerably to an apparent asymmetrical distribution of psychosomatic diagnoses (Kirmayer 1988:71). However, as a consequence of the previously discussed epistemological premises in medical theory and practice, it is literally unthinkable within biomedicine that such asymmetries in the distribution of sickness are constructs.

The fact that physicians possibly or even obviously judge morally rather than medically, could, of course, be related to the "nature" of illnesses falling through the medical grid. What is non-medical and remains unexplained and unnamed within the realm of confirmed knowledge may possibly require judgment different from, and differently grounded than, the majority of other conditions. This argument, however, is not part of the current medical discourse. If an inclusion of other than currently valid aspects were agreed upon, psychosomatic medicine might have an inspiring potential or a corrective influence on the medical view. According to Kirmayer, however, "psychosomatic medicine has had little noticeable impact on the organization of biomedicine." (Kirmayer 1988:71) This is most probably due to the fact that the practitioners of psychosomatic medicine seem to maintain the naturalist position, instead of introducing a different epistemology.

The French physician and philosopher Georges Canguilhem argues that the distinction between the normal and the pathological is normative; the criteria for what is to be considered normal are not scientifically determined but rather morally.[65] A prevailing norm is always a consequence of consensus within a system built upon values, and distinctions between the acceptable and the unacceptable, the normal and the abnormal, the healthy and the sick, are determined in accordance with these values. Looking back in history, this phenomenon is easy to spot; however, if we hope to see how it operates in our own time, we must search consciously for its contemporary manifestations. Those having the right to define the norms also determine the arguments establishing what shall constitute normality. Thus, indirectly, they define the deviant, sick, or insane. As regards the present study, the issue of normative power is of interest in many respects, and certainly with regard to the interrelated notions of gender, sexual integrity, and boundary violation. This has been explored extensively in the feminist critique of the theory, philosophy and practice of both the human and the natural sciences.

Our western culture has a heritage of several thousand years of philosophy and thought relating to the symbolic connections between maleness and form, designating that which has boundaries and is fixed, precise and determinate. The female, on the other hand, is associated with formlessness, that which is without boundaries, is non-specific, imprecise, and vague. This has resulted in maleness becoming the norm for the human being in a tradition of thought which values the fixed above the flexible, the constant above that which is in flux, the linear above the cyclical. Geneviève Lloyd, among others, has examined these relations.[66] The male norm renders the female to be the "other," as outlined by Simone de Beauvoir.[67] Maleness as the norm makes that which thereby is conceptualized as femaleness the deviant. Representing the deviant almost 'by nature,' femaleness is also thereby implicitly conceptualized as a 'natural' origin of abnormality. This provides the epistemological ground for "female" behavior and traits, and it allows seeing the female body and its functions as carrying the greater 'natural' potential for abnormality.

This basic assumption has affected the production of knowledge in the natural sciences as explored by Sandra Harding,[68] in biology and anthropology as discussed by Anne Fausto-Sterling,[69] in genetics as documented by Ruth Habbard and Elija Wald,[70] and in medicine as studied by, among others, Emily Martin,[71] Margaret Lock,[72] Paula Nicolson,[73] and Anne Walker.[74] Anthropologists Sylvia Yanagisako and Carol Delaney, among others, have contributed to an understanding of how the gendered distribution of power has been naturalized. Anthropologist Susan McKinnon has elaborated theories about the culturally constituted views upon father-daughter-incest as contrasted with mother-son-incest. [75] Fausto-Sterling and Habbard & Wald in particular have emphasized the significant influence of "scientific" attributions on the social construction of sexuality and sexual identity in the Western cultures. They also touch upon an area which, until now, has been thematized very little within gender research, namely the conceptualization of difference within the male gender. The Norwegian sociologist Annick Prieur has explored this topic in relation to the concept and practice of homosexuality and transvestism in Mexico City.[76] She has conceptualized a group of criteria which constitute social gender. According to her, gender is constructed both socially and individually; it comprises biology, signs, style, body practice, *and a relationship between power and subordination. Within male gender, ranking may result from situational subordination to symbolic or practical violence expressed in or related to penetration.*

For the purpose of the present study, sexuality as a construction is a very important topic, and aspects of how this construction is constituted and legitimized are highly relevant. Sexuality is practiced among humans in a variety of forms which are embedded into culture and which, as the most intimate of private relations and the most powerful of public interests, represents an interface loaded with meaning and full of impact. Sexuality is practiced with, on or in human bodies, which are the most private arenas of property and may be subject to the most public exposure of social control. Therefore, the "natural object human body" of biomedicine is an untenable starting point for reflections about human bodies, body boundaries, violations that can be inflicted upon them, and the meaning these assaults may attain in the interaction between the individual, the molester, the social context, time and space. The human body and the separate mind, both loaded with covert and rationalized differences in rank between and within genders and races, cannot be reunited by words or concepts. *Consequently, psychosomatic medicine is an impossible project both in an epistemological and philosophical sense. Embedded in dichotomies and guided by gender biased assumptions about body and mind, women and men, health and sickness, the attempt to bridge a gap is doomed to fail, as the gap thus devised, opposes, and ranks according to implicit values.*

I return to Merleau-Ponty and his reflections about the human body as an object/subject. He writes:

> It is particularly true that an object is an object only so far as it can be moved away from me, and ultimately disappear from my field of vision. Its *presence* is such that it entails a possible *absence*. Now the *permanence* of my own body is entirely different in kind: it is not at the extremity of some indefinite exploration; it defies exploration and is always presented to me from the same angle. Its permanence is not a permanence in the world, but a permanence on my part. (Merleau-Ponty 1989:90) (italics mine)

By introducing the phenomenon of the permanence of the perceived body for the perceiving self, Merleau-Ponty names conditions which render the human body a radically different kind of object. The body comprises presence, permanence and absence simultaneously, yet not similar to the usual meaning of absence of objects as opposed to presence. In the terms presence, permanence, and absence, three modalities of *esse*, to be, the simultaneity of a human's being represents not only a plentiful spatiality, a physical being which takes (its) place in a room, but also a plentiful temporality.

In such an understanding, space and time have qualities different from those of the naturalist universe. They have to be conceived of as situational time and situational space, embodied in the human being. As such, they are termed spatiality (lived space), and temporality (lived time). Together with these two, communality or relationality (lived human relation), and corporeality (lived body), make up the four life-world fundamentals in the phenomenological philosophy of the body.

Any human experience can be "opened up" with questions corresponding to these fundamentals. Spatiality designates lived space which represents experienced space, different from the geometrical and numerical measures and qualities of space we must be taught to act and "think with." Lived space as felt space is preverbal. We know it by birth. We relate to it by perceiving it, and ourselves in it. Thus, we can feel small and thoughtful in cathedrals and huge and uneasy in a cot; free and happy in an open landscape or exposed and lost in an open place; comfortable and secure in a little room or caught and frightened in elevators; far but too near to somebody we fear and near but too far from somebody we love. Consequently, space can be felt differently and *be* despair, comfort, anxiety, loneliness, security, excitement and pleasure.

Additionally, humans inhabit space in a simultaneity of places, being "somewhere else" while knowing that they are where they are. Thus, not the memory of a room or place but the memory of how we felt that room or place will be recalled. To put it in extremes: a room of a certain size can be felt like a prison or a shelter. What it feels like, depends on the context, not on the room. Further, a similarly felt room might later recall a feeling of imprisonment or being sheltered. Even objects, sounds or smells from the "shelter" or "prison" may from then on represent the lived space as it was felt.[77] The experience of space has its modalities. Some places feel "good" for reading, thinking and relaxing. *Facilitative* surroundings have their counterpart in *limiting* ones, which create discomfort. Children may experience other modalities of place and space than adults, and these are characteristic of the felt space as it becomes, forever, embodied. Space is culturally structured and defined. A certain distance to strangers feels comfortable, while the same distance, taken by a person we long for, feels intolerable. Spatiality as lived and felt space represents a phenomenon which professionals who focus on human bodies should know.

The notion of corporeality refers to the fact that we are always bodily in the world. Jean Paul Sartre alluded to corporeality when describing how a person, watching someone else through a keyhole, all gaze so to speak, suddenly becomes aware of another person's gaze on him. Immediately, a change of awareness occurs: he "sees himself because somebody else sees him." His totally "absent" body while watching is all of a sudden the object of a gaze. It regains its corporeality by reemerging in awareness, though not only as bent forward with an eye to a keyhole, but as being caught in an indecent activity, evoking an overall feeling of discomfort. [78] This example is relevant to medical settings. Foucault has explored how the detached medical gaze transforms what it looks at into an object.[79] To be objectified changes self-awareness. It may reactivate former experiences of having been objectified, evoking previously integrated feelings of different kind. The physician's objectifying medical gaze influences the patient's corporeality. He or she enters the patient's corporeal sphere, though not deliberately, and partakes in embodied life, though not knowing how and with which kind of impact on the patient's present emotions.

Temporality is lived and embodied time, subjective time. How it feels depends on its lived and experienced context. Embodied time means lived-felt time. Merleau-Ponty, becoming aware of his past when touching the top of a table into which he once has carved his initials, addresses a fundamental epistemological difference between time in western science and medicine, and embodied lifetime as understood in phenomenology. [80] A sensory perception opens the world of meaning present in it, evoking an awareness of presence and past simultaneously. If this present perception were the only awareness, it would not point to itself as being something from the past. If present awareness could not identify the perception as a memory, the past in the present would *be present as presence* only. [81]

If such a perception, perceived as presence only, is presented in a medical setting due to, for example, its painful character, no medical tool or procedure can identify this pain as being the trace of a perception in the past. It is only present in the words of the patient reporting them. If the patient is perceived as being trustworthy, the pain is believed though unverified. If the patient is perceived as being untrustworthy, the words about pain are interpreted as "bodily sensations" of dubious "nature" and probably belonging to the unreal and imagined in the patient's mind. Bodily past, unidentified by the perceiver as past in presence, is an irresolvable problem in the medical world due to the naturalist cosmology upon which medicine rests. When embodied life is present in a pain which is not identified as a memory, health may be impaired and function disturbed without cause, in the medical sense of the term. This does not, however, mean that there is no source or origin of pain, nor that there is no reason to feel it.

The fourth of the fundamental human existentials is relationality or communality, which designates the lived other and embraces the sphere of lived relations. In phenomenology, the Cartesian position that cogito is the only certain ground for the "I" is considered inappropriate. Being as perceiving constitutes the simultaneous ground of subjectivity and transcendence towards others. Though aware of the other(s) in another way than of ourselves, unable ever to perceive them as they do themselves, or as we do ourselves, we can nevertheless be confident of others as being. We know them through our interaction with them in time and space. From our awareness of others, we not only derive our selves, but also a meaning of life and a purpose for living. Consequently, we must consider what we know from experiencing this awareness: others mean something for us. We are born of an other. We are given our lives by others, they provide our living and shape our basics. To encounter others has an impact. They leave traces, they broaden or narrow our abilities, they feed our fear or nourish our dreams. Their importance for us may surpass our own importance for ourselves. Therefore, human relationality is a medical topic.

3.2 HERMENEUTICS

Classical hermeneutics deals with literary texts, but text may also mean discourse[82] and even action.[83]

Psychologist Steinar Kvale[84] points out three central aspects in phenomenological/hermeneutical research:

- open description, which means, "to describe the given as precisely and completely as possible; to describe and not to explain or analyze."
- investigation of essences, which addresses "the transition from the description of separate phenomena to a search for the common essence of the phenomena.",
- phenomenological reduction, concerning, "a suspension of judgment as to the existence or non-existence of the content of an experience." (Kvale 1983)

The process of interpretation represents a gradual unfolding of implicit meaning, a kind of, in principal, infinite spiraling activity. Kvale relates this methodology to focused interview practice by way of seven canons:

- The first canon involves the continuous *back-and-forth process between parts and the whole* which follows from the hermeneutical circle.
- A second canon is that an interpretation of meaning ends when one has reached a *'good Gestalt,'* or the inner unity in the text which is free from logical contradictions.
- A third canon is testing of the part-interpretations against the global meaning and possibly also against other texts.

- A fourth canon is the *autonomy of the text*, it should be understood on the basis of itself.
- A fifth canon is a *knowledge about the theme* of the text one interprets.
- A sixth canon is that interpretation of a text is *not presuppositionless.* The interpreter cannot 'jump outside' the tradition of understanding he lives in.
- A seventh canon states that every interpretation involves innovation and *creativity.* (Kvale 1983)

An interview text is produced within a given situation, not primarily meant for external communication, and as such context bound. It is a product of a cooperation where the one who interprets is involved in generating the "text." Negotiations of preliminary interpretations may go on while the text is taking shape, which means that the text emerges simultaneously with the beginning of its interpretation. Both interviewee and interviewer are involved in this first phase of interpretation. The researcher "brackets" her/his prejudices, and simultaneously draws upon as much pre-knowledge as possible. This requires that the researcher's pre-knowledge is consciously kept present as a tool for the type and degree of sensitive attention required in a dialogical interaction. In the report, however, it must be presented explicitly in terms of epistemological position and theoretical framework.

Role distribution as a precondition for reducing the asymmetry in interview research is a crucial topic. The view of research as reporting information, and the role of interviewees as informants and of interviewers as reporters poses restrictions on the researcher and poses limits to the interpretation of the information. In comparative anthropology this topic has been widely discussed.[85] Another pair of roles is that of actor and advocate. Mishler refers to studies where the researcher takes part in how people experience the aftermath of a catastrophe, and how the researcher's analysis provides arguments for claims in the reconstruction of future security measures. The model has also been applied in studies aiming at enabling people to articulate their concerns regarding work, education, political influence, or personal relations. The model has also been an instrument in addressing silenced topics such as violence against women, sexual harassment and gender discrimination. (Mishler, 1986:131-2)

Kvale's interest is directed toward the interviewer's person and the methodological or even instrumental function he or she assumes by virtue of being the particular person he or she is.[86] In addition, he is concerned with what is going on between the two persons involved in a common endeavor though with different intentions and with not necessarily congruent aims. He views the "reciprocal influence... [as] not primarily a source of error but a strong point of the qualitative research interview" which has to be accounted for adequately. (Kvale 1989:178)

Psychologist Amadeo Giorgi, on the other hand, has reflected upon the very fact that every researcher starts out with a specific background and a specific purpose, which means that the researcher is a specialist. This fact, being a specialist, is what brings her or him into contact with the person(s) who becomes "the other". Giorgi makes explicit, however, that this fact is completely compatible with the fact that the two participants involved are equal as human beings. The frame of the research situation, however, sets a very particular stage,[87] which the participants recognize. According to Giorgi, asymmetry in a research interview is inescapable.

This is not contradicted but rather problematized by Oakley who refers to studies in which the interviewers make explicit how they experienced a gradual change of both frame and interaction which initially was neither intended nor foreseen.[88] The performance of an interview in accordance with the explicit intentions of the researcher and the defined aim of the study depends, certainly, on the design and mode of approach. As frames or conditions can never be perceived of only as "outer," given the decisive way in which they determine what can or cannot be done on the "inner," a distinction between the impact of the setting and the impact of the interaction is neither possible nor fruitful. There is a mutually constitutive and informative interaction between these elements.

The preceding reflections upon interviews as research situations, upon role distribution, and upon the construction of interviews as literary texts, shall now be integrated into an example from the study. The onset of interpretation occurring simultaneously with the starting of an interview shall be demonstrated by quotations from the final interview of the thirty-five interviews recorded, and from one of the early interviews. Interview excerpts are braided into Kvale's description of phases and levels. He writes:

> There exists a continuum between description and interpretation, and below six possible phases - which do not necessarily presuppose each other logically or chronologically - shall be outlined.
>
> 1. The *interviewee describes his life-world...* spontaneously, what he does, feels and thinks about the theme, without any special interpretation of the descriptions from either the interviewer or the interviewee.[89] (Kvale 1983)

The topic is chronic pain, migraine headaches, and its impact, as quoted from page 2-3 in the interview text. ('H' means Hanna, 'A' means the interviewer)

H: My friend supported me in my wanting to find out whether my migraines could disappear. So we chose two weekends when I let myself have an attack without taking medication to get in touch with all the emotions which emerge in such an attack. Very much happens in ... (hesitates a while)...

A: ... in the pain?

H: Yes, it does. I had already acknowledged that it was my uncle. That came through my physiotherapist, during psychomotoric therapy, and she was the one who asked me whether I had been exposed to incest. I didn't think so myself, but she just said: lay here for a while and rest. And then the picture of my uncle came up. Since then I've started to understand and understand and understand ... And then I didn't dare to understand more on my own, alone, any more. Now I've been accepted for treatment at B (name of a psychiatric clinic), I'll stay there for three months to dare to get in touch with it while being free from my job. It's impossible to experience how painful this is, and simultaneously be capable of working. I feel enormously relieved. It's a relief like I've never felt before. I've been in trouble during the last years because I've felt that I have to be brave and stretch myself more and be brave even longer than that ... and then it doesn't work anymore, there isn't anything more, no more teeth to clench, somehow. (pause) And, it feels right.

What we did, my friend and I, without asking anybody, we just found out ourselves, was to invite my uncle to my house one Saturday night, because he sneaks alongside my house every evening at seven, well, he used to so but he doesn't any more now - we invited him to come in. He always wants to come and see me. So we asked him to come in, and that I wanted to ask him about something. Beforehand, we'd placed a video-camera on a little platform above my living room, and then we had two hours of talking which is video-recorded, he and my friend and me talking together. *Probably the video is not as interesting as one could hope, because I showed it to my psychologist, and she was disappointed.* She decided we'd been too harsh with him and scolded him and so on. But for me it was enormously good to really give it to him and bring out all my accusations I now know I have against him. (voice strong, rising, with great emphasis) And I have no migraines any more. That means I haven't had a headache, I haven't taken a pill for nine weeks, and that hasn't happened since I was fourteen, just by having given it a name, all this. (lowering of voice to level before). Last Monday I saw my physiotherapist. She asked: 'How much pain did you feel since we met?' And then I heard myself say: 'I have no migraines anymore.' It's been a plague, almost killing me as it was. Approximately three attacks a week at its worst, and drugs since I was fourteen. It's made me turn to all imaginable places to get help, because the pain was unbearable.

From this example of Phase 1, the speech turns evoked from this statement will be used to illustrate Phase 2, formulated as follows:

2. A second phase would be, that the *interviewee discovers new relations,* sees new meanings in what s/he experiences and does.

The statement in italics was followed up immediately for two reasons. First, it pointed to a central theme of the project, a therapeutic relationship. Next, I felt uncertain about how Hanna had felt about it, as her wording made me wonder whether she felt "let down." Characterizing the video as 'disappointing,' was apparently contradicted by what followed as Hanna seemed content, relieved, almost triumphant. Therefore, the excerpt which followed represents Phase 2, a discovery of new relations.

A: *Why, in your opinion, did your psychologist say so ... how did you react, what did you feel?*

H: I acknowledged that I wouldn't get any further with her after that. I've only seen her once after that.

A: A decisive step. What was your reason?

H: For me it was so important what I'd done, and I hadn't consulted anybody, I felt totally responsible myself. She had advised me, if I was able, to talk to him. And then we had such a nightmare of a weekend, my friend and I, one of those when I didn't take drugs and couldn't sleep, and it's not easy to join a person in migraine. It's something I don't wish on anybody. I used to be torn and adverse. Whenever I was having a migraine night, first comes the flickering and I know that the pain comes, then you're in pain, really, terribly, and then I have such a creative period embracing ideas about ... and they may turn out pretty unrealistic when the attack is over, but I'm so high, I can make the most incredible plans at the end of a migraine attack. This time it turned out that I felt eager to make a film. I felt like documenting it ... because I am afraid that I'll become one of those who has too much fantasy, now ... after all those complaints I've had in my life, since not finding another way, starting to accuse an old person in his seventies of having assaulted me, he who is so humble, as everybody says. I was afraid it would come back at me. So I wanted proof. I'm so furious. So very furious. (long pause)

Now I know a little bit about where my enemy is. I've had the feeling of having enemies around all my life, without being able to grasp them. And then tell him all these things ... I believe it has resulted in my not having migraines any more.

Hanna has discovered a new connection without an interpretation from outside, catching sight of the concrete "enemy." This acknowledgment is a result of distinct steps in a logical sequence: the physiotherapist's remark, the mental images, the awakening memories, the desire for proof, the documented confession, the relief in naming, the disapproving psychotherapist, the recognition and legitimization of anger, and the disappearance of the migraine headaches. Her perception of "hostility all around" creating a rage against something unknown had been transformed into chronic pain which could only be dissolved by facing and naming its source. Hanna expresses precisely how, the very moment she had shown the courage to be in the pain, she had imagined what she had to do. She took responsibility for daring to see and daring to name - and then did it. Despite the subsequent criticism, she experienced the disappearance of her migraine headaches as a validation of the preceding acts. By taking control, she even gained confidence in her own judgment being more adequate than the psychologist's.

The two italicized phrases in the previous excerpts represent Phase 3, an "on the spot validation". In Kvale's words:

> 3. In a third phase *the interviewer during the interview condenses and interprets* the meaning of what the interviewee describes, and may "send" the interpreted meaning back, ... an "on-line interpretation" ... the result is ideally a verification or falsification of the interviewer's interpretations.

An on-line interpretation occurs in yet another excerpt (from page 7-8) linked to the topic of significant relationships. Hanna has talked about her parents' involuntary support of the abuse, and her conflict about telling them.

> **H:** I have problems finding out how to do this, because they've been so fantastic towards me, both of them. I didn't succeed in my marriage and they've built a house for me - they're so fantastically caring. Mainly after I had children they've been careful. But there is a lot of muddiness between them and me. I can't relax in their presence, and I feel like moving from their place. In a way, this is beyond repair. Now I notice - I've become so sensitive - I notice that I go to bed when I come home. I just hide in my bed, sitting there and eating. There is an insecurity out there. I recognize when leaving, for each kilometer I distance myself, it's just like a magnetic field decreasing, I feel better, happier.
>
> **A:** *Does that mean that distancing does good, that you possibly need distance, that it would help?*
>
> **H:** Yes, I'm sure. It might be pushed a bit. But as I have children, one of them is living with me, he's very attached to my parents.
>
> **A:** So you have considerations in all directions?
>
> **H:** Yes. He wouldn't move. It's his security. He hasn't experienced his parents having a good life together, I left his father when he was three years old. I feel I have to be patient until he has finished school and moves. He's a very good boy. So there isn't just misery around me, but I would love to leave right away, because the very moment I acknowledged what had happened to me, I recognized how my lack of strength is linked to that place.

A conflict between a wish to leave and an obligation to stay is addressed. The conflict leads to inadequate behavior. "The place" seems synonymous with "being paralyzed." The ambiguous meaning of the *magnetic field metaphor* evokes the next question. Does Hanna feel caught or rather attracted? The first answer is the son and his bonds. The next is Hanna's guilt. She has deprived him of his father. The "magnet" is not attracting, but absorbing her mental and physical strength, not letting her escape.

> 4. In a fourth phase *the completed and transcribed interview is interpreted by the interviewer alone,* or by another person. One may here broadly distinguish between three levels of interpretation: self-understanding, common sense and theory.
>
>> A. On the first level the interviewer attempts to condense and formulate what the interviewee her/himself understands as the meaning of what s/he describes.

My reading of the completed interview with Hanna on the level of self-understanding according to Kvale is as follows:

Hanna's life has been overshadowed by a chronic, intermittent pain causing disruptions in her work and activities, and by chronic pains elsewhere in her body, leading to many contacts with the health care system, various treatments, and continuous use of medication. She feels she has failed in her marriage, feels guilt toward her children because of the disruption, and she has been in need of help from her parents. In turn, this is obliging her to stay with them despite a strong desire to leave. The place where they live has recently been identified as the place where Hanna was sexually abused by her uncle. She has confronted him in a third person's presence. He has confirmed her memories and confessed. This event has been videotaped in order to serve as proof whenever this might be demanded. Hanna has experienced that to face the abuser and name the abuse has released a great deal of anger; what she sees in retrospect as the true source of her migraine headaches has disappeared. Encouraged by this experience she has dared to question the origin of other bodily pains and chronic sensations. Hanna takes into consideration that more has been done to her than has been confirmed thus far. This assumption causes her both fear and impatience. She wants to know and be certain of her memory and judgment. She hopes to activate the "unknown" which she thinks might be paralyzing her creative abilities.

> B. A second level of analysis implies going beyond what the interviewee her/himself experiences and means about a theme, while remaining on a broad *common sense* level of understanding, ... distinguished between an *object* and a *subject* centered approach, ... one focusing on what an interviewee states about the world, the other on what the statement says about the interviewee. (italics mine)

I shall quote a passage on "the meaning of a migraine," and describe successively an object and subject centered interpretation.

H: So many thoughts are provoked by our way of talking about this. My mother's family suffers from migraines, it's in a way - it has been expressed like this: 'Well, you're doomed to have this too, since we have it as well.' Several of them have it very badly. So it's been so accepted. But (her voice fills with laughter) that's why it was even greater to be allowed to say: 'I haven't got it any more.' It's some kind of a protest against something which has been a family tradition. It's the way they react if something happens. To have a headache is very accepted.

A: You, in a way, went outside the lines of fate, didn't you?

H: Yes. (very glad voice). That's impressive, isn't it?

Taking an *object* centered approach, what does Hanna state about the issue? She has addressed a heritage of bad health, or a certain style, form or kind of pain as apparently inherited, as several members of a family "have it." To inherit a pain expression might be to inherit a bodily place into which a painful experience might be channeled. The bodily place makes the pain legal, as a pain "all of us have." To legitimize a pain due to its place, shape or type as a "usual event in this family" or a "destiny we cannot escape" veils the true source of the pain and makes that source invisible. Adapting an expression of an unnamed pain to one's "heritage" means inscribing it into an existing pattern, which quite effectively will render its true source inaccessible. Any intention or hope the person in pain has to evoke somebody else's interest in where this pain might be coming from will be in vain. The real origin will be transformed into a part of the family narrative about "those of us with migraines." In this way, a pain pattern may seem hereditary, so that children with a need to express a pain can integrate their pain into it. Once that is done, it has found its explanation and therefore the legitimating force of the family narrative will prevent a disclosure of the origin.

Taking a *subject* centered approach, what does this statement say about Hanna? She has reflected a great deal upon her chronic pain and its sudden and surprising disappearance, opening herself to understand connections and meanings hitherto unseen and unthought. The pain had fit into a pattern of "being doomed" to have it. She did not mention whether she had ever before considered her pain to be a reaction to "something," given that she can so easily read meaning into in her relatives' pattern. Had she been able to see this as long as she was "part of the group?" Her own reaction made her too involved in the explanatory system the family had adopted. As long as she was "in migraines," she had no option but to perceive of them as belonging to the destiny of inescapable pain. At a particular moment, however, something changes. When challenged by her physiotherapist, who not only had the ability to "see" but also had the professional skills and courage to address and name what she saw, mental images emerged. When she chose to be in the pain, unmedicated for the first time in her adult life, and let it happen, and as she was courageous enough to open up, she could imagine which way she had to proceed; it was the way outside the line of fate, the step out of the tradition of pain. She left the explanatory system - and by doing so she overcame it and thereby questioned it. That, I suggest, is why she used the term "protest" against a tradition, a pattern of pain and an explanatory system. To me, her triumphant voice expressed the joy of breaking through imposed limitations of both symbolic and real meaning and force. By leaving "the lineage" she had regained self-confidence. And she took pride in this, although she still felt obliged to express herself indirectly by asking me to express agreement with her in seeing the act as impressive and courageous. I even think of her joyful, triumphant laughter as expressing an array of emotions, as holding the promise of a constructive ideation about the next steps to becoming more herself, a more whole person, undivided, healed.

C. On a third level, the interviewer may draw on more *theoretical interpretations*, as e.g. psychoanalytic theory of the individual or Marxist theory of society.

I want to use the previous excerpt once more, this time for a theoretical interpretation, by referring to Nancy Scheper-Hughes and Margaret M. Lock. Hanna's statement of migraine headaches as a family pattern of pain can be interpreted within their theory about *three presentations* of the human body. Since, in most respects, this theory stands in contrast to the biomedical paradigm, it may be appropriate to sketch, very briefly, the dualistic concept of biomedicine: To think with structures of distinction and polarization, and to conceive of all phenomena as defined and opposed, creates many paired expressions denoting mutually excluding opposites. Such pairs of words are, for example, mind/body, culture/nature, magical/rational, seen/unseen, real/unreal, sick/healthy, and so on. Their defined distinctiveness leads to a systematic ordering of things and phenomena as either the one or the other. Dualistic concepts are instruments for sorting and making order. Embedded in such dualism, all thoughts about the human body in biomedicine are structured in dichotomies.

Consequently, scholars of critical medical anthropology have directed their interest mainly to how dualism defines what may be thought about human bodies. The consequences of a concept, as expressed in the theory, structure and practice of the biomedical field, emerge through comparison with other medical concepts. Medical rituals are to be read and interpreted as rituals within a cosmology and ontology, as mirrors rather than as *the only adequate way* of dealing with sickness. Medical taxonomy may be understood as *one possible system for organizing deviance,* and sickness classifications can be read as *one of several possible lists of names for bodily conditions.* As part of a necessary and fruitful exercise in critical reflection upon the basis of biomedicine, analogies, parallels and divergences permit the questioning of that which is taken for granted. When accepting biomedicine as only one of many possible ways to pursue a socially and culturally defined activity of understanding and naming diseases, and the treating or healing of people, relevant questions offer themselves for critical investigation. Lock and Scheper-Hughes anchor their critique of the biomedical paradigm in critical medical anthropology.[90]

What can be known about the human body, health, and illness in a given society is a consequence of social construction. Culture, the whole set of expressions of a cosmology, is simultaneously the frame, the limitation, and the tool for every production of knowledge. How bodies, health and illness are seen, estimated, defined, named, understood, dealt with, accepted and ignored, depends on, and is rationalized within, cultural fundamentals. To the same extent that the knowledge about body, health and illness is culturally constructed, human bodies are conceptualized according to cultural and social premises, and structured by group and rank. Lock and Scheper-Hughes discuss these two fundamental preconditions within a theory of the *simultaneity and mutuality of three bodies* as opposed to the *individual body* in the biomedical paradigm. They write:

> At the first and perhaps most self-evident level is the *individual body, understood in the phenomenological sense of the lived experience of the body-self.*
> At the second level of analysis is the *social body, referring to the representational uses of the body as a natural symbol with which to think about nature, society and culture.*
> At the third level of analysis is the *body politic, referring to the regulation, surveillance, and control of bodies (individual and collective)* in reproduction and sexuality, in work and in leisure, in sickness and other forms of deviance and human difference. (italics mine) (Scheper-Hughes & Lock 1987;20:6-39)

In this sense, the individual experience of being alive as a lived body and as lived experience is not limited to what the individual conceives of; nobody has a mind "empty" of thoughts, images or perceptions when regarding, feeling, and experiencing oneself. This is also so for Hanna. She had learned to conceive of her body as a distinct entity, and to experience her bodily reality as more real than her mental. As such, the pain in her head had its own existence as a bodily disease named, based on its characteristics, migraine headaches. She had learned to accept the name, provided her by physicians, the experts for naming those bodily states which lie outside the spectrum of the normal and acceptable, as the consensus among medical experts defines it at any given time. Located in the head and diagnosed as a disease, this pain had overwhelmed her seemingly at random. She had been treated for years. The disease had not been cured. Its cause had not been known. Hanna herself had apparently never doubted that her disease in the body was caused by her body. She had been the location, the origin and the sufferer, and as such, alone.

At the same time, however, through belonging to a group of people called relatives, she perceived herself as part of an extended body with shared traits and characteristics. She had belonged, but to a diseased collective body. Her pain carried the meaning of a "tradition" in the lineage. In Hanna's family, migraine headaches, though a disease on the individual level, affects individuals as if their genetic heritage "doomed" them to it, in the sense of disease as a group fate. This tradition engenders the acceptance of pain as a cause of intermittent disability, meaning: "I am in pain, I am incapacitated - but it is not my fault." It brings with it a considerable amount of fatalism. The "voice" of the family says: "Yes, you are doomed to have it, as we have it as well." This tautology expresses the core of a "local culture," a family's belief in the simplest and least escapable causality. Fate means no choice, no guilt, and no escape. Additionally, the family as a "body with migraine headaches in a common head" displays migraine headaches as a response. There is a common pattern in what seems to provoke the common answer. Hanna says: it is the way they react when something happens. Although she does not specify the "something," the context of her statement allows an interpretation in the sense of "something probably or certainly adverse." It depends on what individuals or groups consider as adverse or unwanted. A family or subculture is likely to agree on what The Adverse and Unwanted are, based on collective experience, accepted values, or cultural taboos. More probably, a "family body" has as many reasons for reacting to unpleasant things with pain in the head as there are members in pain. In Hanna's family, the issue of individual origins of individual pain had been made irrelevant once and forever by the collective acceptance of its origin, which, as it were, was proven by its reemergence multiply and repeatedly. Therefore, the particular in the general had not to be explored. Hanna's pain had had its cause and explanation simultaneously attributed: something inherent in the lineage and confirmed by the repeating narrative of the "pain in our family." The explanation of heredity in the medical language and of fate in the family language seemed convergent and conclusive.

The very moment Hanna dared question the explanation's accuracy, she recognized it as inconclusive. First then could she feel its personal origin: impotent rage. When recognizing her formerly unfelt rage, and acknowledging it as real and just, Hanna finally gave it an adequate expression. She *did* it. She confronted the person who had caused it. One might ask where and what this sudden, strong and righteous rage had been before. The answer is given in the effect of its display. The rage had been in the pain, or rather, the rage had had to be transformed into pain since pain was accepted but rage was not. Hanna's rage, caused by invasion, molestation, and betrayal, could not be spoken and so was impotent. Her protest against a violation of a social taboo, which had been supported by parents through their refusal to see, could not be acted out. Therefore, it was acted in.

In the reading of Lock and Scheper-Hughes, this is the level of the body politic, the result of social rank and power silencing the violated by rules which assure acceptance and subordination. All members of a society know how authority is ranked and power represented. To be socialized into a group or a culture means to internalize this knowledge and to accept the explicit and implicit rules. Hanna knew the rules guiding her varying roles as child, daughter, girl, divorced woman and mother. Every role imposed on her different stances in regard to changing relations or circumstances deriving from differing rules. However, the impact of rules for her life phases had a common, and, it seems, crucial essence: *limitation.* Likewise, the impact of roles in her life course had a common essence: *dependency.* In accordance with the same conditions which had created these states, any open acting, any protest, had been made unacceptable. Thus her only option was adaptation to the limits, binds and invasions through covert acting. The family pattern of reacting to "something" with pain had offered an accepted expression and simultaneously an explanation. An individual pain called a migraine headache, caused by heredity, made perfect sense in the inner world of the family.

But, as embedded in a culture of dualism and an epistemology of dichotomies, it had made perfect sense also in the outer world of society. Representing an institution of societal power and patriarchal hierarchy, medicine had received a highly ambiguous role and function in the case called "Hanna's migraines." According to the concept of diseases as defects in individual bodies, physicians had diagnosed a disease with reference to scientific knowledge about origins of pain in the head, and labeled it according to a classification system of organic defects or malfunctions. However unintentionally, medicine had nonetheless contributed to the sources of the pain by imposing yet another limitation on her, the role of one diseased, as well as other kinds of dependencies, namely, on the health care system and on drugs. The science of medicine had "re-enacted" the structure which generated a seemingly individual and apparently bodily defect in a young girl. Valid medical knowledge and proper medical practice, providing a diagnosis and an explanation, had shaped what could be called a cover story, a *blinder*. Consequently, *the scientifically grounded knowledge and the correctly applied practice of medicine* took part in a silencing *social* suppression. With the help of medical intervention, what was *socially wrong* was transformed into what was *individually sick*.

Seen through the frame of the theory of three presentations of human bodies in social life, Hanna's pain can be understood as resulting from the body-logical inverted act of adapting to forces which were effective as long as she "adapted." The moment she protested by naming, a spell seemed to break. A fate was, verbally, contradicted. No wonder Hanna's voice sounded triumphant and she herself felt impressed by the courage and will she had harbored without being aware of it for the greatest part of her life. Hanna's statement about the meaning of the migraine headaches, when read in the theoretical frame offered by Lock and Scheper-Hughes, has a "Gestalt," radically challenging the theoretical fundamentals of Western medicine, as precisely expressed by Lock and Scheper-Hughes:

> Sickness is not just an isolated event nor an unfortunate brush with nature. It is a form of communication - the language of the organs - through which *nature, society, and culture speak simultaneously*. The individual body should be seen as the utmost immediate, the proximate terrain where social truths and social contradictions are played out, as well as a locus of personal and social resistance, creativity, and struggle. (italics mine) (Scheper-Hughes & Lock 1987:31)

So far I have shown an interpretation of one interview within one theoretical framework. Now, I return to Kvale's considerations and his last two phases which shall be illustrated in connection with another interview.

5. A fifth phase of interpretation would be a *reinterview*.

I have chosen an interview with a male interviewee that lasted more than three hours. He told about his rather unusual life where a wordless, secret and silenced rape of a little boy held a key to an array of later events, developments, and constellations. While I read and interpreted the text, several themes emerged. In order to elicit their possible impact, I asked the interviewee for a reinterview. We talked together for two more hours, both of us taking notes about the topics we addressed and the questions which arose. Thereby certain phenomena which had puzzled me in the beginning of the study and which had been thematized during interviews preceding the one now under discussion, configured and contributed to later interviews since I actively reintroduced the topics they were related to.

> 6. A possible sixth phase would be, that the continuum of description and interpretation is extended to also involve action. This may mean that the interviewee begins to act from new insights s/he might have gained during the interview.

On the basis of interview and reinterview I formulated ten questions. These became the matrix for a lecture given in post-graduate education of colleagues. The questions gave structure to a dialogue between the interviewee and myself in plenary session, addressing the intersections between the mindful body and trauma memory. Prior to the lecture, the interviewee had prepared his answers which were not elaborated in the setting. This dialogue was tape-recorded and transcribed by me. In this manner, the initial interview had been broadened into a learning activity or educative action. During this collaboration, there emerged an acting, in the sense of Kvale, based on insight the interviewee gained during the process. In the interview, the interviewee told about having been attacked by a man a year before. In the reinterview, he told of having met the offender recently. In our common lecture after this, one question addressed the impact of revictimization. The interviewee's answer to this topic, was later integrated into yet another lecture as follows:

> One year ago, I was attacked by a young man. I protected my head with my arms and was beaten severely. After that, my well known depression returned, and the pain in the arms incapacitated me for half a year. Recently, I ran into him in the market place in our town. He screamed obscene words at me. Then I used an old strategy from my childhood: I shaped a distance, as if putting him under a glass bowl. The week after, I suddenly had a heavy panic attack, the first for many years. I recognize the reaction. I had escaped by dissociating. I acknowledged that I couldn't continue like that. I had to go through it in order to overcome it. Otherwise bad people will always have me in their grip. But I need my autonomy and peace. I'll get it.

In my interpretation, this description holds elements of healing potential: a traumatized person is assaulted again, a depression is reactivated, a bodily pain leads to temporal disability. Dissociation is the immediate response to what feels overwhelming. A hostile confrontation reactivates a strategy which has been invented earlier. The interviewee, by experiencing, telling, and reflecting, acknowledges that this constellation may reappear unless he overcomes and abolishes his strategy of dissociation. A "doing different," as Kvale calls such a development in an interview process, seems possible.

3.3 NARRATIVES

Dialogues about the impact of life-world experiences on individuals include personal memories and reflections. To these, statements or judgments are related, shaped as narrative accounts. In the human sciences, there exists a multidisciplinary agreement that a central part of human communication is embedded in the telling of stories. This is mirrored in the universality of story-telling, and in the grammar structures constituting a linguistic matrix for stories found in all human languages. The story itself resembles a natural psychological unit in emotional life. Such stories present as internally consistent interpretations or reconstructions of presently understood past, experienced present and anticipated future. Merleau-Ponty, reflecting upon how telling relates to living, points to interacting layers of reality which make up the human existence. These reach from the objective level of the physical realm to the linguistic level of expression.

Alasdair MacIntyre formulates the significance of a narrative as follows:

> It is because we all live out narratives in our lives and because we understand our own lives in terms of the narratives we live out that the form of narratives is appropriate for understanding the action of others. Stories are lived before they are told - except in the case of fiction. (MacIntyre 1981:197) [91]

Psychologist Donald Polkinghorne, in stressing the significance of narratives, writes:

> Being human... *is an incarnated or embodied making of meaning* - that is, it is primarily an expressive form of being. Narrative is one of the forms of expressiveness through which life events are conjoined into coherent, meaningful, unified themes. (Polkinghorne 1988:126)

Embodied time has been among the central topics of phenomenologist philosophers. Heidegger conceived of the present as threefold, including a present about the future, which is expectation, a present about the past, which is memory, and a present about the present, which is attention. This threefold present makes time an integrated aspect of human experience. A concentrated attention on the presently ongoing may, however, completely dominate awareness, as may memory and expectation. This means that particular circumstances may accentuate a memory or an expectation so that it may be the attention in the sense of all that which is present.

- The preceding short reflections about discordant time will be useful as a frame for reflecting upon "forgotten" abuse experiences.

The threefold present of memory, attention and expectation may be shared collectively by groups of humans, a prerequisite for realization of common aims. This is true for all endeavors of establishing, maintaining or developing ideas or values, as objectified in institutions, systems, laws, rituals, rules or canons. A collective expectation can blind the common attention to any present sign of jeopardy. Collective mourning can make memory the overall attention, thus preserving the past as a dominating force in the present. A collectively integrated prejudice from the past, present as "tradition," "custom," "right" or "law," can blind the common attention even to obvious present inconsistencies. Collective expectation or memory may be the attention in the sense of all that which is present.

- The preceding short reflections will be useful as a frame for interpreting societal ignorance or silencing of particular boundary violations.

According to Heidegger, the threefold present is transcended at three levels in human consciousness of lived time. The *first level* is awareness of being human, present among the world of objects and in interaction with them. On this level, time is a means to structure our doing and interacting with others. However, human beings know about non-linear time coexisting with the linear time of the natural sciences,[92] representing a simultaneity of imagination and presence. The awareness of co-existing times seems to be overruled when humans have integrated experiences which have been overwhelming and disruptive.[93] Studies give reason to suggest that traumatic experiences can represent forces which may not lend themselves to being ordered in time, thus dominating the present in shorter or longer periods, confusing all relations between the individual and her or his surroundings.[94]

- The preceding short reflections will be useful as a frame for an exploration of the confused times related to assault memories from childhood.

The *second level* of awareness about lived time in Heidegger's terms is historicality, an orientation of the individual within an organization of time flow related to other beings and one's own past.[95] The *third level* in Heidegger's concept is temporality, which means awareness of personal finitude, including the past, present and future as aspects of our existence.

Ricoeur hypothesizes a relationship between the temporal character of human existence and the human narrative activity as non-accidental, but as representing a transcultural form of necessity. To tell stories is the mode of constructing meaning from the various human experiences of time. Time is reflected in actions and plans, in considerations and hesitations. As action becomes interaction, "public" time is woven into the "private" time. Narratives can be understood as the act of telling about acting and interacting, and the act of listening. Listening is part of the structuring of the telling and the structure of the told. Whatever shall be told must be presented so that it stands out from the time flow of continuous or successive "nows." Something must be configured, a person, an event or an intervention, a theme or a point. Ricoeur has introduced his notion of the plot, which he defines as follows:

> A story describes a sequence of actions and the experiences of a number of people, whether real or imaginary. ... To follow a story, then, is to understand the successive actions, thoughts, and feelings as having *a particular directedness*. ... In this sense, the 'conclusion' of the story is the attracting pole of the process. [96] (Ricoeur 1978)

From the point of coherence, the plot, every part may regain a new meaning or new dimensions. And, whenever retelling a story whose plot is known, the elements will be understood according to their condensed meaning. To interpret the past events viewed from the perspective of "their result" has a double function. Telling about events makes experiences a source of learning for the teller and the listener. Narratives communicate common destiny by transmitting past possibilities to present hearers. Thus, narrating enlarges the personal memory into a communal.

- The preceding short reflections will be useful as a frame for the evaluation of memories from abuse experience, and of the meaning these are attributed when narrated.

Some central terms, linked to narratives, and mutually constitutive, are: identity or selfhood, and memory. The question of identity, formulated as "who am I," is culturally shaped, both in its formulation and answer, as anthropology has made evident. Brian Morris, among others, has comprehensively presented different selves as they emerge in the great cultures of our world.[97] The Greek self and the concept of psyche has a ground significance for Western identity and individuality. The Buddhist self and the self of Hinduism have influenced the Eastern view of personhood, morality and duty. The concept of self in Taoism and Confucianism have informed the Chinese perception of the proper place for the one among the others. African philosophy provides the matrix for conceptions of personhood as embedded in nature, and Oceanian people build their relations on concepts of people as social beings.

The word identity has its etymological basis in *ipse* and *idem*, that which is discrete from others (as opposed to others) and that which remains the same (as opposed to change). The apparent opposition of that which designates difference and that which designates sameness, embraces a living tension and represents a dynamic balance. What is "discrete from others and yet same to oneself" constitutes the identity of an individual, undergoing changes through time but still her/himself, and as such recognizable to others and to the self. The characteristics of "discrete from others" and "same to self" have been explored with regard to their essence and their expressions. Answers have been linked to a bodily identity, always in permanent and profound transformation, or a memory identity, similarly in constant transition.[98]

With regard to identity, it seems obvious that humans respond to other humans as "mindful bodies," to use a term from Scheper-Hughes and Lock. As mindful bodies, humans act and behave in accordance with their perceptions, abilities, intentions or interpretations. According to Merleau-Ponty, the source, the reason and the ground of identity are one: I am given to myself, born as vision and knowledge, both being in and of the world, both mind and matter, all my experiences, but irreducible to them or to any one of them. This is the knowledge, strength, and consciousness which makes humans certain of their unique self.

Humans tell their lives to others, in different ways, at differing times, and for varying purposes. The self as narrative is, perhaps, the most unambiguous of selfhoods, as it is the self her/himself who creates it by telling. And it is, perhaps, the most abstract of selfhoods, as the narrative may be continued by others when the self as her/himself is defined. The process of "configuring of personal events into a unity" (Polkinghorne 1988:150) is influenced by a variety of phenomena. In cultures which are grounded in a lunar mythology, analogical phases and circularity are the main structuring elements, while solar mythologies provide a matrix for dichotomies and linearity.[99] Members of a Christian community will, in their reflections upon meaning, be guided by a patriarchal death mythology, which offers the notions of sin, guilt, condemnation, and grace as explanatory symbols. Persons of Western societies from this last century will not be able to figure out meanings of human interaction experiences without "thinking within Freud," or of primarily bodily felt experiences without "thinking within Darwin" or within an established biological model of heredity, hormonal function or cellular activity.

Persons who have related to societal institutions over time will tend to configure their thinking according to how these institutions have defined the nature of the person's experiences. A. Jamie Saris has pointed out how narratives of a chronic condition are deeply embedded within various institutional structures that influence its production as a story. This is presented as follows:

> I am defining institutions, then, as bundles of technologies, narrative styles, modes of discourse, and, as importantly, *erasures and silences.* Culturally and historically situated subjects produce and reproduce these knowledges, practices, and silences as a condition of being within the orbit of the institution. This definition of institution focuses our attention on the ability of the institution to define and constitute as well as on the *silences and erasures that provide the persuasive force for such definitions.* (Saris 1995) (italics mine)

By analysis of one single dialogue, Saris traces the plot of the story, its temporal frame, and the narrator's style. The story is about a personal experience of being marginalized, and is embedded in a multilayered "institutional topography" consisting of psychiatry, the Catholic church and the local health bureaucracy. The narrator is guided by the triad of mystery, knowledge and authority present in each institution, mirrored in describing himself in their words, and judging himself according to their norm. In Saris' analysis, the narrator's *"institutional voicelessness"* is made evident as a result of the systematic erasures by institutions of authority. By describing how the narrative takes shape in the tension between the narrator and that which is silenced by psychiatry, Saris points to two different, yet interlinked kinds of silence, with different but related consequences:

First , the *silence and erasure practiced by institutions* as a force for constituting and maintaining power. That which is not and never will be a topic of discussion, is thereby designated the institutional property or the terrain of expert knowledge. Whoever makes what is institutionally silenced or erased into a topic, displays inappropriateness, disobedience or insanity, which is transformed into a double proof: the wrongness of the speaker, and the rightness of experts. To disobey by voicing what is silenced is, in psychiatry, transformed into a proof of the patient's lack of insight into her or his own conditions, which confirms insanity. In somatic medicine, disobedience is termed non-compliance.[100]

Second, the *silence and erasure practiced by society* as a force for constituting and maintaining societal power. That which neither now nor ever will be a topic is thereby declared as nonexistent in this society. Whoever makes a topic of what is socially silenced or erased, behaves amorally, shamelessly and indecently, which translates into proof of two things: the wrongness of the speaker, and the rightness of the order of society. Disobedience through voicing the silenced is met with a variety of societal reactions in a continuum from ignoring and ridiculing to judicial conviction or compulsory medical treatment.[101] Here, obviously, institutional silence and societal silencing as tools of control meet, which in turn renders an individual voiceless in all respects.

- The preceding short reflections on the silencing power of society and institutions are useful as a frame by which to evaluate the impact of societal denial and institutional ignorance of actual sexual violation.

Whoever enters the world of narratives will have to be aware of silence and erasure as a matrix providing the tacit background, with that which is to be heard as foreground. Provided concepts through which to think and to formulate, humans can tell, reflect, revise, or oppose, contradict, dispute. However, the situation is radically different when there are neither concepts with which to think nor a topic to which to relate. This is the case in experiences belonging to the culturally silenced phenomena.

Until now, very few realms of human science, with the exception primarily of the fields of psychology and anthropology, have addressed problems linked to silenced experiences. The unending enterprise of maintaining one's identity through a continuous integration of experiences, acknowledgments, recollections and revisions, is obviously related to and dependent on it being possible to tell one's life.

If the telling of experience is a prerequisite for creating or confirming meaning, not telling must interfere with the construction of meaning. If this continual process of creating and confirming meaning is a prerequisite to the establishment or maintenance of identity or selfhood, then any obstacle to the telling must have an effect. If the telling is neither listened to nor believed, is neither shared nor confirmed, leaving the story "as-if-not-told," then selfhood must be affected adversely.

- The preceding short reflections about the possible impact of not being able or allowed to tell, or not being listened to or believed, are useful as a frame for the exploration of the impact of untold violation on selfhood and on the personal memory of the one who is violated.[102]

Human memory is conceived of as fallible, and as such reduced to an individual and psychological phenomenon. It is considered a problematic source of information in research. However, in a study of human life-world experiences, this "fallible" source is the only one possible. Experience is, by definition, something having happened in the past, even if this past is immediate. Only when experienced, only as past, can it be reflected upon or recalled. A recollection does not need to be verbal telling. It can be communicated without a word. A pale, white face, widening eyes, a holding of the breath, a turning away of the whole body, a visible trembling, a repeated swallowing, hands hiding the face, an apparently shrinking body, a rash on the skin, a laugh and a sigh, and other non-verbal communications have narrative impact.

Linguistic communications are inseparable from non-verbal; however, as soon as a dialogue is transformed from recorded speech into a written text, the two modes are separated. This represents a central difficulty for interpretation: splitting context from text renders invisible or incomprehensible why the dialogue flowed as it did. Actually, this transformation means even more than dividing two modes of communication. While verbal and non-verbal communication both have an aspect of content and of relationship, these two aspects differ in the way they are associated with the two communication forms: "verbal" is much more strongly associated with content and "non-verbal" much more strongly with relationship. Consequently, separating the two modes results in a communication from which has been removed more of the information concerning *relationship*. This is so regardless of how the transcribing is performed. [103]

- The preceding short reflections about communication are useful in the evaluation of the interviews when elaborating how the "unspeakable" is narrated.

In research into areas of a private, intimate or even stigmatizing nature, such as alcohol or drug consumption, sexual violence, or criminal activities, the fallibility of memory, its unreliability as regards precision and exactitude, has been an integral part of the discussion about both the reliability and the generalizability of findings. Likewise, the issue of recall bias has been central in research into sexual boundary violation.[104] As it is impossible to have direct access to other people's experiences, only indirect access may be had.

This fact confronts the researcher with a dilemma, forcing her or him to make a choice between two options: to *rely on observations* and to *interpret what is registered* according to an accepted theory; or, to *rely on people* and to *interpret what is communicated* within appropriate frames of theory.[105] The radically different preconditions characterizing these two approaches imply that the researcher's reliance and interpretation must also, and necessarily, be anchored and grounded differently.

According to Ricoeur, research into human experience has to consider some fundamentals, namely: that, "we make our history and are historical beings," and that, "the form of life in which the *speaking of narrative is a part, is our historical condition itself.*" (Ricoeur 1978) Stories about what happened and how it was experienced, what it led to, how it influenced certain successive phases of life, how it informed particular views of later events, how it became an integral part of a special pattern of reactions or a cluster of feelings - all this may be accessible to researchers. However, the main precondition for access is that introspection and retrospection are given legitimacy in a research setting.

Asking people to tell about their life is asking for *responses, and not replies,* to use a distinction underlined by Goffman. These responses may involve the interviewer in characteristic ways, demonstrated in his or her connectedness. Merleau-Ponty has introduced a term, unusual but rich in connotations, to grasp the qualities of listening. Instead of *Empfänger* (receptor/receiver) he coins the word *Empfinder* (perceiver). (Merleau-Ponty 1989:73) This term is a linguistic equivalent of what is going on between two people during a discourse about life-world experiences, and it makes comprehensible why, indeed, it is impossible for the researcher not to become involved.

We know both from life and from social research that to ask people to tell but not to respond while they do so, stops the flow of speech. Lack of any response carries a message. Whether this message might be unintended, or is definitely unintended, it relates nonetheless to the basic conditions for communication among humans beings, as thematized by Watzlawick and his co-authors. They write:

> ...if it is accepted that all behavior in an interactional situation has message value, i.e. is communication, it follows that no matter how one may try, one cannot *not* communicate. (Watzlawick et al. 1967:48-9)

The non-verbal, corporally expressed and perceived communication, however ambiguous, is interpreted according to rules which are to be found in the implicit social knowledge of members of groups or cultures. In fact, when research of human experience neglects these fundamental conditions of communication, the research itself may estrange the study "object" and make the study situation an alienating one. This very fact of estrangement by lack of response, as it has been observed by sociolinguists in a variety of human societies, seems to reflect a genuine necessity for an "other" to whom the telling is directed. The German word "mitteilen" and the Norwegian word "meddele," meaning "to share with," express this necessity. Telling, mitteilen or meddele, means opening for the "other" and for signs of connectedness, dissent, wondering, doubt, mistrust or agreement, which influence the flow, wording, focus, intensity, and consistency of how the teller tells.

Certainly, not only the interviewer must respond. The interviewee, too, must have the freedom to do so by modulating the interviewer's questions, by not answering them, or by answering to what as yet has not been asked but is of relevance for how the interviewee wants to confirm or correct an interviewer assumption implicit in the phrasing of a question. Consequently, narratives are of necessity joint products.

- The preceding short reflections about the structuring elements of narratives, are useful as a frame for the evaluation of the interaction between interviewee and interviewer.

All narratives are to be approached analytically via the linguistic triad of: the syntax, asking for the structure of what is said; the semantics, asking for the meaning of what is said in this language; and, the pragmatics, asking for the interactional context which defines what is meant here and now. This analytic triad, established by Charles Morris and Rudolf Carnap in 1938, represents a tool. It's validity is grounded in the fact that all shared information presupposes a semantic convention. Regarding the area of pragmatics, a primary methodological concern in the present project, the interactional context deserves special attention. The role and influence of the interviewer is among the most central aspects focused upon, whenever validity and reliability of research interviews have been discussed. Since "the interviewer's presence and form of involvement is integral to a respondent's account," (Mishler 1986:82) the solution cannot be to make her or him "invisible or inaudible."

Kvale addresses another topic of criticism linked to the influence of the interviewer, namely, "the question of the leading question." The phenomenon is of concern and of methodological interest, mainly in research which seeks to avoid any interaction and to promote distance and non-involvement. The primary issue regarding the "question of the leading question" is that it indicates a presupposition. Those who ask this question are seen as introducing a measure from one area of research and one methodology into another. As explored earlier, any evaluation of a research enterprise with measures from a different research tradition is seen to include an epistemological error, and, as such, to be unscientific. At the same time, such an approach reveals the presupposition of a norm wherein validity is expressed in one, and only one methodology, regardless of differences in the theoretical framework within which the object of evaluation is placed. This means, that the very question of the leading question is *not evaluative,* but rather normative.[106]

Anthropologist Renato Rosaldo has addressed the topic of the involved researcher in a reflection about the observer role in ethnography, its method being participant observation. He writes:

> In my view, social analysts can rarely, if ever, become detached observers. There is no Archimedian point from which to remove oneself from the mutual conditioning of social relations and human knowledge. Cultures and their 'positioned subjects' are laced with power, and power in turn is shaped by cultural forms. Like form and feeling, culture and power are inextricably intertwined. (Rosaldo 1989:169)

Although this argument is part of an ongoing discourse about validation of knowledge in the field of ethnography, its core is the central problem in all kinds of social research endeavors. Humans' situatedness in a web of cultural preconditions and personal presuppositions is simultaneously an inevitable fact and the topic of the research question at hand. (Geertz 1973) The researcher, in his or her efforts to be as distanced as possible in order to remove observer bias, can never overcome the fundamentals of being an emotionally, cognitively and morally informed subject and not a blank slate with a kind of zero position towards others or other. The myth of the detached observer as "innocent" with regard to his or her view of the observed has, both in the social and in the natural sciences, gradually been overruled, although from different points of departure. Consequently, neither in the natural nor in the social sciences, is the researcher any longer granted the status of one who is uninvolved.

Giorgi has argued in favor of redefining the central terms of research evaluation, that of validity and reliability, in phenomenological studies. Phenomenological validity is different from experimental, and so is phenomenological reliability. In his words, in phenomenological studies of human life-world experience, these terms are connected in the following way:

> If the essential description truly captures the intuited essence, one has validity in a phenomenological sense. This means that one adequately describes the general essence that is given to the consciousness of the researcher. If one can use this essential description consistently, one has reliability. (Giorgi 1988)

Psychologist Marcia Salner argues that, "the term 'validity' is bounded by the context of empirical epistemology," and that, "from the perspective of human science epistemology, it makes more sense to talk in terms of *defensible knowledge claims* than of validity per se." (Salner 1986:127-8) Kvale emphasizes that to validate is to question, to investigate, as well as to theorize in a critical and continuously ongoing process, communicating about the theory and the aspects of explorations into the field of human experience.[107] He refers to Rorty's argument that the basic condition of communicative validation is, "conversation as the ultimate context within which knowledge is to be understood."

Sociologist Frigga Haug objects to the hermeneutical criterion of coherence, which means freedom from any inner contradictions, as an untenable position when facing the fact of the contradictory nature of social reality in itself. Haug argues that methods aiming for consensus are invalid tools to explore contradictory social worlds. (Kvale 1989)

By definition, social science research enters a world of contradictions, conflicts and ambiguities. These fundamentals of social reality cannot be "corrected for" by any method, practice or procedure without generating consequences which counteract the intentions of the scientific approach. Thus, the approach, by necessity, must be estimated and evaluated in accordance to these basic preconditions. Marcia Salner has reflected upon the problems of evaluating research with inappropriate tools: it not only leads to the mismeasuring of what one intends to measure, as there might not be something "measurable" in the sense of a degree, quantity, proportion or rank. It even might produce a rather impossible "dialogue" from different levels about differently coded and typed topics. Therefore, she argues:

> The human science researcher often makes the mistake at this point in the discussion of assuming that the defense of the validity of his or her methodology rests on an adequate defense of subjectivity to counteract the empiricist's obsession with objectivity. This is a tactical and a logical error. (Salner 1989:48)

In the human life-world there exist facts which are socially constituted and as such, in a certain way, objective, true for all experiences within this social frame, and neither subjectively "invented" nor non-objectifiable. Human experiences are embedded in and informed by social practices which are "collectively created over time and, as a result, they transcend the capacity of any single individual to shape them. They are neither objective nor subjective in the empiricist's definition of these terms; rather, they constitute human experience." (Salner 1989:48) In other words, human science researchers operate in the world of intersubjective social meanings which are based on the ideas and norms constituting the common property of the society. Individual members of this society may share or oppose these norms, but they cannot configure their individual beliefs and attitudes without referring to the constitutional elements. The socially constructed meaningful practices that make up human experience are invisible from the perspectives inherent in the theoretical opposition between objective and subjective. Salner claims that the primary validity criteria in the human sciences are those aspects which make competing and fallible knowledge claims distinguishable. Furthermore, an evaluation has to acknowledge the interaction between the observer and the observed, rendering both part of a "unit" in which each must be defined with reference to the other.

In connection with the evaluation of the "involved researcher," there is one more issue to be explored. This is the notion of empathy. Geyla Frank's reflections upon "becoming the other" address con-subjectivity as part of a common human subjectivity different from the notion of intersubjectivity.[108] It denotes an immediacy of sensing a feeling, in accordance with the position of Edith Stein. She opposes the prevailing views of her time, which regarded empathy as the ability to grasp another's feeling by representing it in an analogy with one's own experiences, which means to understand another's feelings through a manipulation of symbols. Stein countered by advocating that empathy with another person's feelings is a direct intuition, an experience all of its own, not needing symbols, as expressions of feelings are self-revelatory. As such, empathy is a direct apprehension of the other's state. Stein claimed that analogies such as the one implicit in the question: "how would it be if that had happened to me?" are not part of genuine empathy.

Neither identification nor comparison are the entrance into the other's emotional sphere. Stein stressed the complex states of empathy which enable us to feel with other people some of the most enduring aspects of their being. To be sensible (able to sense) does not merely mean to *operate a sensory apparatus*, but rather to *be able to apply an imaginative approach* to what can be perceived. "Grasping another person's range of values is at the heart of the empathic act... As a method, empathy cannot be merely talked about; it must be practiced - tried and tried again - to be comprehended, for it involves an act." (Frank 1985:196-9).

- The preceding short reflections about the criteria for validity in research grounded in phenomenology as differing from those in the objectivist sciences, are useful as a frame for an evaluation of the research process, the hypotheses which are generated, and the conclusions which are derived from the project.

4. METHODOLOGICAL ASPECTS

4.1 DESIGN

Aiming at deeper insight and a more profound understanding of the existential impact of hidden sexual boundary violation, I planned to speak to people who had experienced such violations. In order to learn, I was willing to listen. I intended to give space to the whole spectrum of themes, events and emotions connected to these experiences. I was prepared to utilize my clinical experience from many years as a primary care doctor. I would meet my respondents in an exploratory manner, encourage them to forge possible connections, and to formulate untold experiences. I would by no means try to hide my reactions, although being aware that it would influence the dialogue, either in a facilitative, or perhaps even in an inhibiting way. In addition, I was prepared to hear recounted experiences, and to be told of suffering and humiliations beyond my imagination.

In cooperation with two Norwegian incest centers, from among 72 people volunteering for a study interview about experiences of encounters with health care professionals, 42 subjects, 34 women and 8 men (all of whom volunteered) were chosen. My selection of interviewees was guided by a desire to achieve the widest possible variety of life circumstances as regards age, family background, civil state, abuse experience and health problems. The 34 women chosen represented an age span from 16 to 70 years. They lived singly, in cohabitation, in marriage, were separated or divorced, some of them were remarried, and a few were widowed. Most of the non-single women lived in heterosexual relationships, and a few in homosexual relationships. They were childless or had biological, foster or adopted children in various combinations, several of whom were known to have been abused as well, in a few cases by the same perpetrator as had abused the mother. The family background of these women included unbroken families, broken homes but where the child had lived with one of the biological parents into late adolescence, foster homes, adoptive homes, and a childhood or adolescence in an institution. Concerning the experience of the abuse, there was both intra- and extra-familial abuse, solitary events, repeated abuse by the same or by different perpetrators, continuous abuse over many years by the same or several perpetrators, some of whom had been sentenced, some known or suspected to have abused several persons. Some women were known to have had violent relationships in adult life including sexual and physical violence. Several women were known to suffer from chronic health problems with varying degrees of somatic and psychiatric predominance; some had been in psychiatric institutions but never in somatic hospitals; a few had been treated frequently in somatic wards but had never been admitted to psychiatric institutions.

Of the 34 women chosen, 2 dropped out due to moving, as a result of recent divorce, or due to admission to a psychiatric ward. One woman met me in such a marginal state of functioning that the interview could not be performed in accordance with the interview guide. This interview is reported as an example of different interview situations and is also described in "Impressions."

Of the 8 men chosen, 4 could not be interviewed. This was due to very special circumstances which I want to describe as they seem to illustrate an aspect of the problems researchers in the field of sexual assault may encounter. Man A had been violently assaulted and, injured and in a state of psychosis, had been immediately admitted to a hospital some days before our scheduled meeting; a friend of his told me this when I phoned in order to confirm our appointment. Man A had been abused from early childhood, and he had recently given indications of being an abuser himself. Man B disappeared from his home and could not be found for weeks for unknown reasons. He lived alone and had been abused from early to mid-adolescence. Man C was, at that particular time, involved in police interrogations at his working place, an institution for the disabled. A colleague of his had been suspected of having abused one of the clients, and man C felt overwhelmed by the current offensive by the press and the presence of the police. He, therefore, withdrew. Man D had received hormone medication in anticipation of a transsexual surgery. Though he lived with his parents, they had not been informed about the ongoing treatment. However, when accidentally finding out that he was scheduled for an operation fairly soon, they ordered him to leave their home immediately, otherwise they would take action to have him forcibly referred to a psychiatric ward. He canceled the appointment and went underground.

4.2 PARTICIPANTS

The variety of age, life circumstances, abuse and sickness history among the 32 women who met for an interview, was as previously described.

The four men who met for interviews were between 27 and 56 years of age, three of them living alone, two of these after a divorce. One had never been married but defined himself as heterosexual, while two had, in periods, lived in homosexual relationships. Two of the men had children. None of these had, according to their fathers' knowledge, ever been sexually or physically abused. The four men's childhoods had been varying. One had lived in his eldest sister's home after their parents' death. Here, he was abused by his brother-in-law for several years. The second man had lived with his grandmother after his parents' divorce. There, he had been abused by neighbors for four years. One had been brought up by his family and had been abused by en elderly acquaintance. The fourth had lived in a boarding school, and had been abused by a male staff member. All four of these men had had frequent encounters with different representatives of the health care system, both in somatic and psychiatric institutions.

Information concerning sexual abuse, systematically elicited in the beginning of every interview, can be summarized as follows:

Of these thirty-four adults sexually abused during childhood, eighteen have been molested by several persons. Two of these are men. All offenders are male, apart from in the cases of two of the women, who were abused by both men and women. Twenty-two persons in the group have experienced sexual abuse during adulthood, several of them in combination with physical abuse. At least eight of these twenty-two have had very many different sexual partners and describe themselves, in certain periods of their lives, as either without any sexual boundaries, or as prostitutes in early or late adolescence, or, quite a few of them, in early adulthood. In all cases of sexual abuse in adulthood occurring during an "out-acting" or prostitution, the other persons were men. All cases of forced and unwanted sexuality were also initiated by men. None of these was a stranger, but rather a member of the near or extended family, a partner, a colleague at work, a neighbor, or someone who was trusted, such as a professional.

Concerning life-time illness and health care utilization, systematically elicited in the beginning of every interview, and related to childhood, adolescence and adulthood, the following brief summary may be given:

The predominant health problem in all three phases was pain. This pain, however, clustered differently in the different periods. Childhood pain was mainly named as pain in the stomach, the throat, the head, the ears, the back, and the shoulders. In adolescence, the "localization" changed to the head, the stomach, the pelvic area, the body (everywhere or in large areas). The women described their pain in adulthood as being localized in the muscles, the joints, the pelvis, the body and the head, while the men had pain localized in the chest, the lower back, the head and the knees.

The predominant acute sickness in both childhood and adolescence among the women was urinary tract infections, accompanied in later adolescence by infectious states in the pelvis.

The predominant general problems in childhood were related to breath, teeth, eating, nausea, urinating, eczema, sleep, and anxiety. In adolescence, eating disturbances and anxiety remained, and were accompanied by abuse of alcohol or drugs, self-destructive acts, disrupted school performance, loneliness and/or drifting in a "gang," and first contacts with psychiatry. In adult life, all interviewees reported strained relationships and disturbed sexuality. A majority had suffered from depression and/or anxiety, and more than half of the interviewees had attempted to commit suicide at least once. Twenty-six had been admitted to psychiatric wards at least once due to depression, anxiety, substance abuse or psychotic states. The above mentioned conditions led to contacts with psychiatrists, psychologists, social workers, speech therapists and special education experts.

As regards somatic health, the problem of pain, either localized or generalized, caused most contacts with the health care system in all three life-phases. The medical encounters related to pain as the main symptom included contacts with a variety of medical specialties, and the therapeutic measures taken comprised a range from the general approach of physiotherapy to the specific approach of surgery. Among the women, surgical interventions were directed primarily toward the pelvic and abdominal area, but more than a third of all surgical interventions were non-gynecological. Twenty persons among the interviewees had, during their life time, at least one period of severe eating disturbances of some kind, and twenty-two had at least once had injuries of various origin which demanded medical treatment. The somatic health problems had led to contacts with representatives of all medical specialties except geriatrics, and an array of medical therapies such as physiotherapy, psychomotoric physiotherapy, chiropractics, and acupuncture. Apart from these, several of the interviewees also used non-medical treatment such as foot zone therapy, aroma therapy, healing, hypnosis, yoga, autogenic training, and irido-diagnostics.

Among the thirty women in the group, only two lesbian women and two young women had never been pregnant. Of the twenty-six women ever having been pregnant, ten had experienced terminated pregnancies, the majority through miscarriages. Twenty-three had given birth, and the group had a total of fifty-eight living children. All these women reported problems related to either pregnancy or delivery in at least one of those pregnancies which led to a birth. During pregnancy, the predominant phenomenon mentioned was "distance." This could be expressed as not feeling pregnant at all, not looking forward to having a child and not preparing for its arrival, as not perceiving the movements of the child although having them confirmed by ultrasound or others' hands, and not relating to the abdominal and pelvic region of their own body. This distance was often combined with depression and a feeling of unreality. The bodily complaints were: bleeding through pregnancy; nausea; pelvic pain; and very late or repeated spontaneous abortions. The problems related to delivery were, mentioned in the order of their reported frequency; very slow progress of birth; pre- or dysmature infants; frequent surgical interventions; loss of blood requiring transfusion; psychosis; post-partum depressions; and, bonding problems.

In those cases in which the interview had brought to light specific incidents linked to treatment at a hospital, I asked the person to sign a release form to have a copy of their hospital records sent to me. All respondents I asked gave me their permission. Some people had been referred frequently to different hospitals for different problems and thus could not say with any confidence at which hospital the event(s) of interest had occurred. The maximum total number of hospital admissions for a single person could not be established properly, nor could the total number of individual contacts with different specialties. The largest number of different hospitals one single person had been admitted to was ten. Several of the interviewees had been in hospitals many times, a few in the same hospital each time. As a group, the people I spoke with had been in contact with all the major medical specialties, apart from gerontology, and with several subspecialties.

Their contacts with other counseling or therapeutic professions included psychologists, Gestalt therapists, social workers, chiropractors, acupuncturists, traditional, manual and psychomotor physiotherapists, speech therapists, drama or art therapists, and alternative therapists such as healers, homeopathic practitioners, aroma therapists, foot zone therapists, iris diagnostic practitioners and yoga teachers. Every respondent had had contacts with several such professionals and therapists.

The majority of my respondents stated explicitly that it felt meaningful to speak to an experienced physician about former meetings with doctors and therapists, and about their previous encounters with the health care system. Although aware of the difficulties of speaking about suffering, the fact that the researcher who invited these dialogues was a doctor constituted their explicit and primary motive for participation.

Participants

Pseudonym	Child Abuse	Main abuser/s	Adult Abuse
Oda	4 to16	uncle+friends/nurses	acquaintances
Laura	3 to ?	father/brother/neigh.	
Christine	7 to ?	acquaintances	strangers
Karla	4 to 14	father/acquaintances	strangers
Gisela	2 to 14	father/acquaintances	
Mary	5 to 27	father/brother/neigh.	strangers
Annabella	4 to 19*	mother/uncle/acqu.	strangers
Gunhild	6 to 23*	stepfather+friends	professional
Fredric	10 to15	neighbors	
Bjarne	10 to ?	professional/strangers	
Nora	5 to 11	father	husband
Hedvig	7 to 13	stepfather	
Ruth	5 to 6	acquaintance	
Berit	8 to 14	mother's partners	professional
Synnøve	12 to16	mother's partner	
Beate	12 to ?	uncle/acquaintances	
Line	8 to 25*	father	father/husband
Runa	12 to 15	father	neigh./colleague
Susanne	12 to 17	brother-in-law	acquaintances
Pia	12 to 15	stepfather	partner/acqu.
Veronika	3 to 20*	father/uncles	father
Fanny	5 to 19*	uncle	uncle
Tom	4 to 7	father's friend	strangers
John	13 to 16	brother-in-law	
Eva Maria	7 to 13	father	husband
Jessica	7 to 16	stepfather	acquaintances
Elisabeth	1 to 17	father/stepfather	
Annika	11 to 17	brother	partner
Barbara	10 to 13	uncle	friend
Grethe	11 to 15	father	
Judith	11 to ?	neighbor+friends	strangers
Tanja	4 to 11	uncle/acquaintances	acquaintances
Camilla	5 to 10	neighbors/friend	professional
Hanna	6 to 15	uncle	professional

(*)the same abuser from childhood until adulthood
"strangers" denotes, in this connection, prostitutive relationships
"acquaintances" comprises peers, gang members, colleagues, employers

4.3 INTERVIEWS

My first interview session consisted of seven interviews. When listening to the tapes of the first interviews at the end of the first day, I almost panicked. Hardly any sound had been picked up by the recorder's microphone during long passages of the interviews. Use of earphones reduced the problem only slightly. Interpreting this as a technical question only, I hoped to compensate by turning the recording microphone to its maximum during the interviews of the next day. This apparently technical problem, however, was evidence of something I had witnessed during the interviews, without reflecting upon its technical impact or its profound meaning. I have tried to describe these "fading voices" in the section "Impressions." [109]

Three interview settings shall be described in detail, representing a few of an array of situations where my previous theoretical considerations about the roles and rules in an interview were overridden by a combination of circumstances. The descriptions may illustrate what I have called "confused roles," "balanced interchange," and "marginal presence." The first and second interviews were completely transcribed verbatim. The third was not transcribed, and is consequently not one of the 34 interviews which are analyzed, although the interviewee and I spent almost two hours together and the setting provided valuable experience of an extremely challenging communication.

The first of these three situations, exemplifying a confusion of roles, occurred during the very first interview at the northern incest center. I became aware of a challenge I had not anticipated, and for which I therefore was mentally unprepared. Although my own mother tongue is German, I had been living in Norway for twenty-two years at that time, and my knowledge of Norwegian had continued to increase. Professional contacts and personal friendships with people from all regions of Norway had gradually made a variety of Norwegian dialects sound quite familiar to me. Now, however, I had to admit that this sense of familiarity did not apply, although I was aware that the dialect spoken in this particular northern county was considered particularly difficult, and its northernmost variants even more so. The first of these interviewees came from such a village in the north of the county. She spoke in a dialect so unfamiliar to me that, in the beginning of our dialogue, I had to ask her, again and again, to repeat what she had said, to slow the flow of her speech, or to explain the meaning of certain words. This "incompatibility" between my ears and her pronunciation became a very disturbing and extremely non-facilitative frame for our dialogue. Through the previous interviews I had learned to adapt to the lowering of voices in passages of the narratives about the hidden. I had also understood that certain acts of abuse had no words from any accepted vocabulary. I had conceptualized the main obstacles to communications about abuse to be the silenced, the forbidden, the taboo. Suddenly, I was confronted with an even more effective and most disturbing obstacle, which was not a semantic or pragmatic problem, not residing in the technical equipment, but residing in me, resulting from my own lack of ability and my naive underestimation of language diversity.[110]

This first interview situation in northern Norway profoundly challenged all concepts of a research interview as a defined frame for special communications. I was a guest in a house where the woman to whom I spoke felt familiar. We were seated at a table in a setting indicating a conversation between persons of equal right and rank. I was the one expressing my gratitude for her being willing to meet me and let me hear her experiences. I was the receiver and she the giver. Still everything became influenced by the limitation to my understanding, and by the verbal "breaks" which I literally and repeatedly placed on the flow of her speech, interrupting her thoughts, associations or sequence of mental pictures. Initially, the woman became confused by my facial expressions of what she obviously interpreted to be signs that I doubted her statements about the context of abuse. I had to admit time and again that I had difficulties to understand her words. I had to assure her that that was by no means an indication of any disbelief in what she was telling me. She had to recall what she had said previously, and hesitated several times to assure herself that I was able to follow her. However unintentionally, I had continuously to disturb her concentrating on remembering since she repeatedly was forced to concentrate on me.

During the first half hour, she obviously felt responsible for giving me the possibility to accompany her in her recollections and reflections; that was clearly counterproductive to the purpose of our meeting, and was certainly the opposite of my intentions. Instead of me facilitating her recollecting ambiguous or conflict laden events, she had to facilitate my recognition and comprehension of what she said. Because the interviewee was very careful to speak slowly and distinctly, little by little, my ears and mind adapted. It became obvious that she realized my increasing ease with listening and responding, and this allowed her gradually to shift the locus of her concentration away from me and onto herself. On the way toward "forgetting" me more and more, she also "forgot to help me" more and more. This was expressed in her eyes no longer looking for signs from me that I either "understood" or was "confused." It also showed in her voice becoming more hers and less "for me." And it was mirrored in her speech following the rhythm of her own breath more than that of my interruptions and breaks. Gradually, the feeling of getting lost once again arose from my diaphragm, interfering with my attempts to listen actively. She must have sensed what was going on, perhaps because my paraverbal utterances of, "I am with you, go on," had become less frequent. She hesitated, her gaze returning from afar to zoom in on me. She smiled with a sigh, as if apologizing for having lost me on the way into her mind. This made my diaphragm find its proper place again - and as she continued I must have forgotten about how strange her dialect sounded to my ears. I became aware that both of us were in her story on her premises.

The second interview situation I want to describe represented an example of what I perceived to be a balanced interchange. The setting as such was characterized by certain deviations from the majority of the interviews in that it took place in the office of the respondent, at the school where he worked, in a little town one hour outside Oslo. The interviewee, one of the four men with whom I spoke, had invited me to meet him after school hours. I reached the town by train, walked to the school, and, when entering the building, heard choir music coming from one of the classrooms. The voices of children performing an elaborate hymn were guided by a strong male voice, evidently that of a trained singer. The chorus filled the hall where I had been asked to take a seat with joyful music which immediately engaged me emotionally and cognitively, and I searched my mind to identify the origin of the composition. When the bell rang, the pupils left the room, most of them singing, laughing and obviously still "in tune" with what they had been singing. The teacher, a bearded man in his fifties dressed in black, was the last to leave the room, and, when seeing me, hesitated a second and then approached me and presented himself as the person I had come to meet. On the way to his office, we began talking about singing and very soon recognized in each other an interest in classical vocal music. Thus, when starting the interview, we were already aware that we shared a passion and that both of us were influenced by the music I had heard and he had sung. That we knew something about each other already contributed to an ease of talking, clearly facilitating our entrée into the interview and the flow of speech.

The teacher proved to be a trained narrator, one accustomed to shaping thoughts and impressions in general, and explicitly aware of the details of his experience of having been abused as a child. At the same time, as a professional care-taker of children, he was engaged in and felt obligated to make all sorts of contributions to assure that the silenced topic of child sexual abuse be made an issue of public concern. He told me immediately how, after having listened to a radio-interview with a woman he had known in their childhood, he had regained his memory of having been raped as a little schoolboy. Until then, he had been unable to recall the abuse. When he had heard her speak about her own sexual abuse experience in a boarding school in Japan, however, he could suddenly remember what had been done to him. He then contacted the woman, and together they had succeeded in collecting adults, all of whom were children of Norwegian missionaries in Asia before and after the Second World War, who had known each other during a certain period of time. The group members met frequently for more than a year, helping each other to recall and to speak out about the harassment and sexual abuse several of them had experienced at the same boarding school. The members of the group had agreed to write about their experiences and engage in political activities in order to influence the public. They also had agreed upon taking initiatives at their places of work appropriate to the local conditions, and to oblige the leaders of the Mission Congregation to discuss both the topic and preventive measures to protect children presently living abroad and separated from their parents. This was a man who had reflected explicitly about the various aspects of childhood sexual abuse as a social phenomenon and as a personal experience.[111]

While talking, we discovered that I was familiar with one of the places where he and his family had lived and worked after their return to Norway, and, what is more, he recognized my family name as familiar through relatives of my husband whom he had met and spoken with. As he told his biography, these details in the flow of information which revealed links to my private life and my husband's family created an atmosphere of equality between us. A balance of roles was established which overrode the fact that I had initiated the meeting as researcher and physician. This was true despite the fact that a greater part of the interview concentrated on reports of medical treatments and therapeutic encounters, and upon various and detailed descriptions of bodily perceptions and sensory phenomena. He thematized connections between childhood memories and adulthood emotions which revealed a particular body logic which I had not heard in the previous interviews. In addition, although capable of a high degree of abstraction in the description of assaultive events and in reflections upon their life long impact, he was an authentic and visibly emotionally engaged narrator. This was especially true when touching upon crucial events or occasions. It was also the case when he spoke of the powerful influence of music during his life which expression was obviously motivated by our already acknowledged shared passion for music and his assumption of my familiarity with the music he mentioned.

The third interview situation which might fill out the picture of the variety as regards the distribution of roles, stands as an example of what I perceived to be the marginal presence of an interviewee, an extreme situation which occurred only once. One might, therefore, object that it is not an example representative of a group of interview settings. In one way, this is correct. However, several other interviews held elements similar to this one as regards changes in the mental presence of the interviewee and the complementary change of the impact of the role of interviewer. This interview resulted not in an interview text but in an "Impression," meaning that the only way I could grasp this marginal situation was not through textual analysis and interpretation but by "associative and perceptive writing" on the basis of notes. The interviewee, a 26-year old woman, had made an appointment for the interview several weeks in advance of our meeting, as did all the other respondents. She belonged to the group I had been scheduled to interview on the second day of one of my visits to the center. I had seen her arrive in a police car late the previous afternoon, as I was returning from the home of another interviewee. She had been accompanied by two policemen who, walking beside her, almost had carried her into the house; she seemed unable to stand upright. She had been met by a member of the staff who helped her upstairs to one of the guest rooms where it was intended for her to stay the night. Before leaving for the night, the staff member informed me briefly of the woman's situation and asked me to be attentive. The woman had left her home, assisted by the police, after her husband had battered her severely. She had refused to be taken to a local doctor and had insisted on being given shelter for some days at the center where she had come on other similar occasions. Regardless how concerned the staff was about her impaired condition, she would not allow them to call upon a friend or relative to stay with her. Therefore, she and I were alone at the center during the night, although I did not see her as she had asked to be left undisturbed in her room in order to sleep.

The next day, I did not expect her to meet for the interview, but she came to my room, although she needed help to walk and be seated, at the scheduled time. She looked very depressed and as if absent, but assured me that she wanted to talk to me. I could not establish eye contact with her as she avoided looking at me. Her voice was low and hoarse, as if she had not spoken for quite a while. I had to do my utmost to understand what she said, and I repeatedly asked her not to feel obliged to keep our appointment. She whispered her insistence on staying, saying that she had something she needed urgently to tell because it sometimes overwhelmed her, and this was one of those times. Consequently, I decided to accept her decision. I presumed, however, that I would write out the information form since she was obviously unable to write herself. She agreed and answered slowly, hesitatingly, her voice almost tuneless and the intervals between her utterances continuing to increase. Occasionally, it seemed that she had not heard my question, so, after some time, I repeated it. Then, her body would react on one side only, her eyes would not meet mine, and she would strive visibly to concentrate upon her answer, which consisted mostly of a few words, parts of a sentence or just a nod of her head. She sat as if twisted, her head slightly bowed forward, her eyes fixed on the floor in front of her feet, her left arm hanging along the side of her chair as if not belonging to her, and with her left leg likewise appearing lifeless. She was so extremely withdrawn that I felt nearly forced into the role of the examiner performing an inquiry and repeating my questions almost insistently, although this was exactly the kind of interaction I had specifically rejected as being neither supportive nor facilitative. Still, the interviewee had indicated that she wanted most urgently to express something, although she almost disappeared from our setting between every attempt to answer. I felt that I more or less had to hold her in a state of consciousness by my continuously focused, intense and listening attention, otherwise she might fade. Sometimes I felt she had left the room mentally, which provoked physical discomfort in me, and which made me move closer as if trying to keep her present. Sometimes I had to read her lips to know what she had said, and was forced to repeat what I believed to have understood in order to ask for her confirmation.[112]

The information I received concerning her abuse history revealed extended abuse from the age of five by four persons, two men and two women, successively. Two of the abusers were members of the family, one was an acquaintance and the fourth was a stranger. The interviewee could not state when the last of these abusive relationships had terminated, but she indicated it might have been late in adolescence. Since she was married, she had been abused by her husband, predominantly physically, through threats and terrorization. Sexual abuse in her marriage was not brought out. Concerning her sickness history, she could recall various health problems and frequent injuries during childhood, some accidental but most self-afflicted. She had continued to afflict harm on herself ever since, and had done so recently. She had been involved in several therapeutic relationships, of various duration, with psychologists ever since late childhood, and had been referred to a psychiatric ward at the age of just over twenty. Since early adolescence, she had suffered from eating disturbances, alternating between phases of anorexia and periods predominantly of bulimia. She had been pregnant four times, the first resulting in a miscarriage when she was 17, and two others terminating in abortion. She had given birth to one child in a very complicated delivery a few years ago.

All this information was elicited in the manner described above. No complete sentence was spoken, no narrative unfolded. The setting was characterized by marginal communication in all respects, a struggle to understand, to keep contact, to answer and to remain present. No dialogue was possible, and most of what I received as answers was non-verbal and physical, movements as if frozen, creating the impression of watching a slow-motion film through an unfocused and therefore distorting lens. The predominance of physical communication, however, made it possible to grasp the horror which the interviewee tried to express, mirrored partly in facial expressions of disgust, and partly in gestures and interruptions of breath, plus some words. What she tried to verbalize was the sensation of experiencing one half of her body as dead and rotten, so that she had to turn away from it in dismay. At some occasions, however, she would experience a rather threatening terror of floating electric shocks seeming to fill the left side of her body with painful, pulsating streams. When they ended, these would leave her exhausted and tormented - and again as if the left side of her were dead.

She gave me a sign of agreement when I told her what I presumed to have understood. She only nodded when I asked her whether she thought I would know how this felt and what might be the cause of these sensations. Still she avoided my eyes and maintained her position. I told her that, according to my knowledge, there could not be any medical explanation for these intermittent sensations affecting half of her body. I also said that such painful phenomena had been mentioned to me in previous interviews, but not exactly as she had indicated, as splitting her body and rendering one half of her almost a strange and repulsive object to herself. She seemed to accept that I had neither an explanation nor a conclusion, and she "left" the room mentally, as she had done before. After a while she "returned," looking at me for the first time, hastily and as if ashamed when she met my eyes. She turned her eyes away and stood. On her way to the door, speaking away from me, she said goodbye and, "Thank you," before she left the room.[113]

4.4 TRANSCRIPTION

As a point of departure for describing the process behind the interview texts, the transcribing of the taped interviews, I once again refer to Mishler and his reflections on transcription in general:

> The complexity of the task of transcribing speech into written text merits brief comment. There are many ways to prepare a transcript and each is only a partial representation of speech. Further, and most important, each

> representation is also a transformation. That is, each transcript includes some and excludes some other features of speech and rearranges the flow of speech into lines of text within the limits of a page. Some features of speech, such as rapid changes in pitch, stress, volume, and rate, seem almost impossible to represent adequately while at the same time retaining the legibility of the text. Adding another complexity are the nonlinguistic features of any speech situation, such as gestures, facial expressions, body movements, that are not captured on audio tape recordings and are difficult to describe and record from observations or videotapes. Lastly, it must be born in mind that the initial record - audio- or videotape or running observation - is itself only a partial representation of what 'actually' occurred. (Mishler 1986:47-8)

Mishler emphasizes, furthermore, that no universal form is adequate for all settings. The mode of transcription adopted should be sensitive to an array of theoretical and practical conditions which are peculiar to that particular research project. (Mishler 1986:49) When applying these considerations for the present project and its main aims, I had to make consistent choices. Due to the intimate character of the topics at hand, the first decision was to do all the transcribing myself.[114] This decision, although resulting in approximately 550 hours of typing, lent a structuring effect to the whole interview period. Since I tried to have all the interviews collected at one visit to the incest centers transcribed before holding the next group of interviews, the writing process was enhanced by the freshness of my recollections of each interview and its circumstances. Furthermore, the intense process of listening to the tapes while writing kept the central messages, points or questions from previous interviews vividly present, and as such they were available to re-enter later interviews as questions, associations or analogies presented to the interviewees.

The next choice concerning the mode of transcription was guided by the main focus of the study. The linguistic interaction of interviewer and interviewee was not the central focus of interest, but rather the world of the subjective, personal experiences linked to past events. Therefore no linguistic analysis on a micro-level was aimed for; instead, the aim was an understanding of patterns and constellations as they were described and verbalized according to how they had been experienced or perceived, and how they were recalled, transformed or interpreted by the narrator. The transcript would not, as a consequence of this choice, contain paraverbal features or an exact representation of overlaps of speech. Nor should the length of pauses be noted as a measure of hesitation or silence. This ought not, however, preclude the possibility for making marks in brackets regarding non-verbal communications and utterances such as sobbing, crying, deep sighs, laughter, a sudden rise in volume or tone, a marked accent on words or sentences, a break in the voice, a whisper, a sudden stuttering or a stumbling. As an example and in order to illustrate how this was done, an excerpt from an interview text concerning the circumstances of a suicide attempt will be quoted in translation. The text in brackets, given here in italics, is in Norwegian and English:

> But the line had remained open and my name had been mentioned. I may have had a little wish. But I was waking up at the intensive care ward the next day, I was still in this rage. Lying there, naked, tied to a bed (hiver etter pusten/struggling to breathe) with plugs here and there and tubes and everything - it's only this - what I can remember is that I was so furious that I arched my back and tried to get free. I didn't want to be in life. I was back in life but didn't want to be. This was the most serious attempt I've made ... and it was my last attempt to commit suicide ... because I was so ... I've never felt such rage ... (stemmen blir lav, hes og nesten forundret/ voice becomes low, harsh, and almost amazed) (pause/interval)
>
> But I didn't succeed in going ahead with the rage in my therapy, because then there were new medications ...

In the same interview text, the following remarks are similarly to be found added to the transcripts and may illustrate what I registered and in which connection. The interviewee speaks about his recollection of how he became totally sexualized after having been raped (stemmen går opp i siste setning/voice rises in the last sentence); he recalls how memories of his childhood nanny came back to him, as if she herself had come back to him (hans stemme blir gråtkvalt - hviskende/his voice is choked with weeping - whimpering); he talks about how he was declared cured of homosexuality when he announced his marriage (vi ler begge meget hjertelig og lenge/both laugh heartily and for some time); he recalls the rape and its circumstances (dyp pust/deep breath), (nøler/hesitates), (stammer nesten/almost stuttering), (stemmen tetner/his voice thickens), (stemmen blir toneløs, han gråter/his voice becomes monotone, he weeps), (stemmen nesten dør, lang pause/ his voice nearly dies out, long interval); he talks about the coincidence of a rape, a death and a wonderful concert (stemmen svikter/voice fails), (helt stemmeløs/completely without a voice), (lang pause/long pause), (dyp pust/deep breath); he recalls a perception while having being unconscious (han er tydelig oppstemt/he is clearly geared up); he recognizes an experience from an analogy (han holder pusten litt/he holds his breath a little).

All these remarks in parentheses are, of course, only a shorthand, signs for words which indicate interpretations of certain, selected nonverbal features which could be heard on the tape.[115] They represent some of what Mishler pointed to as, "what seems almost impossible to represent adequately while at the same time retaining the legibility of the text." The remarks in brackets, however, may serve a documentary function, making them a key to an understanding of a certain train of thought or associations which otherwise might remain inaccessible, as was the case in the analysis of Hanna's triumph when stepping out of the limits of fate. The remarks about her voice (stemmen blir lattermild, glad/ her voice fills with laughter, happy) (veldig glad i stemmen/very happy voice) make clear both the meaning of the term, "protest," and the hopeful expectation of understanding phenomena which were not yet comprehended.

Of course, a critical voice might argue that, when the eyes and the ears of the interviewer provide the constitutive premises for the ears of the one who transcribes, only those paraverbal features most consistent with the interviewer's presuppositions or impressions will emerge in the final interview text. From there they will go on to acquire a quasi-documentary force which could very effectively direct the reader's eyes towards certain elements and away from others. This is certainly a valid argument for the present situation in which the generator of the project's central question, the designer of the project, the interviewer, the transcriber, the analyst, the interpreter and the reporter are one and the same person. On the other hand, dividing the implicit influence of the interconnected processes of wondering, questioning, planning, choosing, interpreting and naming among several people would introduce an array of other elements which are arguably more difficult to know and to take into consideration when evaluating a particular step in the study or the study process as a whole. No matter what, the fact that there is one mind, preference or understanding throughout the entire endeavor of a phenomenological study of human experiences, will be reflected in the transcribing of audio taped dialogues where the transcriber has thus taken part in the original dialogues.

Since the guiding principle in the present study was a hermeneutical/phenomenological approach to hidden events, a problem of language was anticipated regarding the difficulties of speaking about events from the realm of sexuality, due to a lack of words or expressions, or due to a hesitation to use "forbidden" words or terms. Consequently, all passages revealing this kind of difficulty, a struggle to find the appropriate word or an attempt to overcome the barrier to speaking a taboo word, were to be transcribed as precisely as possible, even if that would affect the legibility of the text. The following excerpt from an interview, concerning the recollection of the first abusive event experienced, may serve as a clarifying example. The number "1" stands for the interviewee, "2" for the interviewer:

2 What did your father do to you?

1 He fondled me, he touched my breasts and my genitals, and he forced me ...to...forced me to ...forced me so I would...(long intervals between each attempt)

2 This is difficult to say?

1 He forced me to suck him. I had to suck him. Yes. (voice fades) But I didn't want to, but he forced me.

Apart from the "resistance made visible" in the form of dots between repeated words to represent the unmeasured length of hesitations and pauses, the non-transcribable struggle to speak what is taboo and unspoken remained unwritten in the interview text, as a text. I repeat that experiences such as witnessing the re-triggering of disgust, written across a facial expression, had to be channeled into an "Impression."

In addition to those principles of transcription, which had been chosen or defined in advance and in accordance with the epistemological frame, aim and focus of the study, others were introduced during the process of transcribing. These added procedures could be the result of a growing insight into the dynamics of the interviews. I used two approaches in the transcripts of the recollection of life-time diseases: either a summary in the first-person singular, simply referring the facts given to me as facts and not developed as dialogue; or transcribed in the form of dialogue from the very start.

The following is an excerpt from an interview text demonstrating the first approach, a summary told in the first-person singular:

1 I am 28 years old, and I was almost ten when the abuse started which continued for four to five years. There were two perpetrators, the eldest had received his driver's license when I was ten, which means he was 18, the other was two years younger, that means he was 16.

I got extra lessons at school, and lessons in speaking, but not with a speech therapist. I've been admitted for two hernia operations, and then I had to be operated on because of phimosis; for that I was hospitalized a few days. I had adenoids which were removed when I was little. I've never had a fracture, and I've gotten small injuries in sports activities. As an adult I've been, this is some years ago that I suffered from epileptic seizures. I was hospitalized for that once. The first time it hit me really hard. Then I had a rather thorough check-up. They didn't find any cause. The next time I was examined at an out-patient clinic, they performed an EEG and ...I don't remember what else. The third time, I wasn't examined. Another time I was offered an in-patient examination, but I refused. I had conferred with my doctor and we decided that there was no point in it. We didn't want to do the same thing all over again. The first time they took a sort of no-sleep test, I'd been awake for 24 hours and they observed me until the next morning, then they performed the EEG and some other tests ...which I don't recall right now.

2 Did you go through the kind of brain scan using contrast dye?

1 No. I can't recall that.

2 Did you have a computer tomography?

1 No. I didn't.

Terminated registering questionnaire responses. Interview starts.

Other kinds of inventions were sometimes forced upon me by technical conditions beyond my control, such as was mentioned earlier regarding the fading voices. As a result of that situation, the transcripts of the first seven interviews all contain some passages marked with an asterisk indicating that that passage is a summary of what had been talked about. I recalled the content from fragments of sentences and from my memory of the details. These passages, however, do not constitute a major part of the dialogues. An excerpt from one of these interview texts illustrates how this was done. In the beginning, I noted the following:

(She talks in a very low and almost flat voice, nearly without modulation, and the recording is very faint. She talks slowly and hesitatingly, with long pauses. Often, she searches for words. Most of the time she is sitting with her head bowed down and her eyelids lowered. She speaks as if only to herself. During long passages her voice is unclear, hardly audible. Thus, in parts of the interview the woman's narrative is supplied by what I remember from our dialogue. These parts are marked with an asterisk (*).

The excerpt from the interview text is linked to her recollections of "mad or strange" survival strategies, such as, for example, wanting to donate her tears to the clouds in order to save them there so the clouds might cry on her behalf. This act of donation of tears to clouds is the reason why she loved rain and ran outside whenever it was raining. She herself could not cry since her weeping made her father angry, and when he became angry, he abused her. The text is as follows:

2 I find this moving, it seems completely comprehensible and not mad or strange at all. It's possible that your sensitivity could only have been preserved in this way.

1 Now I think of this the same way myself. As an adult, I can see that to donate your tears to the clouds and to talk to the moon for hours - I loved moonlit nights, then I couldn't afford to sleep. I sat by the window or went out. This was my real life. (stress on each word)

***** She tells about how often she was punished for being wet, for having damaged her clothes or her shoes out in the rain, or how often she was terribly tired after having spent the night talking to the moon, and consequently fell asleep at school. She remembers a great number of comments from both adults and children upon how special, foolish or mad she was.

2 I find it fascinating to think how inventive children can be in protecting themselves. How you managed to preserve good pictures about yourself inside you and good thoughts and feelings for things outside you. That cannot be an indication of madness.

1 No, I've taught myself to look upon this as right, I've learned that such behaviors were pure survival thoughts and pure protection against my soul dying.

* She recalls how she tried to keep a picture of herself as somebody able to do something. Secretly, she performed tasks in order to feel that even she could succeed at something. She remembers how powerful she felt after having proven her strength by cracking a glass which had been said to be unbreakable even by adults. Someone had brought it home from a trip. It was a little glass ball with water in it and a landscape, and whenever you turned it upside down, snow fell on the mountains and trees. She cracked it. Therefore she knew she was strong, possibly even stronger than her father.

An asterisk as a mark of summarized passages represents "the meaning, not the wording," a distinction introduced in the evaluation of interviews by Lazarsfeld and referred to by Mishler was also used in other connections, as, for example, to indicate what was called, according to its function, a "vehicle story." During the interviews, the narrators used stories as something by which to move closer to, or to find the way back to, the core of the topic they wanted to tell about. They themselves often underlined that the story was a vehicle, an instrument to come closer to the essence, and their comments might be, for example: "...this was of course of no importance for what I wanted to tell, I just had to trace back to..., and I couldn't go straight back to ..., but that gave me the association which made me remember ..."

An asterisk marking such a vehicle story might well be linked to yet another kind of constellation, an example of which will be given in what follows. In one interview with a woman, an observation was made revealing a pattern of methodological as well as hermeneutical importance. The interviewee told a long story about her brother, which led to a detailed story about her female friend, after which she continued with a narrative concerning her dead mother. Obviously, she was far afield from the subject of our concern. When I sensed a pattern and asked her whether she used to distract attention from herself by talking about others, as she might feel uneasy when being the focus of another's gaze, she responded: "You certainly hit a point; my therapist has mentioned this several times. Well, that was not what we were going to talk about, was it?" Although I could see her habitual avoidance as a strategy to distract an unwanted gaze or an uncomfortable attention, I chose to concentrate the three stories in the transcript, and mark them with an asterisk as their contents by no means gave access to new information about the interviewee's experiences. Their informative impact, however, resided at the meta-level as it became obvious that they belonged to a pattern of habits, one of an array of possible avoidance strategies and techniques, which constituted a theme emerging frequently, and are found either explicitly or implicitly in every interview, as will be demonstrated later.

The discovery of a habitual pattern could sometimes be the key to an entire interview, as the habit appeared to be special, unusual and "strange". However, this habit proved to be of central interest when interpreted in the context of the life story and the hidden abuse. In order to stress the importance of such a pattern for understanding the interview text, and to make "visible" or "readable" what could not be transcribed, such structures were described in the opening of the interview text. The following text is a complete quotation of such a keynote from one interview:

> The woman answers all questions with a long latency; she watches the interviewer with great concentration, thinks visibly - and answers adequately, fluently and consistently, but after such a long hesitation that the interviewer every time wonders whether she has gone elsewhere in her thoughts. Our dialogue is therefore filled with unusually long pauses. Nonetheless, all her utterances are precise and clear. This creates some bewilderment as the time which passes between question and answer would, in a normal conversation, indicate that one of the partners had abandoned the topic. It takes a while to comprehend this. At some junctures, the interviewer repeats or deepens the question, but this proves to be unnecessary every time; the woman has understood perfectly well, her gaze is steady, her face open. This unusual hesitation without any other kind of sign or utterance causes the interviewer to suspect that the interviewee most probably may appear provocative to people who expect immediate answers - or in situations which demand a quick reaction.

This observation, and the reflection concerning the feeling of impatience which the habit provoked in the interviewer, is thematized in the interview and confirmed by the report of two decisive events among the woman's experiences with therapists. Further, its onset can be traced back in the woman's biography, along with the interview topics. Finally, and even more significantly, its probable origin can be envisioned by combining an array of statements during the interview. Consequently, this initial description of a speech pattern is a highly relevant observation, despite its going far beyond what is recorded. Its relevance renders this short opening passage a necessary accompaniment to the transcript, although it concerns something inaudible in terms of recorded language and unanalyzable as speech turns, neither of which would mirror the woman's obvious mental presence in the phases of absent speech.

Here I wish to stress that such a central feature, which will be demonstrated later as being a complete "Gestalt of an adaptive strategy turned maladaptive," would never have been grasped if the transcription had been performed by another person than the interviewer. Furthermore, another type of absence of information would have remained, likewise not demonstrable in a transcript, but crucial to comprehending its function as a trigger. This next "absence" in the interaction during the dialogue was only visible and consisted of the interviewee's slight physical withdrawal and change of facial expression to indicate fear every time the interviewer, insistently and with increased stress on the words, repeated a question. This "odd" reaction pointed to a possible experience of invasive inquiry or an expectation of being scolded or blamed for something. Both the hesitation and the withdrawal generated a suspicion of the presence of a meta-narrative linked to talking, the mouth, vocalizing or verbalizing. This gradually guided the interviewer's interest during this particular interview toward narratives concerning the mouth in relation to abuse, as will be shown later, which emerged as a hidden or as a central topic in a considerable number of the interviews.

Finally, the issue of transcribing a dialect must be addressed. There were, of course, different dialects represented in the group as a whole. These are, in general, not transcribable due to phonetic characteristics which are neither possible to transform precisely into script, nor are standardized as regards their transcription. This means, that the dialect had to be translated into the dominant Norwegian speech, which the interviewer masters and which may be considered a common, "generic" base language. There exists, however, a second official, written language in Norway which is much more closely related to many of the Western dialects. However, as the interviewer was incapable of using this language to express the narratives in dialect, all texts have received the same linguistic shape. In all cases where the interviewees used dialect idioms, these were translated according to their meaning. It is highly unlikely that these translations have led to distortions of what the persons wanted to express. As the interviewer herself did not speak a dialect, the interviewees responded to her according to their interpretation of what she said. This blend of language in the interview situations disappeared in the interview texts. However, any misunderstanding or misinterpretation occurring due to a mismatch of language would probably have become apparent based on the speech turns.[116]

5. CRITICAL QUESTIONS

Veronika's interview is one of those which had a special influence on the choice of perspectives in the present study. It generated an arrow of questions which guided my attention and directed my interest, both as regards the conducting of successive interviews and also regarding the analytic approach to all interviews. It also holds keys to a variety of associations, themes and patterns which became central in the analysis of the study material.

Likewise, the interview text illustrates how, within the flow of talk during the interview, statements, questions, and comments from both the interviewer and the interviewee are intertwined. Thus it becomes evident that there is no role distribution such that one person does the asking and another does the responding. The speech turns represent different kinds of interpretations in the sense of Kvale's Phases one, two and three. Apart from this, the speech turns represent reflections upon topics where both interviewee and interviewer contribute. Furthermore, the excerpts from the texts will show that the topics tend to be interrelated: one brings forth another or leads back to a preceding topic.

Because of the impression Veronika made on me and of the impact the interview made, her main characteristic ought to be presented in a more detailed way. Her pseudonym is an analogy to a film by the Polish director, Krzyszof Kieslowski, because she, as the main character in one of his films named "Veronika's Two Lives," has always lived two parallel lives. Veronika's two lives have been incompatible, and she has been formed by this incompatibility which has blurred boundaries and distinctions, and made truth appear to her as the opposite of reality. To master her two lives the interviewee Veronika had to use all her vitality, strength and creativity. Her efforts to adapt to incompatible realities found their expressions and outer manifestation in particular health problems, complaints and a conduct which professionals in the health care system interpreted as indicating somatic and mental diseases. Her sickness history, its most salient aspects documented in hospital records which Veronika herself had collected, as well as her abuse history both showed a deeply intertwined intrinsic mutuality.[117]

She was the mother of three children, studying toward a post-graduate degree in education when we met. After having mentioned her childhood diseases and periods with headaches and abdominal pain before the age of ten, she continued as follows in a quotation from the interview (V means Veronika, A means Anna Luise):

V: ...and then I had this hearing loss which started when I was around ten-years old.

A: Hearing loss? Do you know what caused it?

V: No, I don't. But it became so serious that I had to use a kind of hearing-aid. I used hearing-aids in both ears for several years. When I went to grammar school I had to use two. But it wasn't a result of infections in my ears - which I never have had. The loss lasted until I was twenty.

A: Did you develop problems with your speech since your hearing was impaired? Did you need help from a speech therapist?

V: No, my hearing decreased slowly. I became very good at lip-reading.

A: You said your hearing loss lasted from when you were ten until you were twenty years old. What happened during the years in between?

> **V:** I had to visit the specialist's clinic for check-ups once or twice every year. The curves always got worse. But nobody could say what happened and why it happened. Of course I was given a diagnosis. But it didn't mean anything to me. I left home when I was twenty, then I moved to L (she mentions a southern University town). My hearing was restored pretty soon. I realized that myself. And the last time I had my ears tested the curves were totally normal. They were compared to those I'd had taken earlier in the out-patient clinics at home. Then I gave back my hearing-aids.

From records and copies of audiograms which Veronika had received and collected during the years of her deafness, I could reconstruct the following sickness history: after several years with steadily increasing hearing difficulties, Veronika, at that time attending primary school, had been sent to an out-patient clinic for ear, nose, and throat diseases to have audiograms performed. The first note on her records read as follows:

> The first audiogram from July this year showed hypacusis neurogenis bilateralis. We advised her to use hearing-aids in case of difficulties at school. Today she presents problems linked to reduced hearing at school. She has tried to sit in different places in her classroom without any positive results. She is able to hear provided she can see the one who is talking, but has general difficulty following a few of the teachers. As things seem today, she has problems at school and should be helped to use hearing-aids.

At this point, Veronika added that her school had never received this information, and that some teachers had doubted her hearing loss. Repeated check-ups in the following years resulted in audiograms with continually dropping curves. Apart from what her general practitioner had noted regarding her hearing were several consultations due to an almost chronic headache which would not respond to analgesics. A note in her records five years later read as follows:

> To our knowledge, the patient's great-grandfather and grandfather suffered from deafness at a young age. The patient herself has been examined here for many years. Her latest audiograms showed increasing impairment as compared to former registrations. The diagnosis is Hypacusis neurogenis hereditaria bilateralis. In cases of such impairment, it's impossible to predict whether, and if so, when, further progression will take place. Therefore it seems of utmost importance that she receives her theoretical education as early as possible.

By one year later, Veronika had moved to a University town and been examined at the out-patient clinic of the University hospital. The record notes and a letter to her former specialist from this date said the following:

> Your patient NN has been referred to us due to a possible allergic reaction. We learned that she had been treated in your ward. I have received copies of the records and the audiograms. She informed us that, in the meantime, she

> has been healed by prayers in a sect. This happened here in our town. After that event she has not used her hearing-aids, and her hearing is completely restored, which will be confirmed by the audiograms included. I don't have any other explanation than that it must have been psychogenic, but I presumed that you too might be interested in knowing this.

Veronika's comment to this letter was: "What the doctor called a sect was the Christian Student-Union at the University. One of the leaders had asked me to permit them to pray for my improvement, and I accepted. At that time, that was the only explanation I could give."

Veronika had been chronically disabled for ten years. But her sickness history had most obviously not developed according to the expectations described in the above records. Nor had any of the usual information concerning a cause been given, or any suggestion for a treatment. Mention had been made of ancestors and a possible family disposition for impaired hearing at a young age, but this hypothesis hadn't been developed further, perhaps due to lack of knowledge about the health in adulthood of the two men mentioned. In addition, the sickness history had been terminated in a rather unusual way. The letter from the specialist to his colleague gave no clues about his real reflections on this case. Had a long-term medical follow-up of a chronic disease found a medically reasonable solution? How could a physician from a third line institution so easily conclude that the restitution was proof that the former chronic state had a psychosocial origin? And how had medicine been involved? There had been a patient with a defect which could be measured with technical tools. The defect had received a medical name. There had been no verified cause, no specified medical treatment, only technical measurement and compensation. Then, all of a sudden, the same patient had presented a normality, demonstrated by the same technical tool as before and directly comparable. The proof of the normality had been constituted in precisely the same way as the proof of the defect. A cause for the normalization had not been found. The physician, when admitting that the hearing capacity was clearly restored, wrote that, "I don't have any other explanation than that it must have been psychogenic, but I presumed that you too might be interested in knowing this." How he might explain the causal relationship between an objectively proven reduction and a subjectively felt origin remains unknown. If the proof of the normality could finally explain the defect as having been caused by emotions, why had the proof of the defect not been presumed to mirror a world of subjective emotions? And in case both proofs had been equally correct, what had made the suggested psychic origin disappear? That this question had not been raised, however, might be interpreted as indicating that a legitimate object of years of specialists' interventions had become devalued by the probability of the ailment having an emotional source. What kind of emotional burden might have been the true origin of the defect was of no interest even to specialists in the field. Had the diagnostic basis been insufficient? Probably not, provided that audiograms are acceptable as a means of objectifying the subjective capability of hearing.

Chronically impaired hearing had become normalized without any medical intervention. How could this happen? How many theoretical problems of greatest relevance, neither mentioned nor discussed, had remained unformulated in the final report? Could it have been fruitful to think of this case as exemplifying a problem which is unmanageable for the most part within medicine? Might this have been an expression of structures, bearing witness to the way in which life in human bodies may remain incomprehensible to the field of medicine because of medical concepts of the meaning of "body" and "human"? In Veronika's case, the biomedical frame of reference had nothing to offer in the way of a consistent interpretation of a disease and a recovery from it.

Could a key for such an interpretation be found in Veronika's biography? The restitution had occurred quickly, the year after she had left her family and moved to the University town. She herself had presented these two events as closely related to each other, as if she perceived their sequence as carrying a particular meaning. What had made it a meaningful connection, capable of providing an entrée into understanding the reversal of a chronic defect? Her answer had been: "When I moved from my father." How had that subjective attribution made sense of the change in her sickness? If moving from her father had had an impact on her hearing, might the proximity to her father then have been related functionally to the failing of her hearing? Here the shortcomings of the medical nomenclature and the focus on defect had become obvious. There had been no need to ask what had caused the defect, only which function the reduced hearing had had. It would, perhaps, have been more fruitful to ask: against what did deafness shield her? She herself gave the answer when talking about how she, as a child, had often been awakened during the night by the sense of somebody in her bed, fondling her. Very soon, she had learned that to pretend to be asleep and let "it" happen shortened its duration. She would stop herself from sensing the adverse touches and sounds. And apart from that, she had experienced repeatedly that her resistance made the nightly visits worse and longer. She would also be punished the next day for some banal transgression, which Veronika interpreted an obvious, delayed reaction to her resistance, a punishment in disguise, so to speak, which only she had been aware of. Her father had been her main abuser from early childhood. An uncle had abused her as well during her visits to the home of those relatives.

She had been made non-resistant, had been silenced, and she had numbed herself gradually. Later, her father had abused her during the day as well. What he said and asked for had been unbearable. She had refused to listen, hear, and understand. Her protection had been, among other things, to lose the capacity to hear, a measurable defect. Was Veronika's deafness a disease, then, according to the criteria for diagnosis and classification, or rather an incredibly creative act to protect health, dignity and personal integrity? Did it testify to her lack of capability, or rather express her vitality, her will to defend herself despite all violation, her potential for resisting destruction, and her courage in maintaining dignity despite all humiliation? Could her deafness be interpreted as an incomprehensible potential instead of a measurable failure? Veronika had to cope with her own, hidden reality. According to social rules, this reality was both non-real, in the sense of non-existent, and invisible to others. Her individual perception of and reaction to this "private" reality, which apparently was non-existent to everyone but herself, had thus the outer appearance of a bodily defect coming from within. It thereby became a task for medicine, where bodily defects are conceived of as diseases provided that their substrate can be made visible by the tools which are currently available, and which are acknowledged by professional consensus to have a documentary potential. Documentation in medicine means making visible. The tools for making the substrate of diseases visible are expanded eyes. What is made visible is made real; it is both realized and verified.

Could Veronika's deafness be a field for the exploration of interfaces, of layers upon layers of mutually constitutive phenomena, as the hidden and the visible, the non-real and the realized, the individually experienced and the objectively shown, the sensory perception and the bodily reaction, the world of particular sensations and the arena of generalized proofs? Was Veronika's deafness a possible doorway into a field of obvious polarities, as they are conceived of by members of dualistic cultures? Could her sensory "defect" be read as "potential" and could this reading be a challenge to the concept of either/or as expressed in the opposing concepts of either mind or body, either active or passive, either action or reaction, either sick or healthy? Might the shield against the sounds of invasion and the voices of intrusion, expressed in her reduced communicative will/ability, represent a key concept of perception and memory? Was there a chain, reaching from what, by socio-cultural consent, cannot be real (sexual molestation of a daughter by her father), to what is realized by professio-cultural consent (deviating curves in audiograms)?

In that case, what were the parts of it and how were they linked? Or was "chain" a rather misleading metaphor for what had been at stake? Could "reading," stepwise, her denial/inability to communicate with an unacceptable reality give entrance to what appeared rather to have been untranslatable communications about incompatible realities and incongruent worlds? The abuse as such, forbidden and hidden, had not been communicated to anyone other than the abused girl. The girl had denied all communication by becoming deaf, numbed and silent. This reaction had been interpreted as indicating a sensory dysfunction. That sense, as located in the body and linked to organs, made the dysfunction a bodily symptom. As such, it entered the arena of the medical profession. The medical frame of reference and the professional language of pathology turned the dysfunction into a bodily defect. The defect as localized in the auditory sensory system was rendered visible by instruments designed to prove what is deviant.

Following the stepwise transition of this phenomenon, it did not seem "translated" in the sense of a mediation of meaning from one system of symbols to another. Had it been transformed instead of translated? Obviously not, as the outer form, the appearance, had remained unchanged. Could one think of a process of "trans-meaning," a change of meaning on its way from origin to naming? Might Veronika's proven deafness be interpreted as an expression of a social structuring, influenced by forces in the sense of society and informed by institutionalized concepts of power? What had happened?

Provided that Veronika's memory of being abused by her father for years was true, the abuse, the core of a mad human relationship, had become a medical diagnosis, the individualized defect of the abused person. The failure had become hers, residing in organs of her body, proven instrumentally and explained in terms of heredity. If this was a correct reading, what was the underlying rationale, the hidden intention, the driving force? Veronika's deafness had been registered by her surroundings. It had been a making public of something, attracting other's eyes to her. Focusing on her had led to calling upon the medical gaze. The medical gaze found the something in her. Thereby, a subject's reaction had become the medically verified cause of an object's defect. No other origin needed to be considered as the 'deafness' had found its medical explanation; could this have been guided by hidden interests? And, in that case, which?

Had medicine contributed according to the way the medical profession ought, in other words, for the best of the patient? Had Veronika received help? Had artificial hearing-aids improved her health? How had Veronika experienced the examinations and interventions? I quote from the text after she had told about how she had behaved during the acts of abuse as described above:

> **A:** After having listened to you I must wonder if to "become deaf" and thereby create a room which was yours alone - which nobody else could enter - was what kept you alive, what let you survive?
>
> **V:** Yes, I've thought so myself. What you say now is what I've thought. And, also, I read. I read whatever I could get and stayed in the library for hours and hours.
>
> **A:** I consider your 'becoming deaf' as very inventive.
>
> **V:** It just happened, it wasn't conscious, it was guided by something else. Later, there was a couple in my neighborhood, both physicians. We met often, and they were close during that period when things turned up. I kept in touch, and when I told them that I'd experienced incest, that it was a part of me as well - suddenly the puzzle pieces they had recognized found their place. The missing pieces were there.
>
> **A:** For your doctors, your life had a lot of blanks, and what was needed in order to understand was lacking. They didn't reach it. And you are on your way to filling the holes. Because now you have acknowledged that the curves went down the greater your troubles at home (as referred to before). Like a seismograph for family conflicts. Incredible.
>
> **V:** I was very tense every time they took a test, whether the curves had gone up. But they never did, they always continued to go down. Further and further down. I checked every time how they looked. Because I had the feeling of them going up and down, that they now and then had jumped up again, so I might be unmasked, but I couldn't hear more of those sounds anyhow. I switched off, I didn't understand what they were doing, it was something or other about being found out, but I didn't know exactly how I'd be found out, but I was scared to be seen through, that someone would understand something.
>
> **A:** To be unmasked by means of an instrument which you knew was regarded as a kind of truth-detector, was it that? As it's used to tell the truth - and the truth matched what you wished for, namely that you didn't want to hear ...
>
> **V:** Yes. Yes, it was like that. It was in my mind all the time, and I became divided into two parts. I had to hide the truth. I mean the real truth.

What kind of confusion of terms did Veronika express, and what kind of confusion of realities had she to face and to balance? Did her perception of reality, which she had to hide from others, correspond to the reality of the curves, brought about by a machine which materialized what was actually a response to a hidden truth? Veronika, in her conflict between a need to hide and a wish to be seen, experienced the examinations both as a threat and a hopeful possibility. The machine produced curves which both gave witness to and hid, depending on whose eyes looked at them. Veronika "read" them with eyes both wishful and fearful, as a way of proving that she did not hear what she did not want to, and as a possible way of disclosing her unreal deafness since she did not really know how the curves were produced. The physicians "read" them as a concrete, direct and unmistakable representation of an equal degree of reduced functioning on both sides of the auditory system.

What are the valid criteria for medical proofs? What kind of reality do they mirror? Which realities do physicians access with the tools they apply? According to which concept of the human body and senses are these tools constructed? How do they, and the concepts which have informed their shape, influence medical interpretation of a persons' actions? What are physicians prevented from acknowledging about human life via these tools? Which consensus about normality is built into them? What is, consequently, abnormal? On which definition do physicians base their perception of rationality? And what will appear irrational as a consequence?

Veronika's sickness history had been commented upon in a short letter from a doctor to a colleague. There, the sudden disappearance of her chronic deafness was presented as retrospective proof of its non-somatic nature. The sudden restoration of function without any medical assistance turned a disease into a non-disease. The term psychogenic was the last step in the process of a non-translation. Now, a chronic defect had not only disappeared, but had proven itself not to be in the body at all, which equals it not really having been verified, despite its visibility. Had the disappearance falsified the curves? Had the truth of the diagnosis disappeared by means of the same instrument which, for ten years had shown something which now was shown to have been nothing? How could physicians relate to and rely on such contradictory concepts of what forms the legitimate basis for a defined disease calling for a diagnosis in order to be classified properly? By decoding the disease and re-coding it as an emotional disorder, the label from the ICD, International Classification System of Diseases, was removed and a label from the DSM, Diagnostic System of Mental Disorders, was indicated. The new label, hysterical deafness, suddenly seemed more appropriate. Had Veronika simply suffered from hysterical deafness? And in that case, did this indicate, as opposed to the diagnosis of hypacusis, that the cause of the disorder had not been within her? Were outer origins considered now? Possibly, though not expressed. The shift of label, mirroring a shift from one explanatory system to another, did not initiate a broadening of the medical perspective. Rather the opposite seemed to have occurred: the case was not a case any more. The physician was even slightly uncertain whether his communication of the "healing" of a chronic disease really was of interest for those colleagues who had diagnosed, measured and treated it for ten years.

Did this mean that Veronika was perceived of as both the locus and the cause of a disease which revealed itself as madness by its disappearance? What kind of logic was expressed in this shift of perspective in a medical document, decoding a condition from having been a somatic fact and re-coding it into a mental non-fact? Was the obvious lack of curiosity concerning this rather unexplained case, as it was demonstrated in the final remark, an expression of a lack of awareness about theoretical inconsistencies in the field of medicine? The shift from somatic to psychic could hardly be a shift of field, indicating the limit for medical terrain and responsibility. However, the shift was not only a shift, but a degradation of a somatic problem to a madness.

The Norwegian psychiatrist, Tormod Huseby,[118] reported a case from his psychiatric ward concerning a female patient who, from early childhood, suffered from an aggressive arthritis. Twenty-three years old, she was totally disabled by chronic pain, had had several rheumasurgical interventions, and a strumectomy. After her first delivery at the age of twenty-five, she became psychotic, with visual and auditory hallucinations, thus unable to take care of her child. When apsychotic again, she insisted on being sterilized. After one year without treatment, she had to be admitted again to a psychiatric ward as she was suicidal, psychotic and losing consciousness repeatedly. Tests for epilepsy or an intracranial tumor were negative. During the following four years she was admitted twelve times. After her sixth admission she received the diagnosis, "schizophrenia, paranoid type, chronic." At twenty-nine years old she was referred to the long-term ward because of a therapy resistance and a bad prognosis. As it happened, she was the only patient available for the training of a young doctor. This new relationship brought forth totally new aspects. Reviewing her file, it became apparent that, whenever admitted to the psychiatric ward, the woman had talked about having been raped. This story had been interpreted, again and again, as "symbolic of the patient's life situation." [119]

But now, sexual abuse in childhood and a forcible rape in late adolescence were gradually uncovered. One week after the rapist was identified, the woman was apsychotic. Her nightmares disappeared, her sleep improved, her depression faded. Seven months later she was released to her home with a referral to a local support network.

Huseby describes how the visual and auditory hallucinations become the doorway to the woman's hidden world of experience. The very moment she feels confident that what she tells will be met with an open mind, a detailed description of her visions, intrusive smells, sounds and tastes is brought about. She describes her rapist whom she does not know, but whom she has talked about and named each time she was admitted, so that her mother is able to identify the man. When connecting this man to rape, a discrepancy in time appears, which, in turn, indicates abusive events prior to this rape. By and by, this is confirmed in cooperation between the patient, her mother and the therapist. The rapist and the childhood molester are identified and the patient improves substantially within a few days of each step in the process of recalling, telling and confirming.[120]

Huseby concludes: "this case demonstrates how long-term psychotic symptoms may hide severe psychic trauma contributing to failure of function and chronification." Thus a history of suffering becomes almost transparent in the light of knowledge about hidden violation. I would like to focus on another quotation from the article: "From her very first admission, the women had talked about having been raped. This story had been interpreted, again and again, as symbolic of the patient's life situation." In a reformulation, linking those two quotations and thereby accentuating their obviously related meaning, one might say: 'The case demonstrates how severe traumata which the patient herself has mentioned again and again, become visible in her psychotic state. As they are interpreted according to psychiatric theory, however, they are rendered false. This results in an ever-increasing failure to function and a chronification of unrecognized violation.' As long as psychiatry defines the individual to be the place and the source of its madness, psychiatry steers its eyes away from the madness within human relations, and certainly away from social taboos and the asymmetry of power, wherein some people take so much space that others become deformed by their impact. Is it probable that a dogmatic attitude towards what must not be true, forces psychiatrists to ignore relevant information in order to maintain a classification system of individualized mental deficiency? [121]

This question leads back to Veronika, who, by being healed of a disease was presumed to have had a psychogenic disorder. The physician who made the suggestion represented a somatic specialty in medicine. But, did Veronika ever have her mental state classified according to the psychiatric diagnostic system and theory? She had been admitted to the psychiatric ward twice, the first time for eight days, the next time for more than three months. What did she refer from these admissions?

V: I was in the ward for one week, because I couldn't stand it any more. I was referred to the acute ward. But there they didn't want to talk about incest at all. They told me that I had my psychologist to talk to after my release, but at the ward they wouldn't talk about it. They were afraid I'd act out, and besides, it (incest) used to be such a common occurrence that it wasn't anything special for someone my age, and this was the reason why they didn't want me to talk about that in the ward. But I was in trouble anyhow.

A: It was made unreal for you. You really had not experienced anything special.

V: Yes. That's how it felt. It was very tough to be there. I had no idea what to talk to them about. I had no idea why I was there. I didn't understand what they wanted. I started to make drawings instead, but nobody would see them.

A: What did you draw?

V: I mixed events as they were in my head, with emotions. With strong emotions. They thought it had relations to a rape. The newspapers had been writing about rape all the week. About a little girl one suspected had been raped on her way home. There was lots of noise about it, mainly the horrible physical things that had happened. And I thought that, of course, the physical part was bad, but in that ward I hadn't been allowed to talk about the other side, which also happened. Everybody was very preoccupied with what had happened to a little girl, but nobody could deal with what had happened to me. Apart from that, I'd recognized they were inexperienced concerning incest. They had a lot of strange excuses. In a psychiatric ward, the staff should be trained to take up things like that.

A: Right you are. But psychiatric understanding is still based on a theory of incest being more often a fantasy than a reality.

V: Yes. Therefore you have to be very tough to dare say that you're an incest survivor. I'd been referred because I was suicidal. After one week, I said I wasn't suicidal any more. I wanted to go home. I didn't want to be there anymore. The very day I myself felt that I was in control again, I asked to be released. But I've been taking medication for more than two years now, after having had this psychomotor therapy. And I see my psychologist regularly. It helps. The voices I heard before, which scared me so much, they don't come so often anymore.

A: You were scared by your own perceptions?

V: Yes, it was scary to be there (a room with voices, referred to earlier) and not be able to leave it. It didn't help to try to leave, they would follow.

A: The medication made this part of your inner world more tolerable?

V: I don't know. But I couldn't have gone on that way. I don't know how it would have been without the pills. I couldn't work. Now I get a limited disability pension. I couldn't continue in my job. I want to go further in my profession, so I have to complete my academic training, which is the reason that I've received the disability grant.

These excerpts concerning Veronika's perception of her encounters with psychiatry are in her own voice. How was the voice of psychiatry concerning this episode? The diagnosis given in Veronika's record was Depressive Neurosis 300.4 and the reasons for referral mentioned, were attempted suicide and depression. From these records, one can learn that the staff of the ward had been well aware of the patient's very traumatic background, of sexual abuse by her father and by several male family-members from the age of three or four, of a dramatic rape at home at the age of fourteen and repeated suicidality and several attempts to commit suicide, whereof one, at the age of eighteen, had almost been successful. The staff members were also aware that Veronika felt burned-out, that her sleep was severely disturbed by intrusive nightmares and that her suicidality felt overwhelming. Veronika had expressed her fear that something serious might happen as she had no control at all. She was described as silent, withdrawn, unwilling to talk about the sexual assaults, but expressing feelings of guilt. Concerning Veronika's wish to talk about the abuse, which she explicitly addressed in the previous excerpt, the record is as follows, in the words of the ward physician:

> The patient was admitted due to her increasing anxiety over not being able to control her suicidality. According to the staff at the ward, it soon became clear that the patient wished to limit the treatment during this stay to issues regarding her suicidal tendencies. The long-term therapy concerning her incest problems was delegated to her psychologist.

Although syntactically very ambiguously or imprecisely expressed, the staff seems to have met her according to what was perceived as her "wish to limit thc treatment," which meant not to deal with the incest problem but to concentrate on the actual situation, her suicidal tendencies. No mention, however, is made of how the "suicidality" could be met without addressing the underlying emotions and motives for it. Veronika's withdrawal was mentioned, as was her tendency to be on her own in her room, drawing. It was described as though it were voluntary, as a demand which was respected, though it had not been expressed verbally. Mention was also made of her general somatic state, and especially her nausea, her problems with eating and swallowing.

Veronika's own presentation of her stay in the ward, and the hospital's record did not differ as to her condition, conduct or as to other information. What differed was the interpretation of why the issue of incest and sexual molestation was not dealt with, although it was known that she had suffered a great amount of traumatic experiences and had repeatedly attempted suicide from early age. Obviously, the members of the staff had not been able at all to see Veronika's despair and the overall importance of the abuse experience as its matrix, which is even emphasized by Veronika in comments on some events mentioned later in our talk about that week.

> **V:** I talked to one of the others - you find each other in those wards - and she understood my problem in a way, although I hadn't told her exactly. It's strange, but there were three of us in that ward. We walked around

afraid. Because it was just in that week this little girl had been raped. The staff talked about a suspect, and then we were told that person had been sent to this very hospital. We panicked immediately, wondering where he was. We saw the police-cars outside. The staff didn't understand us at all. They said we were hysterical. This happens when people don't understand. But we were scared knowing a rapist was around. We were completely in a panic every time a male member of the staff opened the door to our rooms after bedtime. That was enough. We went to bed fully dressed and with extra clothes, we were constantly on alert, we didn't dare to move, we were scared. We were strange.

A: But it's not strange to react like this when having experienced rape. It would be strange not to react.

V: Yes. I think so too. (long silence)

A: You said 'there were three of us', but you were referred because of being suicidal ...

V: Three of us had been sexually assaulted. We found each other. We recognized the pattern in each other, although one of them hadn't thought of it as a cause for her actual trouble. She hadn't reflected on it. She told us her story and we understood that it was the same. I can't comprehend that people working in such a ward don't understand that people who talk about their abuse have come far already. We recognized a pattern in each other. I don't know exactly how. But I see it pretty quickly.

A: I understand that you have developed a sensitivity to things others don't register and which tell you something which other people could also be able to hear, see or sense if they wanted to. What prevents others from seeing?

V: I guess, in the first place, they have to believe that such things are true. But it may happen that they don't want to go into such thoughts. What's lacking is an ability to imagine what really happens and how.[122] '

The communication between Veronika and the members of the staff in a psychiatric ward had, it seemed, been inadequate. Veronika's recollections and the text of the record correspond as to details of description - but not as to the meaning of what is described. Two worlds seemed to exist side-by-side, their interface consisting of words for describing events or problems. The staff had felt obliged to respond to what they interpreted as being Veronika's expressed wish to be silent about experiences of sexual assault. This made it legitimate - or gave a reason - to concentrate on what was labeled "her suicidality." Veronika, on the other hand, had made it explicit that she needed to talk as what had happened felt omnipresent, too intrusive to ignore and too overwhelming to control. In addition, there was the sheer coincidence of a rape in the outer world, a rapist within the same hospital, and an array of what had been termed "hysterical" and "strange" reactions displayed by three patients whom the staff knew had been raped - none of which was mentioned in the record, obviously regarded as not being of any informative value to any one. Clearly, to mention a patient's withdrawal into isolation with paper and a pencil, is to have seen something. But nobody saw the drawings. To mention a rape experience is to confirm having received a piece of information. Not to see a raped person's panic in the potential presence of a rapist as a sign, but rather misinterpret it as indicating hysteria, is more than simply not to see.

The term, suicidality, seemed to have acquired an array of functions in this story. The language of psychiatry and the praxis of the psychiatric staff made of it an entity, a limited fact, as it were, something to deal with and to concentrate on, almost as if it were a thing or a characteristic trait of the person to whom this word had been attached. By this very practice of naming, the name became a wall behind which a woman tried to express something about why she could not live anymore and why she was afraid she might be unable to resist the destructive forces she had internalized. She herself saw how the wall was established, how she was placed on the other side of it, how she was left alone, literally, in her room, while drawing what she could not say. But no one asked to see her attempts at externalizing what felt too explosive inside her. From Veronika's side, this "wall" gradually revealed its function as a shield behind which the others could hide their inability to meet her, their lack of knowledge and skills, their lack of courage to dare to imagine what it meant and how it was done, this something which nobody wanted to talk about with her, although the issue was also present externally, in the form of actual dramatic and sensational events in the hospital.

Veronika came to the long-term ward of the same hospital a year later. That time she stayed, participating in group settings which used expressive and creative activities as tools for approaching traumatic experiences. She was still described as "being unable to put words to thoughts and reactions," but was communicating through drawings, mainly in individual therapy settings. She was still labeled with diagnostic, psychiatric terms: Panic Neurosis 300.01 and Post-traumatic Stress Disorder 308.31. During her stay, she suffered from pain in her back and was referred to a physiotherapist. A short note in her records indicate some possibly adverse reactions to the examination done by the therapist, which were interpreted as being connected to her abuse experiences. A more detailed description of the effect was not given so the remark concerning Veronika's reaction was not quite clear. In our talk, there came a long passage concerning her physical pain wherein the various kinds of evaluation and treatment she had received during the years were mentioned. Veronika spoke of one particular episode to which the record had referred. An excerpt from the interview text concerning her pain now follows:

V: I had pain in all my limbs, pain everywhere. I believed it was arthritis. I consulted specialists, but they didn't find anything. The same was the case for my head, I always had a headache. I hurt, I had pain all over. I can still feel it now and then, when there are difficulties around. I believe that everything around me can lead to pain in my body.

A: OK, the worse things get 'around you,' the more pain you have in your head and body?

V: Yes. I guess I didn't realize this connection before I was twenty. Then my doctor said that he believed in a connection. By then, I'd gone to him many times during a few months, because of pain in my head and pain in my limbs and in my body and in my heart and everywhere. Then he asked me how I was. He was the first one who had ever asked me how I was. He asked whether there might be a connection between how my life was, and all the complaints in my body.

A: Had he known you for a long time?

V: No. I went to him during a year or two for everything that was wrong with me - after I'd left home. He didn't know my family. But I got angry and left. I didn't want to hear such things. I didn't want to talk about it. My hearing improved very fast. But my pain returns frequently, when I feel tired, when I have a lot of nightmares or ... then it returns, either in my back or somewhere else.

A: Do your nightmares come more often when there is something going on 'around you'?

V: Yes, that's a distinct sign. And then I get this headache so I can hardly walk with my back straight, or it starts in my back, so I can't do a thing. Now, I acknowledge this means 'there is something.' But very often I don't know what it means and where it comes from. And I don't know why it comes still, now, I mean as everything is over. I have a problem making clear what I want to say.

A: I guess I understand. 'That' is over, yes. But the traces, the memory of it, are still in your body, in your head, in your memory, aren't they? And there might be a lot of things in the world which awaken a memory?

V: Yes, that's right. Sounds, smells, especially smells. Yes, smells. But sounds as well. Or what happens. When the curtain in the shower moves (holds her breath for a while). Or if somebody touches a certain place on my back, then... it works very fast ... then my headache returns. I had psychomotoric therapy last year. But I had to stop. It didn't work. She (the therapist) took my head while I was lying on the bench, it was the first time I lay down, suddenly, I didn't understand, everything was ... there was an explosion, it happened ... and then I heard voices ... and then I heard ... I disappeared into a world right beside me ...

A: ... when your head was held in a certain way ...

V: Yes. There suddenly all the other things were back again. Everything mixed up, past and present, just in a turmoil ... something like that had happened before, but not as drastic as this time, but it comes like flashbacks, in periods very often. I had to stop. This contributed to my having to stop. It was too intense. Finally, I didn't understand a thing ...I couldn't function at all, just these voices, these sounds, I knew what it was, but got I scared even though I knew. In the beginning I could cope, the first therapy sessions were acceptable, but the last few times after she had held my head ...I couldn't continue.

A: A great deal of anxiety seems to be linked to your head?

V: Yes, because it was held ...like that ...I didn't hold it myself ...that didn't work.

A: Might there be concrete reasons? Might this say that you've experienced your head being held and a great anxiety being evoked? [123]

V: I got breathless - I don't know where I got air from - although nothing happened there and then - the memories took my breath - yes - and I guess it was my head being held against father - against his body - pushed against a body ...

A: ...when he was lying on you? ...when he forced you to suck him?

V: Yes. The latter. That was it ...I can't eat for a long time afterwards ... then I have this taste in my mouth...

How should a physician relate to this story thematizing the medical terms "persistent or intermittent pain," "panic-anxiety," "auditory hallucination," "psychotic state" and "eating problems"? Of course, Veronika had not used these terms. She had described what she had felt, sensed, experienced more or less frequently, depictions which would activate a physician's search for appropriate medical names: pain everywhere linked to something "around" which could be things, events, sensations or sensory perceptions; panic evoked by a grip around the head when close to another human body; auditory hallucinations consisting of intrusive and intolerable voices; a state of acute psychosis occurring suddenly in a psychomotoric session; eating problems due to a particular taste in the mouth; a headache as a response to a certain movement in the shower curtain. If this last connection should be considered, this certain movement had to mean something very specific which only the one having the pain could know about. How could any physician, being asked about the possible or probable origin of a frequent pain after having a shower ever be able to guess that the person in pain, seeing the curtain move in a certain way, was, by this optical perception, pushed back into a past where this movement meant a demand from somebody outside the shower, an offensive and painful demand without words, signaled by the movement of the curtain? [124] The medical gaze would see nothing: no proof of pain, no origin of the sensation of pain, no damage to tissue, no cause, no measurable impulses in the nerves, no effect of analgesics. But still there would be a person in pain, complaining, unable to work, move, sleep, relax or function. Could pain be grasped, understood and healed by a process of searching for and finding particular embodiments? Could human imagination prove to be a more proper tool than the technical eyes of the medical armamentarium?

Veronika presented yet another challenge to medical concepts in her narrative about the voices. In one way, it may be read as a presentation of an auditory hallucination, and in an other way, it may be seen as an icon. What she tried to tell and what she gradually recalled was nothing less than a breakdown in the distinction of tenses, the past emerging in the present and observed by the one experiencing this breakdown, which in a way can be called a three-fold simultaneity. She stated verbally that past and present were mixed up and that she herself felt that she went into a world or room beside her (see excerpt above), where the voices and sounds were. The breakdown of the tenses had been provoked by a sensation. In a prone position, two hands had held her head against a body, making it impossible for Veronika to hold her own head and to move away from that body. During the interview, by daring to re-imagine the breath-taking anxiety called forth by a particular combination of sensory perceptions (tight grip around her head/ inability to move it/ it being pressed against a body) and provoked by a direct question to which she, apparently, didn't respond directly, Veronika could make an association to her father's body. Still, she could not mention the kind of situation with which these perceptions were identical. Her associative recall, more a rush of pieces than a flow of impressions or a sentence, ended open. She remained in, "...being held against father - against his body - pushed against a body ...," the intonation of her voice remaining raised, indicated as an unfinished sentence, represented with a dotted line in the transcript, and then continued in two, concrete, short questions derived from the image she had alluded by her way of uttering. The answer was clear. She said: "Yes. The latter. That was it..." The answer to the questions emerging from the incomprehensible reaction during a psychomotor therapy session was: forced fellatio.

A very particular mixture of bodily perceptions had shown to be too similar to an abusive experience to be tolerable. They were even so similar that they prohibited maintaining the therapy. They were so intensely similar that they erased the imaginary distinction between the tenses, generating a powerful demonstration of human corporeality. And they were so overwhelmingly similar that they broke down a wall Veronika had established during the time of her protective deafness. They created "...an explosion, it happened...and then I heard voices...and then I heard...I disappeared into a world right beside me ..." In this same passage she had mentioned that she knew what the voices and the sounds were all about. And still, she said, she felt scared. In a later sequence in our dialogue, when talking about the distinction between the true and the real, we returned to the voices and sounds.

A: The problem with what was true and what was real became really complicated, just in the same way as when your head was held and you dropped into this other world. Have you heard what these voices say? Do you recognize something?

V: You mean whether I've heard them before? Yes, indeed. I recognize everything. It's quite distinct, exactly as you talk to me. Just like that.

A: Have you dared to repeat it, speak it aloud?

V: No.

A: Do you block yourself?

V: Yes. (long silence) It's quite distinct. I know what they say. I'm not quite certain who is talking at any one time. I've heard everything before.

The last thread had emerged, the web was complete. Here they were, the sounds and the voices Veronika had deafened herself against, released by a grip of hands around her head, stored in the room of her auditory memory and "forgotten." The entrance to this room could not be words, the access could not be found in the verbal aspect of a psychotherapeutic setting. An entrance was encoded into a sensorimotor perception of forced immobility of the head. The mountain could not open unless the Open Sesame was found, released, by chance, in a situation framed totally differently. Only something which was similar enough to a specific, idiosyncratic pattern could open it up. It would have been impossible to search for it, as nobody could have known what was similar as the similarity could neither be seen nor defined "from outside." What was similar enough could only be felt "from inside."

Could this mean that the events I had been searching for, the hidden assaults of personal, physical and sexual integrity I wanted to encounter, had been embodied in a manner which rendered them inaccessible to words, and unapproachable for thought? Was this another aspect of integrity assault, and in particular of sexual molestations? I had acknowledged the difficulties of speech, voice, words and language before. To tell from unspoken, wordless and unnamed events, perhaps having taken place at an age before language had been acquired, and as such, never having been encoded into words, was obviously difficult. To tell about crime, taboo violation, shame and guilt, humiliation and defeat, was most definitely not easy. But to tell about what is not accessible to words, is impossible. Was this sensorimotoric entrance into the world of trauma a special variant of memory, a kind of experience-embodiment totally and principally different from the experience of the pleasurable, the delightful, the arousing or ecstatic? And if so, how should such an entrance be found by a therapist ignorant of former molestation, forced penetration and abuse?

Could it be, as an analogy, that other "bodily events" might also provide such an entrance, in particular those in which one's body was the locus of the event, and the event was of a kind which demanded other's helping hands, as for example during a delivery? Could Veronika tell anything about that from her three pregnancies and deliveries?

> **V:** I was depressed almost all the time, during each pregnancy. I didn't feel pregnant, wasn't ready for it, made no preparations. There was a distance, especially the last time, when I'd bled at the start and was convinced it (the pregnancy) wouldn't continue. I didn't feel pregnant, I really wasn't. But it continued even afterwards. I didn't feel that I'd given birth. During the deliveries I had bled heavily, my uterus wouldn't contract, they said. So I was rather weak. And I became quite anemic.

This information had been given in summary during the first part of the interview when talking about the sickness history. Veronika had had a long recuperative period after every pregnancy, several months of depression, and very little interest in her surroundings and in everyday life. Later, we came back to this topic in relation to pregnancy examinations, conducted by different doctors because she moved from one place to another.

> **A:** You still kept the door closed, and the abuse was still hidden. Was that what made you react so intensely after every pregnancy?
>
> **V:** Yes. Especially the first time. The midwife placed herself on my stomach ... she had to press the baby out ... in the beginning I just faded, then I collapsed - the baby was well, I mean it turned out all right, but precisely that (strong accent) was too hard.
>
> **A:** You moved out, you left the place when somebody lay on your stomach ...
>
> **V:** I had problems pushing, I had no idea how the child ought to come out, I didn't understand what was going on. I had ... this is also something I don't understand ... I had problems during every delivery, I had no idea how to push, I had no idea which muscles there are in my stomach, how I could reach them, I had no connection that way.

Not feeling pregnant; not preparing for a child; not being mentally directed towards delivery; not having contact with certain parts of the body; not being able to use muscles; losing consciousness and collapsing the moment someone lies on her stomach; not reacting adequately "inside," not feeling having given birth, moving away, keeping distance during the process, and later - reacting with depression, disinterest and withdrawal. Could all this be a consequence of a previously acquired technique of "leaving," of keeping a distance from acts aiming for the body or parts of it, of having one's body alienated so what was going on with or in it could not be tolerated unless one "moved away?" Dissociation as adaptation to inescapable forces? [125] If so, this presented a threat to mother and child. Veronika's survival strategies from abuse, reactivated by events in her body, meant both a threat to the child and herself. Could withdrawal and distance even reach the endocrine system so it would not know its timing during the process of delivery, leaving the uterus atonic and rendering the woman at risk to blood loss? Veronika had needed blood transfusions every time, creating an emergency which led to her being given the advice, after the second delivery, that, if there were a future pregnancy she ought to warn the staff of the delivery ward in advance of this tendency to hemorrhage.[126]

As a conclusion, the interview with Veronika had presented several themes which evoked curiosity, created associations, directed attention and represented challenges. It generated most of the radically critical questions concerning biomedical theory which were explored throughout the present research project.

Chapter 2

UNFOLDING THE IMPACT OF SEXUAL VIOLATIONS

INTRODUCTION

From the very beginning, hermeneutical activities were integrated into every phase of the bringing forth of empirical material. Every narrative account was grounded in the teller's previous interpretations; every interview situation shaped a particular context; every dialogue was a joint activity of understanding and validating; the process of successively interviewing a group of individuals meant a continuous process of learning, searching and configuring; the steps transforming a recollection into a narrative, a talk, a transcript, and an analysis, encompassed a flow of mutually informing acts of interpretation.

In every single interview situation, and during the whole process of interviewing and interpreting, I was guided by the basic questions: *"What does this person talk about? What do these people thematize? How does s/he address personal experiences? What are these people's main concerns?"* These questions informed my listening, asking and speculating during the conversations and during the analysis, as I sought a doorway into experiential areas on an individual and a group level. My attention was directed to the themes or messages emerging from the various interviewee statements, as was demonstrated in the description of Veronika's interview. Whenever I surmised I was encountering a "theme" which was new to me, I tried to explore it immediately, which I shall show in the analysis of Fredric's interview in what follows.[127] When I surmised I was encountering a theme previously addressed, however, I sought to validate both my supposition and, simultaneously, if possible, the theme itself by reintroducing it into the flow of talk, as demonstrated in

Veronika's interview in connection with the breakdown of the sense of time in a therapeutic session.

The variety of themes which had presented themselves, so to speak, in the flow of speech and the flow of the interviews, were finally checked out as to their consistency or discreteness. After having performed thirty-five, and transcribed thirty-four interviews, I let my further interpreting be guided by a few principles which I presumed would allow new analytical steps. First, previously encountered themes were given a preliminary name or title according to the formulations in the interviews. A frequently thematized field was, for example, "feeling bound, obligated, trapped," another was, "trying to cope with ...," and yet another was "knowing abused people." Together, these themes comprised thirteen associative fields. Next, when trying to verify them as they were presented in the interviews, several overlapped or proved to represent modalities of an overriding topic. Third, while searching for their various presentations in the interviews, some themes proved to have been offered by the interviewee while talking, but had not been grasped by the interviewer. This concerned, for example, three of the sixteen interviews addressing the theme, "having one's mouth abused." Fourth, when again applying the initial question, "what do these people talk about?" some of the preliminary themes appeared subordinate to others. An example might be the theme, "feeling seen," which, in some cases expressed modalities of the overriding topic, "strained relationships," while, in some other narratives, the final conclusion sorted under "unheard messages."

The process of reading as an extension of listening, talking, asking and exploring during the interviews, brought about seven specific sorts of structures and problems formulations which remained consistent during the steps described above. They were addressed by each of the interviewees, and they were thematized in various ways, even, at times, presenting in several modalities within one interview. According to my interpretation of the descriptions and my perception of a confluence, I named these patterns: "lived meanings", "confused judgments," "maladaptive adaptations," strained relationships," "recognized memories," "unheard messages," and "reactivated experiences."

In the following seven chapters, these patterns shall be exemplified and configured. Their various modalities and the breadth of their variation will be shown. Interview excerpts will be used to demonstrate both how a particular context influences the interpretation, and to show how the theory and methodology chosen are interlinked. Likewise, the relevance of theoretical considerations for the interview practice shall be explored. Finally, the interview situations shall be addressed as an important and influential part of the empirical field in the present research project.

1. LIVED MEANINGS

1.1 FREDRIC

From the age of ten and for approximately five years, Fredric had been abused by two young men. When 21 years old, he was referred to a neurology ward due to seizures observed by several people, among those his father and a ten-year old boy. According to the witnesses, the seizures appeared epileptic, warranting an immediate referral to a tertiary-care unit. In the interview, this topic is addressed as follows:

A: These epileptic attacks are what you have experienced as health problems, nothing else of a serious nature?

F: No, only these attacks. The first two were truly provoked by *hard labor*, on those occasions I'd done *hard labor*. All of them took place outside. And it was warm weather, the first two times. It happened so fast that I didn't understand a thing. I became rather dizzy, I slept some hours afterwards.

A: Do you remember these attacks, can you describe how they felt?

F: Well, it happened very quickly, very suddenly, I'd talked to somebody I worked with, all of a sudden I felt strange, it happened so fast, no warnings at all, no warnings really. It was just one, two, three, like that.

A: Some people say they smell something strange or become hypersensitive to light or feel a trembling in their body ... and then begins what you can't control.

F: I felt *a little trembling* in my body, but it started here (pointing with both index fingers to his temples, placing the hands along the sides of his face.) [128] ... which means I went right down. It was more like a little, a kind of *intense pain, perhaps a little, as inside the head, and so I lost consciousness*.

A: Were you alone?

F: The two first times I had somebody beside me, the third time I was alone.

A: So, somebody found you immediately, did something immediately?

F: Yes.

A: Did they tell you later?

F: Yes, they told me that I, I had such typical epileptic seizures, all the symptoms, I was lying there having a seizure, frothing at the mouth, I don't remember whether I bit my tongue, I bit my teeth, I do remember. (italics mine, and will be commented upon later).

Some aspects of theoretical and methodological nature may be glimpsed, mirrored in the shape of this text concerning the interviewee's major illness. The excerpt contains many speech turns making it look more like an interrogation than a narrative about a particular sickness experience. The obviously probing style seems to be influenced by an underlying agenda which steers the questions, the common center of which are the immediate circumstances at the onset of the seizure attacks. Their insistence on detailed recollection seems to be guided by certain motives not as yet revealed, and of associations not brought into the dialogue so far. This passage, interviewer questions in a short and frequent turn alternating with interviewee answers, expresses an interviewer presupposition. Indeed, there are two presuppositions, nourished by what might be called two different "informative fields." They literally "are in the room," partly as a material fact, partly as an imagined presence. The materialized precondition is visible, but inaudible. The imagined presence is both invisible and inaudible. Both will be addressed during the flow of the interview. Consequently, this excerpt, and the following, as will be seen, represent a covert attempt at an on-the-spot validation in the sense of Kvale.

The materialized "informative field" requires a description of the setting. Fredric had asked me to visit him and to hold the interview in his flat. We were seated in a one-room apartment, where the major part of the walls were covered with sports trophies of all kinds, from various sports activities, but mainly bicycling. One short glance was enough to see that Fredric had been active in this sport for many years on a very high, amateur level. In addition, an obviously expensive and carefully polished racing bicycle stood along one of the walls, occupying a lot of space in the flat, and evidently of central importance in the life of its inhabitant. These surroundings gave the impression of top training and fitness, stamina and good physical condition. This was not really contrasted or contradicted by Fredric's first statement in the interview, that the attacks had been provoked by hard labor and warm weather. There was, however, in one way or another, a tension between this statement and the informational field in the room. How could this "room impression" of hard, physical exercise fit together with the "word impression," formulated as if the interviewee were suggesting something about an interrelatedness of hard physical labor and seizures. Although the feeling of discrepancy was not addressed explicitly, it steered the questions concerning the nature of the attacks and the circumstances of their occurrence. As to a suggested trigger or origin of the epileptic seizures, training hard apparently meant something different than working hard. The occurrence of the seizures had obviously not reduced his sports activities, although they probably engendered a comparable physical strain. This meant that the interviewee regarded them as different in a particular way, and that a difference in effect had proven their difference in nature as regards an unknown but important aspect. Thus, the on-the-spot validation had led to one, preliminary conclusion. The initially felt tension between the "room impression" and the "word impression" had been validated. This discrepancy or difference proved to "make a difference," in Gregory Bateson's sense: valid information was to be found. However, how this difference was conceived of, and what it was made of, was not yet to be spotted.

Leaving the topic of influences on the interview talk as mirrored in the interview text, I return to the topic of epileptic seizures. Thus far, these have been presented as being evoked by physical, outdoors labor together with others, and as inaugurated by a brief sensation of painful trembling in both temples, followed by generalized seizures and loss of consciousness. The dialogue continues as follows:

A: Those who watched you must have been scared.

F: Yes, my father said he got rather scared, and the first time there was a little boy as well - he was almost ten years old - he was pretty scared. He had been standing right beside me.

A: You said your father was there in both occasions?

F: Yes, the first two times, but really all three of times, but the third time he didn't find me. We were in the forest, searching for sheep. There were several of us, so we separated. He passed by at the foot of the hill, I crossed it higher up. I was just about to turn. Then I understood that something was about to happen. It started to ... something like ... *precisely like a bursting against my temples, so I collapsed.* I awoke by ... the last time it happened I regained consciousness while walking ... then I'd fallen and gotten up and walked through the forest and climbed over some fences without any problems and then I'd walked along the white stripe in the middle of a heavily trafficked road.

A: You had no idea how you came to be there?

F: Not from when I was on the hillside in the forest until I was down on the road. I woke up walking along the white line. The distance is ... I don't remember the hour ... perhaps twenty minutes I must have been on my way without realizing it. It was about one kilometer. But I knew the area, I knew exactly where to go ...

A: In a way you had switched on your auto-pilot. Rather incredible. But your head was not with you.

F: No, not the first part of the route, which I can't remember at all. (italics mine, and will be commented upon later)

Fredric describes a state of mind called, in the medical terminology, dissociation. The term comprises a group of mental states, personal traits or special behaviors representing a discontinuity, a rupture of unity, a plurality of being, or a simultaneity of different levels of consciousness though not within the awareness of the person who is or has dissociated. The interest within psychiatry in this group of "disorders" is enormous, mirrored in an ever-expanding literature and body of research, and in journals concerned exclusively with this topic. The earliest studies of these phenomena were performed by the French psychiatrist and philosopher Pierre Janet, a contemporary of Sigmund Freud. His detailed work of description, analysis, hypothesis generation and theory building were overshadowed by the work of Freud and the eventual predominance of Freudian psychoanalysis in most countries of the Western world. Quite recently, his merits regarding an understanding of dissociative states have been rediscovered by neurophysiologists and neuroendocrinologists.[129]

His being rediscovered is connected to the intensive research concerning the impact of trauma, provoked by observations of a group of phenomena in people having been exposed to different kinds of inescapable threat. Although described by doctors treating soldiers in the American and European wars, the mental aftermath of trauma was not studied systematically prior to the First World War. From then on, however, its exploration intensified, increasing even more after the Second World War. The various observations among soldiers were supported by and differentiated in studies of concentration camp survivors, of Jewish people having escaped from extermination camps, and, most extensively, during the last decades, on American veterans from the Vietnam war. Gradually, the field of traumatology formulated the notion of a post-traumatic stress syndrome, comprising various phenomena in an array of patterns with depression, self-destructive behavior, recurrent invasive thoughts or pictures, grades of amnesia and dissociative states as dominant features, though variously presented and expressed.[130]

The diagnoses are defined by nineteen diagnostic criteria in six groups according to the International classification of mental diseases, DSM IV.[131] Recently, dissociative states or techniques have been described in connection with the long-term impact of sexual trauma as well. In an ever-increasing body of literature from different medical disciplines, ranging from gynecology, neurology, neurophysiology and psychiatry to torture rehabilitation, dissociation has been described as associated with sexual trauma experience, either in childhood or in adulthood.[132]

In the international literature on the topic of dissociation, the process is conceptualized as a compartementalization of experience, indicating "experiences which are not integrated into a unitary whole, but which are stored in memory as isolated fragments consisting of sensory perceptions or affective states".[133] The term is also used to describe four distinct, but interrelated phenomena, which are: the sensory and emotional fragmentation of experience; derealisation and depersonalization at the moment of trauma; "spacing out" in the sense of an on-going depersonalization during everyday life; and, containing the trauma memories in distinct parts of the self, conceived of as separate personalities. According to

Pierre Janet, who is referred to in several articles by van der Kolk and co-workers, memory traces of trauma linger as what he called "unconscious fixed ideas" that cannot be "liquidated" as long *as they have not been translated into a personal narrative.* Failure to organize the memory into a narrative leads to the intrusion of elements of the trauma into consciousness as terrifying perceptions, obsessive preoccupations and as somatic flashbacks, such as anxiety reactions.

Returning to Fredric's report of his disease, the appropriate question now is: was there something which ought to be told, which had not been translated into a personal narrative and therefore interfered with normal functioning and mental balance? Were his seizures a "dissociation," created by a conflict between the desire to tell and the dread of doing so? And if something should be told, to whom? The interview continues as follows:

A: At all three occasions you were outdoors, there was no warning, but you said you were exhausted. There was nothing else which was similar, you weren't upset?

F: There was nothing special, perhaps I was a little upset but not really ... (hesitating) ... I don't believe I was upset, and if I was, it was because of the job, I'd done physical labor the first time. The last time I'd walked and climbed in the hillside. *I was exhausted.*

A: But you are used to it, I mean I can see that you've practiced sports -, all this ... which is hanging here on the walls - merely to be physically exhausted shouldn't be sufficient.

F: No. I can stand a lot.[134] To be physically exhausted is not the whole explanation. Something must have been there in addition. I have no idea.

A: I can't help but be curious. Once I talked to a young, abused woman who had attacks such as yours - without a verified reason - she had them whenever she felt squeezed, when she felt a conflict. The seizures started when she was fifteen. ... Does this give you any idea about your attacks?[135]

F: Yes, what they had in common was that I was exhausted and that my father was near by, I mean once he was very close, twice a little in the distance. But we were together.

A: He holds a key for understanding? Your relationship to your father is all right?

F: Well, up and down. We have perhaps a little ... we don't make such good contact, but after I told about the abuse, he's understood why I've been the way I have. He understands more about my behavior. When we were together last summer, he made more contact than he usually does, he's just not the type you talk to so easily. Apart from that, I'd rather he would talk than me having to ask all the time. I don't believe he has - if I started to ask things, I'm not sure he has the answer ...

A: What would you have asked him ...

F: I would have asked ... how much he had comprehended, what he'd understood ...

A: Do you suppose he knew something?

F: Yes, I think there might have been something. At least the one of them, he was so strange, he, he was a little... perhaps a little retarded, he had a very strange way of behaving, the oldest of those two.

A: You think that somebody should have registered that he abused you?

F: Well, not exactly that, but that somebody had reacted to the way he behaved. It's possible my father knew him. It happened during a period when I had almost no contact with my father, because I was born and brought up in my grandmother's house. He was my grandmother's neighbor. I lived with her until I was almost fifteen. Then I moved to my father's. That was the end of the abuse.

Yes, there was indeed something which should be told to somebody. Something should be told to a father who, very often or rather usually, was remote and "not easy to talk to." But outdoors, when working together, there might have been a possibility. The relationship and the situation, however, seem to have been loaded with conflicting interests or impulses. When being presented with an example of a history of seizures triggered by conflicts, the interviewee grasped immediately the core of it as what he was familiar with and what, obviously, was at stake in the context of his own attacks of seizures. His conflict was, on the one side, a need to let his father know what had been done to him when he was a boy, and to ask his father whether he had known or sensed something. Perhaps this constellation already created conflicting interests. Had the father known - why did he let it happen? Had he not known - why had he been so unaware and disinterested? The abuser was a very odd young man; the adults around should have kept an eye on him; they should have been able to imagine that this man might do improper things. On the other hand, perhaps Fredrik did not dare to tell, possibly out of shame or of fear of being blamed, but equally possibly out of the fear of indirectly accusing his father, whom he wanted to be close to, of not to having taken good enough care of his only son. Still, there were several more questions than answers concerning the origin of the seizures and the timing of their initiation. Could any additional information be found in the records of his hospital referral? The following text is an excerpt from the hospital admission records after his second attack:

> During heavy labor ... the patient registered *a pressure in his head after having been standing bent forward for a while.* When he stood erect he lost consciousness; according to eye-witnesses, he was unconscious for some minutes, there were tonic cramps in his upper limbs, and cyanosis in the face and tongue; after the attack he was sleepy. No discharge of excrement, but he bit his tongue. Three years ago *he had a similar attack also during hard labor, and likewise after having been standing bent forward for some time.* (italics mine)

After an array of examinations, among them, a standard EEG, an EEG after deprivation of food and deprivation of sleep, and a cerebral computer tomogram, the conclusion was that no epilepsy could be verified. The tentative diagnosis was Syncope 782.59 (ICD-9). No course of medication was initiated. I have marked the repetition of a description of the bodily position preceding both attacks, as it is slightly more precise than mentioned in the interview. The double presentation of a special posture implies that Fredric himself, or someone else, must have mentioned it specifically, and that it seems to have been given importance by the physician writing the record. It was, however, not to be found noted in the records of the last contact with the neurological out-patient clinic, where Fredric was examined after the third attack. Those notes contain a detailed report of the circumstances as Fredric described them in the interview, and particular mention was made of that "he immediately before the attack felt a sudden pressure in both temples." After this third episode, the physician decided to start the complete examination anew, as epilepsy apparently could not be excluded after all. Fredric had mentioned initially, however, that he had made an agreement with his general practitioner not to have any further examinations.

In the interview, we returned to the topic of cramps after having talked about the family structure and the central relationships, which proved to be rather complex. The interviewee reported a family history of sexual abuse experience in at least two generations, revealed by his recollections and research. He mentioned doing memory work with a Gestalt-therapist after which, three years prior to our interview, he had finally dared to tell about the abuse. This was provoked in part, and helpfully supported, by his former wife who had sensed that he was under the influence of experiences about which she did not know, and of which he had not told her, but which she felt were crucial. From the succeeding phase of the interview, the excerpt is as follows:

A: Have you ever had attacks after you started to talk about the abuse?

F: No.

A: Do you feel you have overcome it?

F: You mean the attacks? Yes, in a way. What happens with my head, that my head in a way *feels squeezed*, or that there is *something inside my head that moves*, I felt that once more when walking along a road. I stopped, I calmed myself down completely, then it disappeared. But that's all I have registered since I started in my therapy. It happened right after I'd spoken about it for the first time. Since then I haven't registered a single symptom.

A: Where do you get this feeling of having overcome it?

F: From what I observe in my everyday life. Before, when I received a lot of information or *lots of things came in*, when I was studying something, I always reached a point where it stopped. *A block came. Then I couldn't take in any more and my head felt like bursting.* It often turned out to be half of what I should have taken in. But now I can take the double amount and more than that. I don't feel *overwhelmed.*

A: I feel tempted to follow one of the threads: the one is your feeling that something happens with your head. Does this relate concretely to the abuse, when your head was held?

F: Yes, on several occasions my head perhaps was . . . I was about to say used for different purposes, when I didn't want to continue or when I said no, then my head was *grasped and pushed against a wall.* And he tried *to force oral sex on me,* then he held my head like this with spread fingers (shows this by placing his hands alongside his head)[136] around my ears, and it's as if, well, what should I say, *in this zone* I felt the pain. It is *precisely* in this zone. (italics mine, and will be commented upon below)

Here, a shift of level is indicated allowing comments on the flow of speech and the content of talk. The last statement, offering a description of how the abuse had been performed and experienced, was a reply. The question concerned a possible relation between the sensations in the head associated with the seizures, and the abusive acts as they had been practiced. This question was generated by an array of formulations, all of which had been told and expressed in the preceding excerpts (see my italics). These remarks concerned the place and nature of the painful, trembling, moving or bursting sensation inaugurating the seizures. Something very diffuse and ambiguous was depicted in this sequence of fragmentary clues. When I suddenly imagined what these scattered remarks might be the code for, a probing question was engendered. It was validated on the spot, both as to its character as a code, and to the meaning of its parts as belonging to an imagery of sensory perception turned into logical trauma embodiment.

Once again, a shift of level has to be introduced. This question, as it was generated and as it was asked, represents a methodological problem: the question of the leading question as dealt with in chapter 1.3 is relevant here. Also, the significant difference between a decontextualized and preformulated question in a standardized research interview setting, and a context-generated, not previously considered question in a focused interview, is demonstrated here. The question was neither planned nor thought out. Its core was not derived from a theoretical presupposition. It condensed from a flow of clues during two hours of talking in a particular environment. Although neither planned nor conceptualized in advance, was it nonetheless instrumental, in the sense of being a means to elicit a confirmation of an a priori opinion or suggestion? Clearly not. The clues making up its matrix, the converging imagery, had emerged in the process of the interviewee's attempting repeatedly to describe a sensory perception, a cluster of bodily sensations, his corporeality. It could not have been known before he had formulated it himself, the way he, as a little boy, captured in an abusive grip, had perceived the details and the totality of experience, with its repeated threats, terrorization and secrecy. However, it is exactly this process of formulating, of speaking about the abusive act, of verbalizing the forbidden knowledge, which crosses the boundary into what is socially silenced, what is improper to say and shall not be spoken. It will most probably not be said unless explicitly asked for in an appropriate context. Consequently, such an explicit question represents both a methodological problem and a necessary precondition in research on socially tabooed knowledge.

The relevance of the question regarding was validated immediately; an ideation about a concrete relationship was confirmed. The phenomenological frame had provided an adequate way to understand a health problem so that it made sense - a corporeal, temporal, spatial and relational sense. This health problem had not been, and could not be, understood in terms of biomedicine, where "nothing" could be explained, and "nothing" could be verified, although this conclusion (of it being "nothing") had been invalidated by the reemergence of the problem which seemed to persist. Indeed, the embodiment of forced oral penetration, involving a grip around the head, a bursting, a sensation of something moving in the head, and of being overwhelmed, was the answer. This answer even allowed expanding the perspective to include those situations where the seizures had occurred, when hard labor in a forward, bent position, had generated a pulsating pressure in the forehead. *Here was what felt too similar to be tolerable,* as already discussed in Veronika's case. Loss of consciousness was the corporeal response to a reactivation of sensory memories of forced fellatio under very particular circumstances.

The role of the father's presence in these scenarios has already been reflected upon and elucidated. There still remains a missing piece, however, a presence which had been mentioned but which had not been scrutinized adequately. This presence concerned the little boy whom the interviewee had mentioned at the beginning of the second excerpt. A little boy of nearly ten had been standing very close to him during the first attack. He was not mentioned again during the interview. Yet his presence had been important, as was revealed during a reinterview. Fredric had received a copy of the interview transcript for confirmation. He had been asked to feel free to make any comments upon it. He phoned back and wanted to make several additions which he had found relevant when reading the text and asked me to add his notes to the original. While reading his own description of the situation during the first attack of seizures, he had recalled the image of the little boy standing very close to him while he worked in the earlier described body position. Seeing the little face in front of him, expressing eagerness and pleasure to help and be near, he suddenly caught a glimpse of himself as a ten-year old boy. In the sense of Polanyi, a shift from "from" to "to" occurred. For the first time in adult life Fredric became aware of how small a ten-year old boy is, how trusting, and how vulnerable. He faced himself at that age, the age of onset of the abuse, and he saw how powerless he had been towards his two far older abusers. For the first time since the abuse, he had been able to recognize that he had been overpowered, that he, as a boy, had had no choice, that the crime and its consequent terror had been inflicted upon him violently and intentionally. All of a sudden, he comprehended that he was not guilty, and what had been done to him had not been his fault or responsibility. To see himself in the mirror of the boy's face meant a sudden shifting of position and a gaining of salient insight. Seeing himself as a little boy, and knowing his father was close by, intensified his latent conflict about the experience. Simultaneously, he had been standing in a similar body posture as his abusers often forced him to assume, sensing a heavy pulsation inside his forehead, a pressure against his temples like a bursting pain. This complex simultaneity could not be borne. "Spacing out" was his only option, a technique from the time of abuse, where to "leave the head" and literally leave it to the abuser, had been invented as an adaptation to what was inescapable. The technique of "leaving" can be read directly from the wording of the answer. The interviewee said:

> Yes, on several occasions my head perhaps was ... I was about to say used for different purposes, when I didn't want to continue or when I said no, then my head was *grasped and pushed against a wall.* And he tried *to force oral sex on me*, then he held my head like this with spread fingers ... around the ears ...

The head had been made an object, the place of terror and humiliation, of threat, pain and invasion. This was told as if talking about some randomly chosen thing, an object among others, but not a head, and not one's own. To have one's own head pushed against a wall, to have cooperation demanded, to have the mouth forcibly invaded by a penis, and have it "used for different purposes," is to have one's head alienated, made into something else and possessed by someone else. This can only be tolerated in a kind of absence. And it can only be described in a state and style of distance. The passive mode and the expressions of instrumentalization place the hands doing all this in the central place. The focus on the abusive hands, doing what they please with the objectified head, represents a verbal, syntactical and emotional abstraction.[137] As a protection, the strategy of dissociating had once been adaptive to a particular situation. When that situation ceased, however, this "habit," reactivated in an everyday situation of work, had made no sense. It appeared to be a bodily defect and seemed to indicate a somatic disease. In its present appearance, no one could "read" it as a logical consequence of past sensory imprints from perceptions during sexual abuse.

The international literature concerning studies of a possible relationship between epilepsy or seizures and sexual abuse in childhood or adulthood is not very extensive. It was reviewed in a study of six cases in 1979.[138] This relationship was also mentioned in a letter to the editor of the British Medical Journal in 1990. The authors reported having observed a group of women with particularly poorly controlled psychogenic pseudoseizures, all of whom had proven to have been sexually assaulted.[139] Likewise in 1990, a study among adult female borderline patients reported a co-variance of sexual abuse and seizure disorder.[140] In 1992, neurologists and neurophysiologist at the University of Birmingham confirmed an association between some types of seizure presentation and sexual abuse experience.[141] They also emphasized the dissociative nature of the attacks which were viewed as related to sexual assault experience. Traumatologists write about this phenomenon:

> People who have learned to cope with trauma by dissociating are vulnerable to continue to do so in response to minor stresses. The continued use of dissociation as a way of coping with stresses interferes with the capacity to fully attend to life's on-going challenges. (van der Kolk & Fisler 1995)

This means that "spacing out" means leaving, either partly or completely. This means also that seizures may represent the only possible and only available narrative about abuse experience as long as the experience is not recalled, and as long as it is not narrated explicitly, which means verbalized, told, and thereby integrated into the life history. This may even mean that seizures without a verifiable cause in a medical sense ought to be hypothesized as testimony to some form of traumatic time disruption.

1.2 BERIT

Another kind of perception remaining from abusive acts was thematized in the interview with Berit. From the age of eight until fourteen she had been abused by three adult men successively, all of whom were her mother's lovers. Her interview was tape recorded from the beginning such that the text includes the recall of childhood and adolescent sickness as follows:

> **B:** As a child I once was seriously ill. I had yellow staphylococces in my throat. It was so serious that I had to be in a hospital for three weeks. Apart from that I had lots of trouble with my teeth when I was around nine or ten years old. They said I had too many teeth. But it may have

had something to do with the abuse, since I had this bleeding from my gums, and there were things they couldn't explain.

A: Do you think this serious throat infection was inflicted on you?

B: I've experienced so many strange things, and every time I'm mentally down or involved in something, I easily get pain in my throat or get the flu or get sick, physically sick. When I was nine, I met my worst abuser. So I guess I'd had several shocks that winter before I was admitted in the summer. This is how I think and how I see it as an adult, in retrospect.

A: I think of the type of infection, which is rare to have in your throat. Is there any relationship to oral sex ? Were you forced to suck?

B: There is a thing I - (pause) - ... no, not so early.

A: That came later?

B: Now I'm in a foggy landscape where I've pushed away a good deal. It was this and the next abuser I had - there is a kind of a border ... everything else I've agreed to ... but I have a lot of problems with my throat, I produce a lot of slime and the like, when I start to think about the violations then I (swallows audibly, insistently, as if trying to prevent vomiting) then I get like this immediately. That something has been done? Yes. But emotionally I'm not quite ready to accept this yet.

A: Your throat is still a difficult area?

B: Right. Just when it comes, something contracts there (clears her throat several times, breathes deeply)

In the following, the interviewee tells about other health problems, among these a pain in her arm and a treatment which helped her to regain both the strength in her arm and the sensitivity in her skin. We talk about pregnancies and deliveries, and her conflicts with her mother who was unaware of her daughter having been sexually abused by all three of the men she had taken into their home. The excerpt is as follows:

> I've had the whole register, the curious perpetrator, the violent one who put a knife to my neck when I was fairly little, there we are, back at my neck and throat ... (swallows several times) ... well, well, here is something ... but he placed the knife on the side of my neck with the sharp edge into ... and after than, all he needed to do was to place a finger there. He could ... we would have guests for afternoon coffee, they were about to leave, and he would hold me around my shoulders and say 'I'm so fond of Berit' and it would look like as if he only stroked my neck slightly, but what he did was to place a finger ... like this (showing by placing her own finger close to her neck) ... on my neck. Then I knew what this meant: 'now your mother leaves for her job and I'll send your brother out and then I'll take you.' Just like that, and I was standing there, shivering, and saying good bye to the guests and smiling and thinking 'can't you see that I need you?' but then my respect for this pitbull beside me was too strong.

In the field notes, but unfortunately not in the transcript, it is stated that Berit rose from her chair, suddenly, when starting to tell about the knife at the side of her neck. Her heavy swallowing, however, may be heard on the tape, as well as a change in tone when indicating how her abuser put his finger on her neck. The situation during the interview offered a Gestalt, which also is present in the textual form, although not quite so evident: *branded, cut, marked.* The edge of a knife against a child's neck is still there, its effect is still present in the interruption of breathing, the production of saliva, the harshness of voice, the immediate outbreak of an anxiety attack, and the sensation of being caught, generating the impulse to move away and leave. The finger "meant" abuse, powerlessness, and the futility of resistance. The touch of the finger, appearing to every observer as a caress, was the repeated renewal of the owner's mark. She was his property. The mark and its significance might seem almost too concrete and too direct. Could it be so astoundingly simple? The embodied knife was present even in its symbolic substitution, a finger, and, beyond that, is still present in everything else randomly touching her neck, resulting in, among other things, Berit's consistent avoidance of scarves, necklaces, and high-necked clothing.

1.3 JUDITH

The interview with Judith started with a summary of the information concerning the circumstances of the abuse, which began with a vaginal rape by an adult man in the neighborhood when Judith was eleven years old. From then on, she was abused frequently by the friends of this man. In early adolescence, she was physically and sexually abused by a boyfriend. Later, one of her mother's relatives and one of her father's colleagues abused her. At twelve, she was referred to the hospital because of continual abdominal pain, an icterus and a pancreatitis, resulting in a cholecystectomy. The excerpt from the interview after this information had been given, is as follows:

A: How did things turn out after that operation? Had the problem disappeared? No more abdominal pain, nothing?

J: Yes, I've had pain in my stomach and nausea until quite recently. That wasn't operated out of me. But it's only now I've really understood that the nausea and my anxiety belong together - and always have done so.

A: How is that related? What has been done to you?

J: I don't understand this totally. This nausea feels at first as if I'll be strangled first, so I get very anxious I'll throw up. Then I swallow and swallow. I can lie all the night like that.

A: How were you abused?

J: I was raped ... (hesitates)

A: ... anything more?

J: After that, it was oral sex, mainly with these comrades ...

A: There was the nausea and the anxiety. You were held.

J: Yes. (long pause) The first one, for me it seems almost as if it just lasted one second. He pulled off my clothes and ... performed ... but after that I had to suck ...

A: ... and the nausea and anxiety, I've heard others mention it, to be held and forced to swallow ...

J: Uh. Huh.[142] I myself felt a particular kind of discomfort (long pause) I got very bad teeth. And I developed a phobia about dentists ... to repair my teeth became ... and again and again ... my teeth were so damaged ... I also had all this white stuff in my mouth, fungus, I visited a doctor for that, fungus all over my mouth, a crusty, white plaque and blisters. It disappeared later. (long pause) I also have a panic about water. That's because I can't stand to have my mouth under water. Even in the shower, when I'm accidentally standing so that the water runs over my mouth, then I get this nausea. I can't tolerate anything flowing over my mouth.

A: That has concrete reasons.

J: Yes. (long pause) I've tried to participate in swimming courses several times, but as soon as I get under water I ... (holds her breath)

A: You stop breathing...

J: Yes. (long pause) And think of something so impractical as this with the shower, that I have to be careful where I stand and where the water flows. Then I feel like I'm being strangled.

A: Your body remembers it.

J: Yes. (pause) I've been working in a nursing home, and there's an old lady I help frequently, also when she takes a shower, then she's sitting there and enjoying herself under the water and it flows over her face and then... then ...

A: ... you become breathless on her behalf?

J: Yes. Yes, then I don't dare (takes several heavy breaths) and she lets this water go all over and enjoys it and I just have to breathe and breathe ... as an adult, I've had suffered mainly from problems with breathing, not like as if I have too little air, but as if I'm strangling. And the nausea. The nausea and the strangulation.

A: That's not strange.

J: No. It isn't. I stop breathing. Not like somebody breathing quickly ... I stop, very silently.

A: That's the strangulation. That's your mouth, isn't it? Do you have problems with eating, I mean with food?

J: I haven't got anorexia, but the opposite, binge eating. But I don't throw up. That's perhaps to get rid ... not taste ... but perhaps ... when I eat, I comfort myself, then I forget everything, then I feel good, even if I'm nauseated. Then it disappears, especially with sweets, sweet things and good tastes. Sour things are good too. Everything which tastes, different from the taste I have in my mouth, so to speak. Distinct, good tastes. Then the nausea disappears. But whenever I stop, the nausea comes back.

A: Are there things you can't eat at all?

J: Yes, a lot. No yoghurt and white milky things. I haven't thought about that before, but when I think about it now I find a great deal of things. I have problems with boiled things, which I've experienced when dieting, I can't stand them. And drinking yoghurt is impossible. It's so slimy. I love strong tastes, the stronger the better, it helps against my nausea. (pause) I haven't thought about that before (pensively).

Here, another Gestalt of embodiment emerges, which may be named the alienated mouth. The idiosyncratic embodiment of forced fellatio was already addressed in the interview of Veronica. Her embodiment was dominated by the deafness which enabled her to split off a part of her auditory memory when confronted with experiences too similar in feeling to forced fellatio, triggered by in the sensorimotor perception of her head being held captive against a human body. She too had long periods during adolescence and adulthood in which she suffered from eating disturbances, in the form of anorexia. This was mentioned before, but not discussed during the interview. Forced fellatio embodiment, presenting differently from Veronika's, was also revealed in the analysis of Fredric's interview. However, he did not mention any eating disturbances, or any other kind of problems related to the eating and breathing functions of the mouth. In the third excerpt, however, he described a study situation in which he thematized yet another association to the abuse of the mouth. His formulation of "taking in," meaning learning, was obviously connected to the same sensorimotoric sensation of pressure on the forehead, which "meant" to have his head held. The topic of forced fellatio was also mentioned in the initial part of the excerpt from Berit's interview, generating her reaction of avoidance both of the topic and of the imagination of the act. This was mirrored in the syntax of her reply to a direct question regarding the experience of forced fellatio, where fragmentary sentences literally broke the connection of thoughts. It was even reflected in her metaphor of the foggy landscape, not offering, as yet, any clear view.

Judith describes the alienating impact of having her mouth abused frequently by a group of adult men during the course of several years. In the interview, her presentation of problems related to the functions of the mouth begins with nausea, which, in her corporeality, has always been connected directly to anxiety, as she has gradually come to understand. Next, she mentions bad teeth and many consultations for dental repair, resulting in a phobia about dentists. Furthermore, she tells about various practical consequences. She articulates that to perceive "something" flow over the mouth represents the forced intake of semen. The perception of flow is provoked by a concrete flow, for example of water in the shower or of being in water which may reach the region of the mouth. The same perception, however, is also provoked by the sight of "something flowing over a mouth," whomever the mouth might belong to, as, for example an old lady sitting under the shower. Here the phenomenon of the embodiment of a symbolic representation, Berit's "knife which is a finger which might be a scarf," has attained another stage of transformation, or rather acquired a greater degree of abstraction. For Judith it is sufficient to see what would feel like something flowing over "my mouth if I were this person." The sight evokes the sensation of semen running over her mouth. Still. She is rendered breathless instantly by anxiety and by the sensory perception of being strangled and feeling nauseated. She is able to describe her breathing problems, expressing fear of suffocating because of something in her mouth, the literal and concrete plugging of it. Simultaneously, her wording mirrors how she adapted to this concrete threat implicit in the actual abuse situations: "Not like somebody breathing quickly ... I stop, very silently..." which might imply "leaving." Undoubtedly, this description reflects a situation of complete powerlessness, in which the very survival of the individual is close to being at risk.

When the mouth, the passageway for the breath, is not only alienated by means of an act of abuse but literally obstructed in the course of that act, the result is the terror of being suffocated or strangled. The means by which her breath was stopped establishes a "perceptive synonymy" wherein suffocation equals panic equals nausea. In Judith's corporeality, these states of being have the character of different modalities of the same associative field. They merge as a consequence of bodily logic, since the mouth is the passageway for the breath and the passageway for food, representing the most central of means for physical nourishment and biological survival. Based on the intrinsic logic of Judith's idiosyncratic embodiment, a narrative about eating and food ought to be expected. It does come, and is comprised of various modalities in the relationship between food, taste, and consistency. Initially, Judith characterizes her way of eating as, "binge eating." When using a medical term with an implicit judgment of a deviance from a suggested norm, she is applying to herself an outer gaze informed by medicine's focus on pathology. Then, in the next sentence she talks again from inside herself, where eating means comfort and pleasure, forgetting and help against ... what is not immediately easy to spot or easy to say but looks like this: "That's perhaps to get rid ... not taste ... but perhaps ... when I eat, I comfort myself ... (etc., see quotation above)." The narrative is a reflection about what is good, and tastes good, and what tastes distinct and strong enough. Tasty is what tastes different from the taste in her mouth. Then the nausea disappears. Therefore she has to eat continually in order to keep away the taste in her mouth which equals nausea - yet another "perceptive synonymy."

The reflection about taste is woven into reflections about the color and the consistency of food. Thus, "white, milky things" are unacceptable, likewise "boiled things." Judith is not as precise about why color and consistency matter, it seems as if she did not have the same degree of awareness about the origin of her objection to "white milky things" and "boiled things," although she mentions the sliminess of drinking yoghurt as objectionable. In drawing a precise distinction that the tastes are different from what is in her mouth, one might derive an inference about the color and consistency of what she cannot have in her mouth: that which is too similar in either color or consistency to semen and a penis. Here, a complex constellation becomes visible, linked to the notion of the abject in the sense of Kristeva,[143] and the concept of from-to in the sense of Polanyi: Judith has *embodied what is other;* the "from and "to" have *changed their place* inside her. To embody the abject, in this case the taste of semen, has made the basic taste of her own mouth "semen-like." Thus, what is known as other is not only inside, having passed the zone of transgression and as such representing an internalized and ever-present threat to self; it has also become self: the taste of the other has become the self's own taste. Therefore Judith strives to make clear to herself what she wants to get rid of by eating. When other and self are confused in this way, it is not easy to speak "from" oneself, and it is only logical that the place of that which is different, must be in the other. By the intrinsic logic of her embodiment of forced fellatio and the intake of semen, she feels constantly "abject" to herself. To be forced to contain the other, the abject, inside one of her vital orifices, she has internalized the semen and penis-like taste/color/consistency as a constant state of her own being. Due to this corporeality, "a great deal of things" which are considered edible by most humans, have become for Judith inedible, abject. Furthermore, due to this corporeality, the repulsion has taken place quite literally in her mouth, chest and stomach, rendering her constantly nauseated. Only eating non-abject food makes the nausea disappear, as only such food can make her mouth hers, the place and the state from which *she* tastes. Consequently, what by the outer gaze would appear to qualify for the psychiatric diagnosis of compulsory eating disorder, is, as seen from within, the only way of being Judith.

1.4 ELISABETH

The interview text concerning Elisabeth begins with the following statement:

She speaks in a low, but clear voice, almost without moving her mouth, and her teeth are kept together tightly.

After the topic of gynecological examinations was dealt with in the interview, came the following:

E: I've consulted men (doctors), that was all right. I used to prepare for these examination. I used to be psychically prepared for getting up into that chair.[144] When I'm in this chair, I feel like I'm not there.

A: You aren't there in some way?

E: No. I switch off. This is difficult to describe, but I feel like I've survived by switching off in situations where I feel pressed. I'm there but I'm not there. It's very strange, but I can't manage to place myself on such a chair calmly. But I may feel pushed in completely different situations. Then I switch off. At the dentist's, for example. There I feel like this. In order to get into this chair I - I don't know what I do, but I'm there and in a way I'm not there.

A: Does this mean that afterwards you don't remember what has been done and said?

E: Sometimes I do, sometimes I don't remember a thing.

A: How do you deal with these gaps in your memory?

E: Many times I've had to go back and ask what we talked about. (long pause) Well, I have to know what has been said, after all.

A: Have you thought about why you have these problems in your body exactly where they are, in your neck, back, jaws and pelvis?

E: I've had this dentist phobia since I was a child. My stepfather always came with me to the dentist. He would sit outside in the waiting room, and right now I'm uncertain ... whether it connects ... but I've thought ... when I'm at the dentist's ... when I get ...

A: ... things in your mouth ...

E: Yes. Things in my mouth. Then I feel suffocated. Then I feel that I'm suffocating. But I didn't remember this, I haven't thought this before now ... whether it was because he always came with me. He observed me in a way. He rationalized it by saying that he had to drive me.

A: In this way he controlled what you said and did.

E: Uh-huh. (long silence)

A: Were your teeth damaged because you had to suck him?

E: I don't know. But it was this having something pushed into my mouth. Then I panicked. Then it feels as if I'm suffocating.

In the following speech turns, Elisabeth tells about a discussion she had had with her dentist some years earlier about the constant pain in her jaws and her inability to open her mouth for more than two fingers' breadth. The dentist suggested she might benefit from a prosthetic, a rubber bit to wear on her teeth at night. The interview continues as follows:

A: You got this rubber bit, then something was done after all, how did it work? —— Yes, I see. You had to swallow several times.

E: It was horrible, it was brutal. That was what it was. But I tried it.

A: Yes, I can see in your face how it was, you showed some strong reactions now.

E: Yes, I really taste it now, it was this taste, and this soft rubber. I used it for three months, then I gave up.

A: Had the pain disappeared

E: No, it was still there. But the dentist thought it had improved. He judged by how much I could open my mouth. Usually I couldn't open my mouth. That changed during that period. That's what he called improvement.

A: Were you satisfied yourself?

E: I did have this pain before. I did. And this opening - it didn't last. I closed again, it returned. But I work with it, I try ... (she shows how she tries to train herself by pressing some fingers between her teeth) (starts to cry, cannot continue to talk)[145]

A: This means you do not own your mouth as yet. You still need to have it closed. Can you try to express why this feeling is so strong?

E: I haven't worked as much with my mouth as I have with the rest, after all. I haven't. It's been here all the time. (pause) It's why I keep shut. (pause) I keep behind (her voice is tightened by weeping, softly)

A: There's a story behind your teeth which keep something back, is that what you're thinking?

E: Yes. I do. I haven't gone into it. I haven't talked about it.

This interview gives access to quite another path of embodiment and yet another experience of having the mouth alienated. Elisabeth offers an insight into her dissociative states in certain situations or positions which are triggered by what she has to avoid. As she puts it, she cannot be in "the chair," be it the gynecologist's or the dentist's, unless she "spaces out," as van der Kolk has termed the state of on-going dissociation as an avoidance strategy, which will be explored later. She knows when she does it, she registers a kind of simultaneity of consciousness. The "parted" Elisabeth who sits in the chair, exposed or with things in her mouth, is amnesiac. After the examination, the whole Elisabeth has to go back in order to fill the blanks. Her mouth is alien to her to such a degree that personal expressions as talking and laughing, and practical and private necessities as eating and dental hygiene, are severely impaired. She cannot even open her mouth if and when she wants to. She is advised to practice opening her mouth by pushing her fingers between her teeth, which looked as if she were invading, or "breaking in" to her own mouth. To overcome the block in her jaws she is instructed how to "break into" herself. This means literally to inflict upon herself an aspect of the acts of invasion done by others. This must be not only counterproductive, but be deeply self-humiliating,[146] which emerged before my eyes as Elisabeth's pantomimed attempt ended in her desperate weeping. She obeyed not only against her will, but even against her paramount interest in not having her integrity violated further. Is it probable that she will continue to obey, as her final remark in this sequence may indicate, until she dares to "translate her sensory trauma memory into a personal narrative," as Pierre Janet has formulated it? I shall return to this question later.

1.5 TOM, CHRISTINE AND MARY

The alienation of the mouth as a Gestalt presented in a variety of shapes, not all of which were as explicitly and precisely recognizable as in the previously quoted examples. Tom, the third man, had hidden it in the interview talk. I was unaware of it until the second or third reading of his interview text. He was very verbal, and his wordiness must have distracted my ears. He succeeded in guiding my interest away from a salient connection between his mouth and his predominant health problem at the time of the interview, upon which he commented in great detail.[147] On page one, I find the following statement concerning one of his major health problems:

T: With regard to eating and food, I've always been able to eat whatever I wanted whenever I wanted, but I eat very irregularly. Well, I'm a little conceited. Once my weight was 115 kilograms. Now I'm not sure, but I'm below 90. That's fun. I'm very proud. I aim for even more reduction. I have this angina, after all, so I know that it (excess weight) is inconvenient.

The angina started in 1985. I haven't been admitted, it's always just the emergency room. Sometimes I've been forced to - and it's always in the hour of the rats, as I call it, between three and four at night, I get it. The Nitro and Adalat haven't been enough to cope with it, so I've gone to the emergency room. I've gotten almost psychological help there. I'm scared, it's very painful - and I'm afraid to die. It's a little narrow here (points to both sides of his neck), that's why I take the drugs. They've applied for an operation, I call it an operation where they blow up your arteries or what ever it's called. They go in from down below, at your leg. There are narrow vessels, they've taken such an examination. Twice I've been at the hospital to see a cardiologist. And he has admitted me, I mean my general practitioner has recommended surgery. That takes its time nowadays, you can die in this queue nowadays. But I'm not sufficiently sick, there are so many in the hospitals and in the queues.

The next information, the first concerning the abuse of his mouth, is given on page four, concerning the abusive acts in childhood:

A: The abuse, lasting for four years, how was it done?

T: Most of it was dealing with him. So, nowadays, the smell of semen is for me - it's incredibly awful. I haven't brought out everything as yet. But it's taken a lifetime to forget it. It takes time to get it out altogether. I've got out a lot here in the center, mainly last summer when I dared tell ...

The following is a narrative about the time when he prostituted himself in order to get money to finance his need for alcohol. The topic of "the smell of semen" or the impact of the oral abuse is not pursued. In the course of the interview, the health problems related to his heart are returned to. On page thirteen to fourteen the text is as follows:

A: I wonder whether this problem with your heart and your angina will become less severe the more you get out ...

T: That may be right. It (the nightly pain) has returned now. There will be a new breakthrough. The way I live now is tough, it's a mental strain. You sit up, awake, from three, four o'clock at night until you have to go to your job. Certainly there are a lot of thoughts and questions. And this eating, it's reduced. That means now I get this smell of semen and this taste of semen all the time, from time to time I have to push the food to get it down.

This is the final statement, completing the preceding information about his eating habits, obesity, heart problems and, a small fragment of information given in between, the cardiologist's reluctance to operate on him due to his overweight and the probability of complications. He believes that to have an operation, which will reduce the pain in his heart at night, in "the hour of the rats (I was unfortunately not skilled enough to grasp this metaphor which emerged several times, and probe into it)," he has to loose weight. This means reducing his eating. This, in turn, means have continually to smell and taste semen, which, as he says, is "incredibly awful." Tom presents in a very "hidden" way the same phenomenon as Judith did explicitly, that only eating helps to make what has been internalized disappear. When the abject is inside and has to be "pushed down," as Tom puts it, the problem of his weight interferes with the problem of his heart. The logic of abuse embodiment creates an obstacle to an intervention, demanded according to the logic of medicine. In a different reading, however, the apparent necessity of the medical intervention seems to be caused by the same bodily logic. To put it more explicitly: what seems to be a medical problem is not a problem which can be solved by medical measures alone.

Even more disguised, and in yet another shape of the same Gestalt, was the presentation of oral abuse embodiment in the interview with Christine. She had told about her wide-spectrum allergy, her generalized eczema, and her asthma. She had mentioned forcible penetrations, among others oral, that she had been troubled with bad teeth, and that a constant smell of semen went with her everywhere. The text reveals yet another lived meaning than those described until now, near the very end of the interview text, when she is commenting on her different experiences with male and female physicians:

> **C:** When I think of the others, some of them general practitioners, they should have seen that there was more, they should have understood that there was something wrong. That there was no help in fourteen physiotherapy appointments - or whatever it might have been. And such things as scolding you because you don't use your Cortisone spray - when I can't stand the taste of it *and don't manage to spray something into the rear of my mouth.* Or that I don't take my pills, because I don't like to take drugs for all of this as I can't believe they will help what's really the matter. They ought to have understood something, these physicians. I showed them something, I cried. It's in a way a betrayal, not to register things which are visible, or not to do more when really having understood something. (italics mine)

Christine could not tolerate the sensation of the Cortisone spray spurting against her palate, as this felt too similar to the perception of the ejaculating semen.

Finally, the shortest representation of oral abuse embodiment shall be presented. This variation on the theme was told in such a soft voice that a verbatim transcription was impossible. The interviewee, a young woman, was in the first group of responders, who presented the seemingly technical but actually phenomenological problem of a fading voice, mentioned in chapter 1.4. Mary, as I named her, talked at the very end of the interview about the following, which, in accordance with the transcription rules, reads like this:

> * She tells about her problems with sexuality with her husband and about coitus under the influence of alcohol in order to be able to tolerate it, and about not being seen naked. Whenever she goes swimming, she uses a swim suit to hide her stomach, which is covered with scars. In her frequent periods of self-assaultive behavior (from the age of seven, mentioned earlier in the text), she had tried to cut holes in her stomach in order to get out the semen her father had forced her to swallow. And the semen which spurted over her stomach, she'd tried to scratch away with the side of a razor-blade. The cuts were always taped, apart from when she was in institutions, where the cuts were sutured.

In thirteen of the thirty-four interviews, there emerged narratives of the abuse of the mouth. I had not asked for them initially, unable to imagine them and their particular impact. During the interview process, however, impressions from earlier interviews were actively reintroduced in those succeeding them in order to understand the various kinds of impact and the individual forms of embodiment. These forms, varieties of the Gestalt, "alienated mouth," could be grouped, although each person had embodied the oral abuse in a highly idiosyncratic way. They made obvious that talking, expressing verbally, telling, speaking out, learning and thinking could all become impaired.[148] Likewise, tasting, eating, swallowing, digesting, tolerating and enjoying food were broadly affected. So was breathing, relaxing, and feeling free.[149] Self-assaults, avoidance, intolerance, panic, exotic behavior, seizures, dissociation, psychosis, phobia and hallucinations were among the names for the outer appearance of the embodiments. When read with the knowledge of how the oral abuse had been experienced, however, these "behaviors" became transparent.[150]

1.6 EATING AND BREATHING

I would like to concentrate on the two most frequently described types of corporeality influenced by oral abuse, *eating problems* and *breathing problems*, and to reflect upon them from the perspective of the research on the relationship between somatic disorders and childhood sexual abuse. Starting with a phenomenological frame of reference, I immediately encounter a problem. "Eating problems" or "breathing problems" are not research terms. The experience of having problems with eating or breathing in one way or another, has first to be translated into a diagnosis. The nature of the problem with eating or breathing will steer the research into either the International Classification of Diseases ICD (at present in its 10th revision), or into the Diagnostic and Statistical Manual of Mental Diseases (at present in its 4th revision). If it appears most likely that the problem is eating too much, eating and vomiting, eating too little, or both, it is labeled a psychiatric disorder. If the problem is one of eating a lot without vomiting, and having nausea or abdominal pain, the problem is labeled somatic. The situation is the same with breathing: if it appears most likely that the problem is breathing too rapidly and being anxious, it is most probably perceived of as a psychiatric symptom. If it is breathing heavily (even if being anxious simultaneously), it is most probably perceived of as a somatic symptom. Where would Judith be placed were she to enter a research project? If she presented herself as she did at the beginning of the interview, she would have qualified for a study of eating disorders, subgroup compulsory, type bingeing - in a setting guided by the current theoretical understanding of eating disorders valid in psychiatry. Had she stressed her nausea and the sense of release related to intake of food, she might have been selected for a study in gastroenterology, and be suspected of suffering from an ulcus duodeni. Had she talked mainly about her problems with breathing, the feeling of strangulation and the fear of being suffocated, she might possibly have entered a study on allergy and asthma. On the other hand, Christine, would be registered as asthmatic automatically. Tom would have been advised to join group-dieting programs without receiving a diagnosis at all besides his angina, and Veronika would have been considered to be suffering from anorexia and qualify for psychoanalysis or other psychotherapy. These people, all presenting and talking about eating and breathing problems, which, in a phenomenological interpretation make sense as a corporeality in the wake of silenced oral abuse experience, would certainly not all have been found as subjects in the same studies.

One must keep in mind that every study designed to, and aiming for, the exploration of a relationship between eating disorders and a history of sexual abuse, starts at the opposite position from that of the subjective experience. The point of departure is the objective diagnosis, or, in some cases, the observable symptoms. The relevant studies on this topic, though not representing a large body of knowledge, follow objectivist designs. Patients with such disorders are registered, examined and treated in psychiatry and psychology, as eating disorders are, by definition, mental diseases. The development in these fields has been reviewed recently by the American psychiatrist and specialist on eating disorders, Susan C. Wooley. She reflects upon the somewhat astounding resistance within the disciplines mentioned to accepting the ever-increasing evidence of the probability of a certain relationship between these disorders and sexual abuse. According to Wooley, studies which address the topic provoke far more critical, or even uninformed, responses than studies on every other group of disorders. Both eating disorders and sexual abuse are social phenomena most often concerning women. Primarily female researchers have taken an interest in them, and reported possible connections between them. As Wooley suggests, this may create or represent the core of the obstacles to a clearer research picture, some of which seem deeply embedded in the structures guiding research, which are conceived of as scientific though they carry the mark of being a resistance against an elucidation of a socially silenced phenomenon. She writes:

> It seems an open question whether the science of psychology in general, or the field of eating disorders in particular, has a paradigm *so flawed* that it cannot encompass the influence of gender inequities on the subject of studies, or whether there is simply *a profound cultural resistance* to data generated within the paradigm. As we struggle to answer this question, we will be challenging more than the understanding and treatment of eating disorders. We will be confronting a *failure that signals critical weakness* in inclusiveness, communication, and flexibility within our field. One thing is clear: *We have grown too remote from the most persuasive data bases.* (italics mine) (Wooley 1994:196)

Wooley points to a crucial aspect of the research in the field. The research methodology, aiming at objectification as a means of securing valid knowledge production, demands a considerable distance between the researcher and the person having experienced abuse or experiencing eating problems. The life-world of human experiences can simply not be accessed via that methodology. Consequently, "the most persuasive data bases" are out of reach due to theory and method. What can be reached are research objects. They provide only limited information due to the fragmenting theoretical framework and the reductionist methods implicit in the studies, which may contribute to inconsistent results. Consequently, Wooley is right when stating:

> But the truth is that we really have no idea how many eating-disordered patients have been abused. *The silencing of victims is the core phenomenon of abuse - a virtual prerequisite for its occurrence* and the source of many of the most destructive sequels, since victims must rely on primitive and often incapacitating defenses to accomplish the related psychological tasks of repression and concealment. (italics mine) (Wooley 1994:188-9)

Most of the psychiatric studies of the occurrence of sexual abuse among eating disordered patients have been published during the last decade. Earlier reports, including the contributions of Sigmund Freud, lost their influence after he himself changed his theoretical position from one of considering sexual abuse as an impactful actuality to one treating it as an artifact of sexual fantasy.[151] However, no causal relationship between sexual abuse and either anorexia or bulimia has as yet been verified. A frequent co-morbidity with borderline personality disorder or multiple personality disorder, among others, has been observed, however, and described in controlled and non-controlled retrospective studies.[152]

To my knowledge, the first and only study as yet among patients with eating disorders having explicitly explored abuse of the mouth, was published in 1990. The patients, a fairly small group of women, were hospitalized due to severe mental diseases. Concerning the group of patients with eating disorders, the authors write:

> Almost all of the (15) bulimic patients described forced fellatio. Several described flashbacks to the fellatio as precipitating their vomiting episodes. Several were able to control their bulimia by avoiding food that reminded them either of a penis or of semen. (Goodwin, Cheeves & Connell 1990)

Objections have been raised against the hypothesis of a connection between oral abuse and eating disorders or disturbances of other functions of the mouth by referring to the occurrence of ritual fellatio in a few cultures. The critiques have made the point that the interpretation of oral penetration as sexually abusive is culturally imposed; oral penetration, it is argued, need not be abusive and traumatic if occurring within a cultural framework which does not attribute these meanings to the act. Studies of ritual fellatio are scarce, have been performed by anthropologists who have not focused on health, and were not designed to explore possible adverse effects of what is perceived of as ritual, voluntary and legitimate fellatio. The most comprehensive of these studies was provided by Gilbert H. Herdt among the Sambia of Papua New Guinea.[153] This study, exploring ritual fellatio within a consistent theory of the construction of masculinity, makes evident that fellatio within that culture is the major means by which to acquire male growth, strength, virility, fighting spirit and power over women. Not to "drink" semen from the adult males in the community puts boys at risk of remaining small and weak, endangered by other males and, most humiliating of all, overpowered by women. The study provides an insight into a practice which, really, cannot be chosen, as there is no other way to acquire real manhood as defined by an utterly violent and aggressive society. Thus, these boys, who are obligated "to drink semen" frequently and for years, are thoroughly convinced through their socialization *that not doing so would mean choosing "cowardice," which is identical with dishonor, the lowest social rank, and powerlessness. To refuse to do so would be worse.* The study does not offer perspectives into the experiential sphere of this practice due to its main focus on the cosmology in which it is embedded.

As an analogy and for the sake of comparison, one may think of a similar study concerning female genital mutilation, still called female circumcision in many connections, a term "veiling" both the reality of its practice and the consequences. Undoubtedly, one may argue that this cultural practice is both ritualized and legitimized, and that the girls might even agree to be "treated" in that way. They know, however, that there is no other option for acquiring femaleness and honorability, *because not being "treated" would mean choosing dishonesty and the risk of not getting married which, in many cultures even today, would bring vulnerability, poverty and premature death.* Even taking into consideration the lack of choice and the consequences of refusal, girls obviously suffer due to having a legal, sexual practice being inscribed on their bodies. This is easier to perceive when one is willing to remove the "veiling" words and either imagine or actually see documentation of this practice.

Ritual fellatio may be a "veiling" word as well, prohibiting the seeing or imagining of the practice, and oral penetration may just not be as obviously traumatic as genital molestation. Thus, a documentation of its possible adverse impact might be much more difficult to acquire, as it would have to be searched for actively. Certainly, it will never appear as long as the focus of research is on cosmology and practice, and not on experience. Very recently, *female genital mutilation* has been renamed to reflect what it is and what it means.[154] It has been unveiled as violent, suppressive and endangering, legitimized by the law of patriarchy, designed to exact women's lasting acceptance of their subordination and objectification in the name of female honorability. Now, at last, any argumentation on its behalf burdens those who propose this practice in the name of overall cultural tolerance.

The same development could as well occur were one to name the *abuse of the mouth* for what it is and what it means. Then, it too might become visible as degrading, alienating and humiliating, legitimized by the law of patriarchy, designed to elicit men's lasting contempt for femaleness and their cultivation of aggression in the name of male honor. First then could the aforementioned objection to defining oral penetration of children as indisputably abusive and traumatic be met. Only then, too, would the burden of argumentation shift. At that point, it would become permissible to question whose interests are given priority by those who propose, in the name of promoting either cultural tolerance or acceptance for adult sexual preferences, that oral penetration of children may be beneficial, or even a form of caress.[155]

Concerning the other possible paths for the diagnostic registering of "eating problems" as mentioned above, a relationship between sexual abuse and any gastrointestinal disorder had not been mentioned in the literature before Douglas A. Drossman et al. published a study from a tertiary-care unit in 1990.[156] Until then, a convergent body of knowledge had confirmed a very high probability for sexually abused persons to develop somatization disorders. These were, however, never specified as to their bodily localization, presumably because the studies had been performed in psychiatric settings and the psychiatrists were not interested in a specification of somatoform illness. Drossman et al. showed that, among female patients in specialist care:

> ... a history of sexual and physical abuse is a frequent, yet hidden, experience in women seen in referral-based gastroenterology practice and is particularly common in those with functional gastrointestinal disorders. A history of abuse, regardless of diagnosis, is associated with greater risk for symptom reporting and lifetime surgeries. (Drossman et al. 1990)

Drossman and his group, and other researchers in gastroenterology, have continued to explore the relationship, which seems to be verified. A review on the topic was provided in 1995.[157]

Leaving the "eating problems," and proceeding to the "breathing problems" and their possible relationship to an experience of childhood oral sexual abuse, the current literature provides no specific research reports on this topic. As to what is termed "hyperventilation," the phenomenon in all its various presentations is conceived of as an integral part of anxiety disorders. Consequently, problems with breathing have not been studied with regard to abuse embodiment. On the other hand, in the field of somatic breathing problems, to my knowledge, only one single study has explored the occurrence of bronchial asthma among sexually abused women.[158]

Concerning studies of self-injury as related to a history of sexual abuse, the current body of knowledge is consistent. Self-injury among women is regarded as indisputably related to sexual abuse history, and girls victimized in early adolescence seem to be at highest risk.[159] Many studies focusing on mental illness in the wake of sexual trauma have registered self-assaultive and self-destructive activities (drug and alcohol abuse, risk-taking, self-neglect, etc.) among others, in the group of compulsive disorders. In most studies eliciting lifetime sexual abuse experiences among female psychiatric patients, self-destructiveness expressed as suicide and suicide attempts is positively correlated with sexual abuse.[160] Quite a few recent studies have explored this relationship in non-clinical populations. A positive correlation between sexual abuse experience and suicide attempts could be verified in both adolescent boys and girls attending high school,[161] in young men and women attending college,[162] and in young women.[163]

1.7 SUMMARY

I have explored interview excerpts thematizing a variety of lived meanings as a particular aspect of sexual abuse embodiment. Such explorations open perspectives to health problems far beyond the biomedical explanatory system. Theoretical frameworks developed by Janet, Polanyi and Kristeva offer options for a reading "in the sense" of the experiencing person. If one conceptualizes an assaultive act as not merely a violent transgression of boundaries, but possibly a lasting disturbance or permanent intrusion, it becomes evident that this may result in a corporeality of alienation and of internalized threat or danger. The expressions of suffering, however, must not be viewed as symbolic. Their shape is not arbitrary, but informed by the situational logic of an individual's assault perception and interpretation.

2. CONFUSED JUDGMENTS

2.1 BJARNE AND VERONIKA

In his interview, Bjarne mentions how he experiences a confusion of his vision after an anal rape at the age of ten. When trying to clean himself in the bathroom, he suddenly becomes aware of the water in the sink turning red, and, successively, the water running from the faucet and from the shower turning red as well. Simultaneously, everything else in his field of vision turns black-and-white. Later in life, this black-and-white world reappears frequently. He gradually becomes familiar with a connection between the world loosing its colors and his entering a state of depression. To be deeply depressed, then, means to be unable to perceive the colors of the world and unaware of that lack. This transient "color-insensibility," known to him from the time of his first sexual abuse experience, follows him into adulthood.

In the reinterview, he reflects upon this initial confusion of vision by linking it to another experience of change in visual perception, which occurred while, as an adult, he was being taught by artists to paint shadows. He suddenly became aware of not usually "seeing," in the sense of a conscious perception, the color of shadows. To "see" results from attention and analysis. He had literally to be taught how to see the colors, although his eyes were perfectly capable, physiologically, of perceiving them. Prior to this, he had not helped his eyes to do so, meaning that he had not used them that way, as he had not sought out colors. In his apparently normal visual perceptibility, a sense for colors had not been well-developed. How can this be understood? One may presume his color sense to have been normal from birth, as no mention is made of a congenital color-blindness. All water now seemed red in order to normalize what was so horrifying; may this childhood shock of the "red water" have unmade colors themselves as reliable phenomena of the life-world? Bjarne himself interprets the event of having his "eyes opened" for colors as follows:

> Then I understood that my senses could be taught, that I myself could teach my senses to open up for the world, and that I can train myself always to understand a little bit more. Perhaps it was to understand what I did in the bathroom. After having opened up, I could enter those rooms again. I find a connection to what I felt when painting. What I did in the bathroom was to interpret my situation. I believe that's right. I believe that makes sense.

By exploring the Husserlian term of *Sinn-gebung*, "sense-giving" or "-attributing," Merleau-Ponty has reflected upon how humans make meaning out of their experiences. He states that, "to understand is ultimately always to construct, to constitute, to bring about here and now the synthesis of the object." (Merleau-Ponty,1989:428) He not only addresses the fact that humans try to make meaning out of what occurs to them. He also emphasizes that the search for meaning is identical with the attempt to understand. Understanding is impossible without a frame of references. What things and events mean is derived from the frame of ideas within which the person is situated. The act of giving meaning is an act of active attribution, though neither random nor totally deliberate in nature. To attribute significance demands a "from," a reference which may allow choices, which expresses a purpose and is a precondition. Consequently, to understand is to construct and constitute. This depicts a process of *making*, a creative process which may actively reconstruct, maintain or confirm the existing, or question and deconstruct it, thereby shaping and modeling something new. Its opposite is *unmaking*, which designates a process of destroying the meaning once having been made, and which represents situations or states of disintegration. The English literary scholar, Elaine Scarry, has described the unmaking of the objects of the world in connection with the inflicting of pain during torture. Every object, every room can be unmade by torture. Scarry writes:

> In torture, the world is reduced to a single room or a set of rooms. ... The torture room is not just the setting in which the torture occurs; it is not just the space that happens to house the various instruments used for beating and burning and producing electric shock. It is itself literally converted into another weapon, into an agent of pain. All aspects of the basic structure - walls, ceiling, windows, doors - undergo this conversion. ... Just as all aspects of the concrete structure are inevitably assimilated into the process of torture, so too the contents of the room, its furnishings, are converted into weapons ... The room, both in its structure and its content, *is converted into a weapon, reconverted, undone.* (italics mine) (Scarry 1985:40-41)

Scarry makes evident how a wall one has been thrown against, a ceiling from which one has hung, a bathtub where one has been nearly drowned, a bed where one has been exposed to electric shock, will forever be undone as far as what it once meant. It has been made into a weapon, and, after that, a pain. When objects of the world have been unmade by an inflicted pain, meaning is at risk. When the meaning of objects disintegrates, nothing remains as it was. This was what Bjarne was thrown into, what he confronted and tried to reintegrate. The usual references of sensory perception were not a valid measure any more. As fundamentals were unmade, the senses had lost their "from," the person's basis for judgment was invalidated. Vision, the sensory capacity immediately concerned, became flawed. Its directly affected modality, color perception, was unmade. From the vision of the red water to the rediscovery of colors, four decades of an impaired sense of colors evidenced the act of the unmaking. This had not been recognized, however. Bjarne himself, though aware of this particular impairment when at the depths of his depressions, was not able to see it, as it was his "from," his zero-point of vision. As long as he remained in this zero-point, he could not see what it was that he didn't see, so to speak. The impairment cannot even be described adequately due to a lack of its semantic equivalent, being comparable to what, for the tactile sense, is termed "numbing."

Perhaps, the process involved was a "loss of consciousness" which could not be reversed until Bjarne dared to regain his full vision by visualizing what had been done to him.

In Veronika's interview, the imagined presence of her abuser during a certain movement of the shower curtain has been reflected upon. A movement, unmade as such in the past to mean "invasion," still represents this meaning in the present. The response to pain is logical, as is the response to a particular touch on the back. "Meaning unmade" is a key theory in the intrinsic logic of abuse embodiment. That which appears irrational, strange or "exotic," may represent a doorway into the silenced world for those who want to understand and help. In certain cases and constellations, it may be the only available doorway, as was shown in Veronika's interview regarding her deafness and it's restitution. She also presented and was treated for what was termed auditory hallucinations, though no one asked her what she heard. Living in a permanent double reality, she experienced that *"real" contradicted "true,"* and that her perceptions seemed incompatible with others'. She was in the past and the present simultaneously, aware of two worlds co-existing, with her being in both in a very particular way, while knowing this on yet another level of consciousness. She learned that the socially constructed and collectively constituted frame of reference was an invalid tool to make meaning of her situation. Since no one seemed to register that something forbidden happened, she realized her own experiences were uncorroborated. Consequently, she *mistrusted her ability to judge,* mirrored, for example, in her description of taking of the audiograms, or how she explained her sudden recovery when asked by the physician.

Bjarne too mentioned such a state of confused judgment concerning his responsibility for what had been done to him, since his rapist had always been such a "kind man" who certainly could not do evil. Both Bjarne and Veronika talked about an insecurity regarding the meaning of particular events, their confusion concerning *real and unreal,* and their uncertainty concerning *right and wrong.* These confusions influenced their personal memory, leaving many memories disintegrated, and constellating a state of "in search of understanding" which lasted for decades. Their impaired confidence in their own judgments, and their shattered trust in the reliability of their own memory, influenced all situations wherein a position "from" which to start is a prerequisite for reaching "to" something. Their basics and fundamentals, prerequisites for a place from which to speak, compare, judge, see, remember or decide, were no longer reliable after the acts which unmade them. This was expressed in Veronika's inability, among others', to know of and rely on her own body during pregnancies and deliveries.

2.2 MARY

Other modalities of confused judgment appear in the interview with Mary. As a complete narrative, and in many of its elements, it provides rich sources for the exploration of the seven patterns mentioned in the introduction to this chapter. For instance, her self-assaultiveness, offers examples of the rationale behind the apparently irrational behavior, behavior made comprehensible when seeing the corporeality of Mary, at seven years old. Observing that the jelly which her father forces her to swallow does not leave her body as feces or urine, she is afraid it might harm her insides. Consequently, she tries to get it out by cutting holes. Mary reports of several life phases being dominated by confused judgment. Two of those, interconnected, emerge in the following excerpts from the start of the interview.

> I want to start with my first delivery. I had attended a pregnancy course. Perhaps I hadn't grasped everything, because, as it was, when I arrived at the delivery ward I wasn't prepared for being shaved and having an enema and the like, and that they would keep checking the opening of my cervix. They certainly had mentioned this, but I hadn't grasped it. I wouldn't let them do all this. The enema I did myself. I refused the rest. I didn't let them do the examination to measure the opening. So I didn't have it checked. The minute the pain started, my father talked to me the first time. I started to hear voices, and he told me that we were going to have a child together. The minute my husband registered (on the monitor) that our son's heart beat stopped - or got very slow - and he hurried out to report it to the midwife. She just replied that he was nervous and that everything was all right. But we could see on the monitor that it had stopped and he ran out again to tell them, and several people came in and there was a real turmoil all around - because the heart really had stopped beating. They had to pull out the child. But I saw my father approach me, although I knew it was the midwife, but my father came and pulled out the child.
>
> I broke the oxygen dispenser from the wall and threw it towards his head, but it was the midwife. Then I grabbed hold of the railing of the bed - when a doctor stopped me, but I meant to throw it at the head of my father. They managed to get out the child. I heard the midwife ask the doctor: 'can you sew her?' And he answered: 'You'll have to patch her together as best you can.' I'd been torn totally because of high forceps, I learned later when I got clear again. They gave me lots of drugs, I was doped, I fell asleep and slept a long time. When I woke up they had moved me to another ward, it had become evening, and when I opened my eyes my father stood beside my bed and told me that I'd given him a son. Much more happened which I don't recall properly.
>
> The day after, or some days after, a nurse asked me whether I would go to a certain room where a psychologist was expecting me. I went there, and in the room sat an elderly male psychologist. He started to ask me whether I was troubled by something. I said no. I was deadly afraid that he had seen my father too, and that he might have heard my father say he was the father

> of my child and the like. I denied that something was going on. ... Strange things happened. I stayed at the ward for six days, was in my home one night, and then I was taken into the children's ward where my son had been taken care of. I stayed for one month in a single room, and the head midwife visited me often and sat on my bed and talked to me for hours and was very kind and caring. She often asked whether I struggled with something too, whether I was in trouble. I don't know whether they considered me as having been abused, I don't know whether I'd said something in my confusion, but I'd told her that he, father, had been standing there. Right after the delivery, I had a very high fever, probably due to the infection in the wound and an inflammation in one breast. Because of the fever I wasn't able to take care of my son.
>
> During the next delivery they doped me in advance. I don't remember so much. But my father was standing there, and he was to decide the child's name. The same psychologist came to me afterwards. The midwife came as well. She advised me to go see a psychologist after my release. And when visiting my general practitioner for the six-week check-up,[164] he mentioned this too. They had probably written something about me from the hospital. He was very concerned about this. He didn't leave the topic. I was sitting there for an hour, him talking to me. And he asked me directly whether someone had abused me. Because he may have remembered that during all my pregnancy I didn't allow him to touch my skin. I'd also refused a gynecological exam. He had to measure my stomach and to listen to the baby through my T-shirt. I refused to take it off. That was during both pregnancies.

This excerpt contains narratives of two rather extraordinary deliveries, one of which, the first one, was also dramatic. The short résumé from the preceding pregnancies was not quite usual either. Several times, Mary thematizes judgment confusion, which shall be explored in depth within a frame of deliveries as potentially conflictual situations for sexually abused women. What does Mary tell about? In her words, "the minute the pain started, my father talked to me the first time. I started to hear voices, and he told me that we were going to have a child together." At the time of her first delivery, Mary's father had been dead for seven years. He had abused her from the age of five until she was seventeen years old. The abuse terminated when he died. Mary's husband was unaware of this abusive relationship, as was everyone in the family. Now, at the moment of the onset of the initial pain of the delivery, two worlds coexist within a delivery room, carrying a double meaning of pain and creating a simultaneity beyond comprehension. The pain of abuse and the pain of delivery have taken place in Mary's body and confuse the distinction between past and present, informing her corporeality so that they may be indistinguishable for her. This pain means "father" as well as "birth." He is there, announcing his right to be present at the moment of birth of his child by his daughter. She knows him to be present and knows the birth to have started. Her double or rather, blended, consciousness, manifests in a repeated comment about the "blend" of father and midwife. Mary is, in a very particular way, aware of where she is, what she does, and why.

Two worlds coexist on yet another level. The one concerns *Mary's perceptive world,* comprised of the delivery-room, the pain and the appearance of her father, the other concerns *medicine's registrative world,* the monitor for fetal heart activity. The onset of the pain, the appearance of the father, and the drop of the fetal heart rate coincide. Yet again, a simultaneity of meanings beyond comprehension has emerged. Is there ever, in whatever kind of sense, a connection between "father" and "heart-beat," which may parallel the one between "pain" and "father?" Is this sheer coincidence, and are these phenomena merely occurring at the same moment randomly? This might, of course, be so. To conclude that, however, leaves two salient medical questions unanswered which ought to be asked and thoroughly discussed.

The first question is: What caused the emergent danger of fetal intrauterine death in an apparently uncomplicated delivery setting of a young, healthy primapara? The second question is: Could this emergency situation have been prevented? Of course, the answer to the second would depend directly on the answer to the first. No medical records from this delivery were available, therefore one cannot approach the answers using the record made by the staff members present during the delivery. The situation must be explored through Mary's description and the medical measures taken. The emergency generates a sudden and hectic activity. An immediate decision is obviously needed. High forceps without anesthesia is the option the medical staff judges to be necessary, showing that the probability of the risk of fetal death is being weight against the probability of the risk of damaging tissue. The first is evaluated as critical enough to outweigh the other. Imagining the action, how it must have happened and how it would be done, the adjective "brutal" seems adequate. The pain and the strain it causes to mother and child, are beyond doubt. Mary's attention is not in any way on what is being done to her. Her only interest, absorbing all her strength and awareness, is to prevent her father from taking his child. Yet she knows, on one level, that the midwife is standing there, the midwife is the father "in function." By whatever means necessary, Mary must prohibit her child from being born. So desperate is this demand, that she becomes violent, fighting her father (which she had never dared to do before, as she mentions later in the interview), by pulling an instrument fixed to the wall off its mountings and throwing it towards father/midwife. As this does not stop him/her, Mary tries to pull out the railing attached to the side of her bed, but a male doctor stops her. Once more, a medical conclusion must be drawn indirectly. The conclusion must be an unusual one in a case where the sudden drop in fetal heart rate has a maternal origin. This fighting woman, exhibiting so much physical strength and aggressive activity, appears strong and conscious, though she is obviously confused, and is seemingly unaffected by the pain caused by the invasive medical intervention. She fights those who try to save her child. In the eyes of her helpers, this is, of course, an irrational behavior. Consequently, one possible conclusion, yet not a definitive answer to the initial question, is that the maternal bodily state may not have been the most probable cause of fetal distress.

If there were such a connection, Mary is, by necessity, the mediator of yet another kind of connection. She is the only one aware of these coexisting worlds; she has her attention directed towards both, or rather she is in both. Her body is the arena of an insoluble conflict no one else knows about. The child which is about to leave her body is being claimed by her father as his property. Mary knows that the very moment her son is born, the world can see that her father is his rightful father, and the world will acknowledge the immoral relationship which Mary has wrongly partaken in. In the presence of her husband, this child will testify to the everlasting presence of her father in her body. Mary is caught in what she must have experienced as an inescapable trap. Her only option is to prevent the birth of the child. Only through an understanding from within, and based on the background of Mary's corporeality, may this situation reveal its implicit consistency. There is no possible distinction between past and present. The past is present in the form of this child she shall bring into life. Her father is present in the form of his rights. His appearing at the onset of pain unmakes the meaning of the delivery. What is meant to be an act of giving life is turned into an act of unraveling shame. As life cannot be given, without unraveling her shame, Mary has no choice. She has to "act against life" in the most fundamental sense imaginable: the heart of Mary's child stops at the moment of the appearance of Mary's father during the onset of the pain.

This delivery, causing a medical emergency, results in two immediate measures, both of which are non-medical: Mary is separated from her child for six days due to her state of mind and while suffering from a high fever as a result of infections; and, she is asked to meet with a psychologist whom the staff had consulted. The former action has far-reaching, logical, although apparently irrational, consequences, almost five years later. The latter has no consequences at all, according to Mary's presentation. From these measures one can conclude that the medical staff judges the complications during delivery to be, in one way or another, related to or generated by what, in a medical sense, cannot be comprehended. Further attempts to grasp what must have been going on, come in the form of personal approaches to Mary, mainly involving a midwife. Mary experiences her as caring and attentive, but she cannot tell her more than that her father has been present in the delivery room. It cannot be known what conclusions were draw from this about Mary's mental state or life-world. The experience of the delivery, however, leads to "preventive measures" being taken when Mary gives birth to her second child a few years later. She is heavily drugged before the onset of labor, thus having the conscious memory of her daughter's birth taken from her. This is mirrored in her description of the second delivery, which is three sentences long as opposed to the detailed narrative of her first delivery. In a state of multiple consciousnesses, she had been highly aware, active and responsive. None of these words is appropriate to describe the state of impaired presence imposed on her by drugs the next time. Nevertheless, despite the medications she receives, she again sees her father in the delivery room, and again is perceived as being almost psychotic, resulting in a repetition of advice from the staff that she see a psychologist.

2.3 BERIT

The impact of childhood sexual violation on pregnancy and delivery was, in various aspects, addressed by all 26 of the women who had been pregnant, whereof 23 had given birth. The different pathways from which embodiment could be read emerged from the women's narratives. In the interview with Berit, the topic of her deliveries brought forth another modality of confused judgment.

B: During my first delivery, in the expulsion phase, I suddenly stopped. It took me many, many years to understand why. The very moment the baby's head was about to come out - there was the abuse, this pressure of the head, it was - I refused to recognize it, but I remember stopping pushing, it happened several times, every time the head pushed I stopped helping. Then I heard someone order a vacuum extractor. I thought 'that's painful' and came back to myself. I received oxygen since I'd stopped breathing. But when they placed the mask over my face it went, bang! and I woke up and thought: 'Gosh, you're giving birth to a child, what kind of nonsense are you carrying on with. It must come out whether you want it to or not.' Shortly after that, he was born.

A: What was the sensation like?

B: It was so much like the feeling of this big penis invading me, this feeling of being broken, this feeling of too little space. I was too little. Oh, my goodness ... (her voice trembles, she swallows, moves) ... ohhh...

A: Did your body remember something your head had forgotten, which only your body could recall?

B: Uh-huh. It was the first time, I hadn't felt it like that before, and I guess I suppressed it, the very moment it hit me what it was, I suppressed it, so it disappeared again, because this I didn't want to remember.

A: And, of course, you couldn't open up.

B: No. In retrospect I understood that something pushed and wanted to come out while I closed tight and thought 'I don't want to, I don't want to,' but then I understood that it was a delivery I was involved in and woke up. No one asked me what had been going on with me. ...I recall once I opened my eyes, then I saw the midwife and two nurses, then I stopped cooperating, and the next time I looked, there were eight or ten people around, I'm sure. But I was totally absent, I'd switched off. Then I heard my husband's voice shouting that I had to push, but I was very remote, although I wasn't drugged, I'd only received a little bit for the pain.

A: In a way, you had left the ward...?

B: Yes, I'm very glad they got me back, because I know a couple, they have a boy who got too little oxygen during delivery, and I guess it was him who flashed through my head when I understood what I was doing. I feel it was urgent, it was a question of minutes or seconds.

After having talked about "the knife which became a finger which can be a scarf," and the anxiety-triggering function of this particular bodily sensation, the question concerning her deliveries leads Berit directly to a precise description of another kind of judgment confusion in a very particular situation. The pushing of the child's head is too similar to something threatening which she has experienced before. She presents her complex and ambivalent reaction in detail and consistently, expressed as an active and voluntary "stopping." Though she knows that the present meaning of a particular bodily sensation is about giving birth, her corporeality is dominated by the past meaning of the sensation, being sexually abused. The simultaneity extinguishes the distinction of the tenses, just as Mary has described it. Berit perceives of her body in two different ways, but the difference is too little, and the similarity is too dangerous to allow her to remain in the present. In this situation of conflict between the necessity to help the baby out, and the necessity to avoid what feels too similar to vaginal penetration, an acquired defense reemerges. She "leaves," in a way, and stops breathing. In her mental absence she is, nevertheless, aware, she is conscious of the double meaning and of the present activity and its aims. She is suddenly reminded of a retarded child, the image of whom, in retrospective, seems to have given her a strong impulse to open up for her baby, stronger yet than her need to keep closed. Berit is aware of her intermittent dissociative state. And unlike Mary, she is explicit about her active avoidance which, as she also acknowledges, endangers the child profoundly. She herself has comprehended the connection between the bodily sensation recalled, and her response. She can even express herself from the point of view of the little girl's perceptive memory. Once again, she experiences a particular "disproportionality" which has informed her corporeality, and which only a similar sensation can reactivate. Therefore she says: "it was the first time, I hadn't felt it like that before." She had not given birth before, the only situation which, within the modality of "proportion," felt like the penetration during childhood. What she perceives of as too similar is encoded in a relationship, namely, the relationship between the proportions of her body at a particular time in her childhood, and something transgressing its boundaries. Though in the present involving her adult body and that of her baby, this relationship still is the key to Berit's perception, interpretation, and action. The original perception of a disproportionality is reactivated, recognized and responded to in one act: the withdrawal from and avoidance of what is too similar, though belonging within a totally different frame.

An experience of similarity to a known, yet past, perception, may also include positive variants, for example a recognition which reduces the feeling of danger. Being pregnant again and giving birth the second time, she is prepared for the pushing of the head, and for yet another pain as experienced during the first delivery. She recalls how, during her first labor, a physician tried to widen her cervix manually in order to accelerate the delivery. Berit remembers this intervention as horribly painful. She wonders whether the physician had the slightest idea about the intolerable painfulness of such a procedure. However, being in labor for the second time, she recognizes this pain precisely, when feeling her cervix widening. In the recognition, she recognizes its function: this pain means "opening up" and does not feel threatening. Still, she is not prepared for the sight of an instrument, a metallic stick, like a thick needle, which is used to punctuate her amnion in order to accelerate the birth. She becomes afraid when seeing this instrument. When she asks the physician how it will feel, he assures her that she will not feel pain but merely a relief of inside pressure due to the water flowing out. Berit, however, feels a pain so intense that she screams and twists around in the bed, thus getting slightly hurt, and loses consciousness during the intervention. After having regained consciousness, she gives birth actively and to her own great satisfaction, without withdrawing from the pressure. She characterizes her second delivery as perfect, "apart from when they were going to break my water."

2.4 MARY

In general, a delivery is a situation of indeterminate danger. It may be one of the most threatening situations for sexually abused women, as it requires them to be dependent on the help of others and forces them to expose their body and give others access to it. The conflict between the actual need for help and the habitual strategies for avoidance, and the sensory similarities to previous dangerous and threatening experiences, may result in a breakdown of all the foundations for proper judgment. This may lead to actions which appear to give evidence of insanity, and which are then responded to accordingly. With such responses from others, carrying the implication that the insane person is the origin of the problem, no search for the relationship between the present conduct and some other origin of the problem is initiated. One of Mary's narratives shall be used to explore the hidden rationale behind a seemingly irrational act:

> It must have been in October, I'd been admitted to the acute ward as I said when we filled in the list. From there I was sent to the long-term psychiatric ward, BK. I stayed there until December. In the following year, I got psychotic again, and again I was referred to BK, and from there I was referred once again to the acute ward, because I had an imaginary pregnancy. Because Dad was 'living' with me there too, he had 'awakened,' again. And there something crazy did happen, because I took a pregnancy test at BK which must have been positive, because they ordered an abortion on my behalf at the gynecological ward. I'd told one of the psychologists that I'd been raped by my father. He believed that my father was still alive. My father was so real to me. I told him about my father, and he, the psychologist, decided to report him to the police. He contacted the police and the census people, and that's how he found out that Dad had been dead for years.
>
> The senior physician made an appointment at the hospital for me to have the abortion performed, so I was brought to the ward and it was done. But during this abortion they didn't find fetal tissue. And I turned psychotic during the procedure and awoke in an isolation room. I was convinced I'd given birth to a baby. They had let me have a room on the gynecological ward where I stayed for two weeks. One day, I told the nurses that I wanted to go upstairs to the delivery ward. And I went. There, I stole a baby, I took the child I'd given birth to. They stopped me at the exit, I didn't get it out. This created a real turmoil, and I was brought back. I was convinced that I'd given birth. It was so real.

After having given birth to two children, and having been perceived as psychotic and therefore having been separated from the children during the first days after both deliveries, Mary takes for granted that the same thing has happened again, since she is convinced she has given birth. Prior to this, during three months of therapy with a female psychologist, she has for the first time dared to tell about the years of abuse by her father. He is present in her life. This intense presence confuses her imagination and conduct disturbingly, causing yet another referral to the psychiatric ward. She has the delusion that she is pregnant by her father. When she gets a positive result on a pregnancy test, she cannot make sense of it. The situation is further confused by a psychologist's misunderstanding of Mary's presentation of blurred present and past. Nevertheless, her impaired mental state seems to indicate the appropriateness of performing an abortion. After the intervention, Mary is unable to sort out the events connected to her deliveries. She reexperiences the previous separations from her children. Consequently, she decides to find this child and take it back. Her attempt to kidnap a baby, interpreted as proof of a serious mental disorder, results in her being committed to a psychiatric ward for half a year. Her reality perception of having given birth is as concrete and as confused as before. How can she possibly distinguish now between what is real and what is not, when she was not sure about what was real when actually giving birth? In her corporeality, there is a baby which is taken from her. Thus, it is logical to find it and take it back, as it is hers.

To my knowledge, in the research literature about the long-term consequences of sexual abuse in childhood until now, only two authors have reflected upon a possible connection between imaginary pregnancy and childhood sexual abuse experiences, which they explore in a case history.[165] Their frame of understanding, however, is psychiatric theory, and their approach is informed by the concept of a psychiatric diagnosis in the category of conversion disorders.[166]

2.5 ANNIKA

In Annika's interview, another modality of confused judgment, also provoked by a delivery, is thematized. The excerpt is as follows:

> The second pregnancy was very traumatic. I've thought about that a lot. While pregnant with my daughter, my partner, her father, raped me. That started ... it had caught me again ... When I gave birth to her, which was a very traumatic delivery, and if someone had asked me whether I wanted to remove that baby I'd have answered yes, I don't want her. After that I became very depressed. Half a year passed before I could look at her. She was born premature, she looked like a monster, I felt a chill on my back every time I nursed her. That was awful.[167] Since then, I've thought about that. Of course I have photos of her when she was tiny. The baby was quite normal, there was actually nothing wrong with her.

Annika had been abused by her elder brother for years. She had dared to tell her grandmother who alarmed the family. When confronted with the accusation, however, her brother's fiancée made public that he had confessed to her what he had done to Annika, and that she had forgiven him. The family did not concern themselves with the possibility that he might not have told exactly what he had done. And no one seemed to feel a necessity to explore this further in the presence of the "tolerant" young woman and the, as it seemed, remorseful young man. Nor did anyone seem to expect him to ask Annika to forgive him. She disappeared in this scenario of betrayal and became extremely confused about the rightness or wrongness of what to her felt so unbearable and insulting.

Her brother had had vaginal intercourse with her at least twice a week from the time she was eleven years old, and he continued to do so even after his public "confession." Annika was in a state of fundamental insecurity concerning the discrepancy between her own perception of her reality - and the reactions of the adult members of her family. It took her years to discover that no one had known what had really been done to her, as no one had made any serious attempt to find out.

As the meaning of the abuse had been unmade by the family court, Annika's home, where the abuse had taken place and still occurred, did not feel like a home any more. She almost fled into a relationship with a young man she met when seventeen. She needed a rescuer and needed to escape. She did not leave, however. She was abused again. This was what she means when saying that "it had caught me again." She is caught in violence and abuse, her partner repeating or continuing her brother's acts, which she intended to escape. The baby, symbolizing the abuse, appears to her as a monster.[168] Though a result of abuse, it demands nonetheless to be nourished. Once again, an obligation to do what feels completely wrong but seems right to everyone else confuses Annika's sense of justice. It makes her distrust her own ability to judge, which has serious consequences for her life, as shall be explored later.

2.6 TANJA

The confusion of right and wrong, resulting in an impairment of trust in one's own perceptions and interpretations of situations, was also one of the central topics in Tanja's interview. She was 21 years old when the interview took place. Her own summary of the history of the abuse she had experienced is as follows:

> The abuse started when I was four or five years old, and lasted until I was eleven. The abuser was an uncle, an adult. There was also another uncle, in between. But mostly I was abused by the first one. In my teens, I belonged to a group of kids where there was a lot of drugs and alcohol, most of the members being older than me. I let myself be used by several of the boys, which means young men, because they were five to ten years older than I was.

When we started to reconstruct her sickness history, she reported what is recorded in the excerpt directly following the one above:

> **T:** When I was a child, I had a lot of headaches, and apart from that I had problems sleeping. That started long before I entered school. My mother took me to a doctor, I may have been six by then, when I got sleeping pills, which I took. I don't recall how long I took them, but I remember

clearly that I had to go to bed at a precise time because then I had to take my pill, and I slept until the effect was over and then I was awake again. As it was, I didn't want to sleep. I hid. They would find me in the shower, in the bathtub, below the stairs, or... I hid. Sometimes I fell asleep where I'd hidden. There I could sleep, but not in the bed. I could fall asleep in the entrance hall, on the sofa, but not in my bed. I didn't want to sleep there.

A: There it was dangerous?

T: Uh-huh.

A: Where was this uncle who abused you?

T: He lived in K. As a child, I often traveled to their place. The abuse hadn't occurred in my bed at home, but in my aunt's and uncle's home. I was there frequently. They had no children of their own. When I was little, they thought it was cozy to have me with them. In a way, I was their child as well. I spent the weekends with them, and later my holidays.

A: Did you give signals indicating that you didn't like to go there?

T: No, I liked being there, as it was, it was the best place I knew, it was great fun to be there. Every time I got gifts, and when I was there, I was the center of everything.

A: You were in the center, in all ways.

T: Yes.

A: What your uncle did was counter-balanced by something else?

T: Yes. He was the most kind of those two. He was very kind.

A: He was the kind man. And you were caught by that?

T: Yes. I was. It affected my sleep.

This sequence of speech turns invites comments on a theoretical and a methodological level. The methodological level will be commented upon in a footnote.[169] As regards the theoretical level, one might be tempted to entitle the initial sequence, "The Unmade Bed," in the very sense of Elaine Scarry's thoughts about the "unmaking" of places or things. In a detailed, consistent way, Tanja tells about her childhood experience of the bed as the non-place for sleep and rest. She cannot sleep there. Even under influence of drugs, shortly after she has fallen asleep due to the medication, she awakens. Only six years old, she is treated for sleeping disturbances. The adults around her, her father, mother and the doctor, do not seem to wonder why this apparently intractable inability to fall asleep only occurs when Tanja is in her bed. She falls asleep everywhere else. No hiding place is an impossible place to fall asleep. Only her bed. The bed, unmade as a place for safety, sleep and rest, is what she tells about. There the abuse happens. Although it never occurs in her own bed, but in a bed in her aunt's and uncle's home, bed does not mean safety any more. The safety provided in her home by her obviously concerned parents cannot balance the lack of safety of the bed. But Tanja does not tell of pain, threat or horror, which, in the interpretation of Elaine Scarry, is a prerequisite for the process of unmaking. Rather to the contrary, the bed becomes unmade bed in a cozy home where it is fun to be. How does this get so profoundly linked to a lack of safety that it even is "transported" to Tanja's home? Tanja describes the tension and trap of the dangerous bed in an atmosphere of generous abundance. She is in the center of everything, of attention, love, gifts, and interest. She is caught; there is a danger. The danger is generosity and kindness. The abuser is the kinder of the two people of whom she is so fond. She repeats that he was such a kind man. It is he, however, who makes her bed into a dangerous place. A little girl's foundations are confused within this constellation of love and threat represented and practiced by the same person. Both the gifts and the love are provided by the person who abuses her, which she has no means to spell out, name or tell, since this surpasses her cognitive ability. Feeling trapped by kindness means, apparently to be trapped very effectively indeed, as the trap is rendered invisible by the kindness, and the one who is trapped is left bewildered by the ambiguity. Being violated "carefully" seems to direct the violated person effectively as regards the question of responsibility for the violation. Tanja feels in danger in her bed, resulting from having been violated there. The person who violates her in her bed, however, is the kindest person she knows. As he cannot be dangerous, the bed becomes the dangerous place. And, likewise, since the responsibility cannot be his, it must be attributed to someone else. Tanja finds her solution, which shall now be thematized.

Can kindness mean violence if the effect of it is similar to that of violence? Pierre Bourdieu has considered this question in relation to the presence of power in human relationships and the distribution of power attributed to, or inherent in, social roles and positions. He has analyzed the implicit obligations linked to gifts, their meaning and the meaning of their exchange in a variety of contexts.[170] Regarding their symbolic significance, he has developed the concept that a " symbolic violence" may be embedded in gifts, favors or other acts of kindness. To acquire and maintain a grip on another person, "by means of practical or economic commitments which inform, guide and affect moral and emotional commitments," is, according to Bourdieu, to act violently, although the violence is rendered invisible by the outer appearance of what is done. Another person's trust, loyalty, confidence or gratitude can be violated by "kind" acts hiding suppressive, compulsory, captivating and/or exploratory intentions. Thus, apparently unconditional generosity may be intended to establish bonds which endanger the recipient as they mean captivation, and as the means by which the bonds are maintained of necessity constitute or preserve an asymmetry. Such an asymmetry, rendering one of the parties more deeply obligated, be it to feel gratitude or become complicit with the generous "giver," cannot but alienate the recipient. Bourdieu's concept of "symbolic violence" is a means to conceptualize the essence of a social relationship containing confused meanings such as are described above by Tanja. Her relationship to her uncle contains elements which must be defined as mutually exclusive; it blurs what ought to be distinct, and it transgresses borders which ought to be discrete. Her uncle's practices make her experience these polarities as synonymous. Abuser and caretaker are one; violence and devotion occur simultaneously. The distinction between right and wrong, the difference between good and evil, are affected. Her sense of order and her ability to comprehend are disturbed. The "careful violation" not only affects her sleep by unmaking her bed; it also affects her life-world by alienating her within her family. And it affects her self by alienating her from herself. These topics emerge in the following excerpt as Tanja talks about her regained autonomy:

> **T:** I don't use drugs or anything else any more. I'm through that. I have no pain any more, neither in my stomach nor my pelvis. (her voice "smiles," long interval) Now and then I have trouble falling asleep. But I'm used to it. Then I just get up and read. I can live with it. But when I think of the last year I attended school, from eighteen to nineteen, I didn't dare be alone in a house, and even less sleep alone, but now I live alone (strong, glad voice), I have no problems being alone any more, I'm in command. And if I get afraid, I know what to do. I get up, make a phone call, it doesn't scare me any longer, I can think about what has happened before, it doesn't break me anymore. I can feel shocked - sometimes I suddenly recall things I've pushed away, which have been out of mind; this can be terribly disturbing,[171] but it's possible to relate to it.

> **A:** Does this concern primarily you experiences with your uncle, or something else . . . those young men in those years . . . that you meet someone and you think 'he as well' . . . ?

> **T:** Uh-huh. It happens. Lots of strange things. Mainly from the time I messed around. Much of that time I can't bear to remember. I didn't go to school for a complete year. There it feels foggy. The first years at school aren't clear. I don't remember birthdays or anniversaries. But there are also painful and violent events I can't bear to remember. I've tried to make my mother tell me about those years, when I was little, because so much has disappeared from my head. I try to get a picture of myself as a child. It scares me that I remember so very few things with any certainty, that so much has disappeared. I've seen myself on a video from a family occasion, when I was seven, it was very warm, all the children were sitting there in pants or in swim suits, and they played, ten or fifteen kids, and I was seated by myself, completely curled up,[172] far

away from the others. I saw the video at a confirmation in the family, and it was awful to see ...

A: You caught sight of a lonesome, abandoned child ...

T: I felt pain. I've never thought of myself of once having been little. I saw this, all the kids sitting there and playing, and me, I'm totally curled up, sitting among the adults who are having their coffee. It felt terribly bad to see myself like that.

A: You could see that immediately, and simultaneously you could see that the adults around you didn't recognize it.

T: Yes, it felt terrible. I don't believe this to be the only such occasion. I've checked this out by asking other adults of the family ...

A: What happened when you directed your mother's attention to what you saw?

T: She saw it too. She did. She has apologized many times, but that is so painful, she has such a bad conscience about this that she's totally unable to cope with it. Then she starts crying, and then she feels so miserable, she has such a bad conscience ...I choose not to mention it any more. I know, after all, that she was abused too, from when she was four until she was seventeen years old. It was her grandfather. We've been abused in the same way, in precisely the same way. (pause)

Tanja has freed herself from an incapacitating anxiety which made her completely dependent on others' presence and on tranquilizers. After having named what has been done to her, and after having placed the responsibility where it belongs, she gradually opens up to remembering, and she actively tries to retrieve her memories. Simultaneously, she sorts out what she cannot bear to recall and to know. There are years which are veiled in fog, which Berit thematizes also. These years comprise the time of Tanja's acting out and letting herself be used. To have been abused by her uncle is acknowledgeable now; to have let herself be abused by many others is not yet tolerable to deal with. In retrospect, she can see how she has been caught in a trap of kindness, and this has allowed her to free herself from the responsibility of not having resisted and, as such, having been cooperative. She dare not see, as yet, her voluntary participation in her own violation and abuse, however. To meet those men is "terribly" disturbing, and confrontations with witnesses to her self-humiliation must be avoided. Apart from this, the period without boundaries had painful consequences at a crucial point in Tanja's life, thematized in an excerpt and two passages transcribed with an asterisk:

* Tanja tells that she not only received gifts from her uncle but money as well, large amounts of money for a little girl. When he was confronted by his own mother, Tanja's grandmother, with the fact that she knew because Tanja had told her, he offered Tanja a larger sum of money and an expensive holiday trip. We talk about the many signs indicating that

he neither at that moment nor later, during the legal proceedings, seemed to have comprehended what he had done to her.[173]

T: It's difficult to talk about abuse which has been linked to so much "kindness" and gifts and care, it's much easier to tell about force, then it's clear to everyone that what happened was dramatic. Of course, I've survived. I wasn't pitiful, as it seemed. There I came with new clothes and expensive toys. I was lucky to have a nice uncle (intense voice).[174]

A: Was that one of the reasons why the trial ended as it did? You weren't sufficiently molested?

T: Yes. Apart from that, it was a question of time. The accusation came too late. I mean, there were two trials. The first one, when I was twelve years old, the court case ended, the charges were dropped because of lack of evidence. I tried to get justice again when I was eighteen, but then the case was beyond the statute of limitations, even though I referred to the former trial where the charges were dropped. My uncle's second wife witnessed against me, and there were new witnesses ... I don't know what is acceptable, in a procedural sense. *And partly I had myself to thank. Because between twelve and eighteen I behaved in such a way that everyone doubted my trustworthiness. I was easy to discount, which means judged as being "easy."*

* We talk about how both the abuse and the sexualization her uncle inflicted upon her, had made Tanja act out sexually, which in the second court proceeding, was used against her. In other words, her conduct had made even those who earlier had been confident that she had been abused doubtful about her contribution to what had happened.

The statement in italics concerns yet another kind of confusion, namely the wider responsibility for the consequences of childhood sexual abuse. As concluded in the latest reviews on abuse impact, sexualization of the abused children is one of the most frequently and consistently reported short and long-term impacts.[175] Both Tanja and Bjarne mention how they were sexualized such that their relationships to other people, predominantly those of the same sex as the abuser, became characterized as completely lacking in boundaries. Bjarne became what he himself called, "totally sex-fixated," to such a degree that he was like "a flypaper to flies" for other men. Tanja thought of herself as other people's toy, something to be used however one liked, a thing for other's sexual pleasure and an object to be consumed. Her impaired self-esteem, expressed in a grave neglect of her own interests and self-protection, and appearing as a socially unacceptable promiscuity in adolescence, created yet another trap for her. It made of her a teenager with a bad reputation, a shameless person whose trustworthiness as a witness against a kind and modest man was so flawed as to be nonexistent. Her public shamelessness became a major argument for finding her uncle not guilty of the charges for which she had accused him. The primary violation resulted in a self-neglect bordering on self-destructiveness. This conduct, conceptualized as a personal trait of hers, became proof of Tanja's amorality and her uncle's innocence. The symbolic violation turned out to carry a destructive power beyond expectations, ungraspable in its invisibility, and impossible to argue about because of its ambiguity due to a social consensus around the meaning and function of "acts of kindness."

After starting her work to overcome the limiting consequences of abuse, Tanja concentrates primarily on getting a clearer picture of herself as a child. She cannot see herself. In her memory, there is no child Tanja. This lack of biography, a shocking recognition for Tanja, stems from there being blanks created by the confusions in the wake of the violation, led to impairment to her sense of self. Tanja has entered the painful process of facing her own estrangement from her family and herself. It seems as if there has not been a self. She has to reconstruct it, or rather compose it from fragments the adults in her family can offer her, and from documentation representing a gaze upon her from outside, mediated by a camera. There, Tanja can see what she knows as an absence: there was no child Tanja. There was a stranger in the family, not a child among the kids, but a curly something among the adults, and yet not seen by them, not belonging, not one of them either. Tanja puts words to a central impact of having had her foundations confused by the invasive acts and intrusive presence of others: "It scares me that I remember so very few things with any certainty..." Here she addresses what was central in the narratives in so many of the interviews: How can I know with certainty, how can I judge, how can I distinguish, upon what can I rely? Living in incompatible realities, having had one's foundations unmade - the "from," the precondition of structuring, interpreting, knowing and relating - affects one's selfhood profoundly. Tanja tells how selfhood is put at risk when differences are blurred and distinctions extinguished.

The importance of elucidating the "kindly violation" may be made clear by looking at what it means to have one's selfhood endangered nearly to extinction, to the point where no self is anymore, and no reliable "from" exists. Consequently, the final answer to the question whether kindness can mean violence, must, of necessity be: yes, it may mean violence in the sense of destructive power.

2.7 PIA

The interview with Pia offered yet another insight into the confusions created by abuse. In her life and in her corporeality, sensual lust, desire and tension had been tied together with bodily resistance, repulsion and disgust. These melting sensations, perceptions and emotions had evolved to become the soil of irresolvable conflicts as they affected

Pia's most intimate relationships in adulthood. She had been abused by her stepfather from the age of twelve to fifteen, when his sudden death terminated the abuse. The interview begins with the following excerpt:

(Pia is 25 years old and an extraordinarily, almost stunningly, beautiful woman).

P: I was abused by my stepfather. Since then, I've had real problems being close to people, which became very obvious in my relationship to my partner, the man who is the father of my son. Whenever he wanted to embrace me, especially from behind, I reacted. That was difficult. Even if he was careful in his approach after having understood that I pulled away whenever he became intense, it was a problem for me. After I'd given birth, one and a half years after we met, many things changed. He reacted more explicitly when I withdrew as I did, which resulted in more difficulties between us, both about communication and being close to each other. I was really difficult to talk to, very closed, I didn't feel comfortable at all naming things and talking about them - just in general, everyday things and what had happened to me.

A: Did he gradually learn to understand you?

P: Yes, definitely. He was very good at sitting down and asking me questions, but I could never really answer, I just kept sitting there and thinking, I was far away. But he tried, he really did.

A: But did you feel he repeated something?

P: *For example, whenever we had intercourse, it was exactly as if I was seeing my stepfather above me, the contour and the smell, I could see everything so clearly, from when he stood at the door, until he came in, I remembered everything he was wearing. It has come out more and more during the last years. Then I reacted. Either I kicked him, or I hit him, or I scolded and threw things around - I was totally crazy that way. He really reacted, he got scared.* But he coped with it very well. He is a little older than I am, he understood quite a lot.

A: Are you still together?

P: No. Suddenly last summer it ended. In a way, after we had a baby, it was as if I grew away from him, and he got tired. He had plans to join a foreign company, which he'd done before, at the time when I was in the hospital, and that time he left suddenly. And now he thought he could go there again. It was soon after the delivery. He was probably tired of everything, he felt too much pressure. I guess he began to get tired, fed up in a way. It was really demanding.

Pia cannot enjoy physical proximity with the man she has chosen to live with and have a child with. She cannot explain to him why she reacts as she does. Whenever he feels pushed away, however, he actually recognizes that she does not mean him. He cannot comprehend, however, that she is unable to distinguish. Nor can Pia understand this, factually. She cannot spell it out and she cannot explain. She is unable to answer his questions about the core of a resistance directed towards him. He cannot sense that he himself appears to her to conflate with another man whenever he approaches her. He cannot see the conflation taking place within her imagination, informing her corporeality. Pia cannot spell it out as a direct answer to a distinct question, but she presents the two men syntactically in a situational description. Guided by a question about a feeling of repetition, she confirms this on the spot, without the slightest hesitation or preparation, and as if lapsing into the repetition's context. She is in the intercourse, although the choice of past tense might imply a narrative retrieval, and the 'as if' might testify to an awareness of two parallel images; but when mentioning a contour (Pia says 'skyggen', which can signify both shadow and contour) and a smell, she is in her room, her stepfather standing at the door, entering, the picture becoming ever clearer. Then she reacts, and the one she reacts against is her lover, but the "he" in this narrative are both the stepfather and the lover. She cannot see them separately in the particular situation of initiating and proceeding to have sexual intercourse. Pia spells out exactly, not in words but syntactically, how, in a particular context, the *sight* and *smell* of two persons may conflate. Once again, we encounter perceptions too similar to be distinguished, releasing an explosive activity of defense and repulsion. This acting out, which seems beyond the control of the person herself, is not only out of proportion as regards the situation but is also incomprehensible for the one who feels attacked in lieu of another. Pia knows she makes love with a man who is attractive. She does not try to avoid intercourse, she invites him to come close, as if she were quite certain that he is the man she desires, until the very moment she perceives the contour above her as no longer clearly his, nor the smell.

The keys to this process of conflation are two sensory perceptions with specific meaning, which provoke an almost reflexive and explosive repulsion. The contour obviously represents Pia's abuser, his merging with her lover's. Does the smell also represent the abuser; is it the intimate smell of his male body during sexual activity? Is the smell the specific smell of semen which Tom, Judith and Christine mention? How, though, can this smell emerge during the foreplay to intercourse, so that it, together with a contour, causes the sexual act to turn into Pia's violent repulsion in a state of changed consciousness?

The answer is provided when, in a later passage of the interview, she describes the stepfather's violence against Pia's mother. He batters her severely, and Pia has tried to protect her by getting between them, physically. Gradually a pattern of abuse is constituted; it is initiated by the stepfather's verbal provocation of Pia's mother, expanded by Pia's intervention on her mother's behalf, and occasionally terminated by the stepfather's announcement that he'll "take her". Later the same night, he comes to her room. Pia is aware of this "finale" from the very moment he starts to criticize her mother. She is taken; it is unavoidable; she has caused it by interfering; she deserves his special interest and has no hopes for escape because, if she does, he batters her mother. This description is followed by the next excerpt:

A: What did he do?

P: He got me to do things with him, he told me what I had to do, to undress him, fondle him, a lot of this, he used me, my body ... (very long interval) I felt that he ... he intended to have me ... in a way draw me into ... he didn't say it this way 'do this or that', but he explained to me what I should do, precisely as if telling something in detail ... that he had imagined it before he came to my room ...

A: ... that he had his sexual fantasies and that you should realize them?

P: Uh-huh. But every time he stood at the door I said 'go away,' I managed to say these words, in a way, but he said 'I'll just come in for a while,' he said this as if he would try to make me change my mind ... he started very calm and quiet, but he became more and more threatening when I said no ... and finally I couldn't ...

A: It really was a lost game from the start? What did he do with your body?

P: First, he always fondled me, then he licked me ... and ...

A: Was this what you meant when saying that he drew you into it? He tried to seduce you so you would be aroused?

P: Yes. (the voice is completely monotone)

A: Did he succeed now and then?

P: Yes. (the voice drowns in weeping) (very long interval)

A: This link must make sexuality difficult for you. You really had no chance against this experienced man who knew all the tricks. That means a double abuse.

P: He made me mad. I wasn't worth a thing. In the beginning I took part so he would get finished faster,... in a way, gradually it was ... this ... that I went along it ... but when he left, *there was this smell* ... I got finished with him faster when I went along with it ... (voice fading)

A: He linked arousal and abuse - what did this to the woman in you?

P: Every time I have intercourse now I despise myself for my taking part in something which I really ought not ... I feel that *my arousal increases* all the time - but I *feel no arousal* anyway ... but I (long interval) I felt that I had such a need for someone who cares for me, and then I do it. He hasn't really been good at supporting me, the one I live with. Now and then he says that everything is just a performance, because now and then when we're together I have these attacks, that I feel suffocated ... then I lie there, trembling ... and cry... and he says everything is just a performance (she cries loudly) ... and then it happens that afterwards I throw glasses and get completely mad because he says things like that.

A: It's painful to be misunderstood.

P: I don't manage to say what's going on, he asks what's going on and I can't explain it, and so he thinks it's just an act (the rest of the sentence cannot be understood because of crying) ... then I just wait till he gets finished, I can't manage to respond to his body. *I feel I get aroused but I can't enjoy it,* I have such a bad conscience, I feel I'm so... I sicken myself so much I throw up afterwards. *I'm disgusting...*[176]

A: This suffocating - is it because you had to suck your stepfather?

P: Uh-huh. I've tried to tell him, I've told him, but he forgets it in a way. Whenever I've a bad day, and if I sense that an attack is coming, I mean that I feel suffocated, then it's okay, but it's forgotten the day after...

This excerpt offers a possibility for comments on three different levels, the syntactical, the methodological, and the phenomenological, all of which contribute to an understanding of the complexity of the impact of abuse in general, and the idiosyncrasy of embodiment in particular. From the initial passage concerning the merging of men in Pia's imagination and body perception, which is presented syntactically, a visual and olfactory essence emerges. It triggers defense and repulsion. It signals a break in a voluntary participation in sexual intercourse. From a certain moment, *Pia is in conflict,* resulting from a correspondence between the past and the present which Pia cannot name. However, the unnameable once again emerges syntactically. When Pia enters the imagined situation by describing it, her wording and the fragmentary structure of her utterances show her hesitating approach, step-by-step, as if to see. By literally going into it, she presents parts of her current corporeality thus allowing the conclusion that she is again being seduced into feeling sexual arousal during sexual abuse. The imagined is named, and when the word arousal is spoken, the answer is yes - twice - and the voice disappears. It has taken all her voice to confirm this link, the "perceptive synonyma" of arousal and abuse. The emergence of the link leads to its essence, directly to be read from the text, which is a smell. This second conclusion, a particular smell, and the first, Pia's conflict, will be transferred to the phenomenological level after the comment on the methodological level.

Four important methodological questions arise from this excerpt:

1. does the interview text document an interview technique which leads to a construction of a pattern designed to confirm the interviewer's presupposition?
2. do the interviewer's comments and interpretations invade the interviewee's intimate sphere, or disregard her personal boundaries?
3. is the understanding shared and mutual, or does the interviewer alienate the interviewee from her own perceptions and feelings?

4. is the interpretation of the text as exercised by now in accordance with the interviewee's experiences, or is there a pretense of authenticity, achieved by the interviewer's construct of a pseudo-logic?

The point of departure is Pia's explicit mentioning of an inability to tell, name and spell out confusing emotions creating conflicting responses. Quite obviously, she is unaware that what feels unclear and confused and thus cannot be explained to others, is present in her description. Only when daring to enter the abuse situation, can she "know" the fusing of the central elements in her present sexual conflicts. Consequently, when guided to an imaginative description by a question concerning a possible repetition, she sees how two men, her abuser and her lover, conflate. Still, there is more which seems confused, unnameable, and as yet unspoken. When talking again about the abuse situation, but from another perspective, it seemed logical to ask her to follow the same course, to enter the situation and imagine it. Which she does, and her stumbling approach, mirrored syntactically, allows the interviewer to visualize what has happened. Pia gives clues in her fragmentary utterances, pointing to certain meanings of particular acts, so that the pieces "I felt that he ... he intended to have me ... in a way he drew me into ..." merged into a Gestalt with "first he always fondled me, he licked me ... and ..." The Gestalt was her own sexual arousal. The word had to be said by someone else. The inside of the abuse had to be named by another. Pia could only confirm, twice, first the fact and then the repeated occurrence of it. The interviewer could visualize the situation due to Pia's stumbling verbal description, and she could grasp its meaning due to Pia's fading voice, echoing the former interview situations mentioned, and to her desperate weeping.

Thus, the conclusion grew out of probability, empiricism, experience and empathy: the unnameable, the grounds for confusion, are abuse-and-arousal, interlinked. As this is expressed, Pia can speak again on her own, confirming her corporeal situation in the interface between these conflicting perceptions. She immediately applies it to a statement as to her femaleness and present female sexuality. This is given a precise formulation, articulating what she has not been able to mediate before, in a striking contrast to her former attempts to grasp what was confused, and concerning an emotional link between her voluntary participation in intercourse and her disdain for herself. The recognition evokes the next step in comprehension. Unable to mediate her emotional conflict and, consequently accused of performing by her partner, she has acted out her desperation. The interviewer interprets this as expressing the pain of being misunderstood, and Pia can immediately integrate this into a self-referent acknowledgment of her disgust for herself.

The basis for the answers to the four initial questions may now be elaborated. The interview-text does not show a construction of a pattern, but a real effect of judgment confusion resulting from silenced sexual abuse. The comments and interpretations were obviously not experienced as invasive, but seemed rather to facilitate subsequent steps in the process of distinguishing that which was confused. As such, they do not represent a disregard for the interviewee's interests. Here, one might object that questions which tap directly into unexplored conflicts or fields of unacknowledged tension can introduce an element of danger. The objection is valid when considering the factual effect of such comments or interpretations, namely, to initiate a confrontation with that which has been, possibly by all means, avoided. Perhaps the objection concerning a danger implicit in every direct statement during interviewing cannot be met with a generalization. What is, in fact, dangerous, may depend on the setting, the topic, the frame and the actors involved. Thus, the argument, in the present context must be reasoned situationally. Could the interviewer's comments, obviously activating the "stuff" or matrix of a deep tension with a destructive impact and potential, have evoked a reaction leading to a breakdown of balance, an acute psychosis? To my knowledge, such a development within the setting of research into the impact of abuse has not been described, most probably due to the fact that the dominant methodology does not produce such situations. In studies of the way medical settings are influenced by the explicit or implicit presence of violent experience, the adverse effect of direct questions has more often been addressed in terms of the physicians' reluctance to ask them, or even more, their rationalization of avoiding to ask them, than of a concern of danger to the subject.[177]

When considering the theories related to trauma, memory and selfhood, in the sense of Janet, the objection concerning direct questions or comments may be invalid. According to Janet, it is not the speaking out, but the silencing that is the pathogenic agent. To be given a possibility to tell, to reintegrate disintegrated memories, might prevent a psychosis which would be understood rather as a breakdown of adaptability to silenced trauma. This is supported by neurophysiological studies on stress hormone levels by, among others, Pennebaker and colleagues.[178]

As to the third and fourth questions, concerning methodological considerations, the understanding which evolves during Pia's interview is shared and mutual, not one which causes an alienation, but rather one which virtually leads the interviewee to herself, on a level of improved insight. Consequently, the presented interpretation is consistent with the core of the abuse experiences, which created a constant confusion concerning the distinction of, and the difference between, sexual arousal and sexual abuse, and influencing all judgments in adult sexual life.

On the phenomenological level, Pia describes a sexual conflict which turns voluntary sexuality into violent defense. One trigger is a situational merging of past and present through a perception of a merging of her lover with her abuser. A particular sight, the confluent contours of her abuser and lover, however, connects to a smell not specified further, so far. While the significance of the contour is obvious, mirrored in Pia's syntactical merging of two men into one shape, the significance of the smell remains rather ambiguous as it is not described. Which smell is this trigger which breaks the distinction of the tenses? The next interview passage permits the conclusion that the core of the conflict is the link between sexual arousal and sexual abuse, or rather the arousal *by* the abuse, by the manner the abuse has been performed. By being stimulated in a skilled way, Pia is "drawn into" participating sexually, as if voluntarily, by will and out of sexual desire. Although she tries to recall and to reestablish her own rationalization for participating ("I got finished with him faster when I went along with it"), a smell remains when he leaves. Now, the next conclusion concerns an essence of the arousal-abuse confusion. It is the blend of the odor of arousal, both *hers and his*, which become literally conflated in the abusive act. Consequently, the deepest origin of Pia's adult sexual conflict is her idiosyncratic embodiment of a three-layered confluence, consisting of the *merging of abuser and lover*, the congruence of *abuse and arousal*, and the *confluence of male and female odors of arousal*, creating the blend of a particular smell which can only reemerge when Pia is aroused by sexual desire for her lover. In a phenomenological reading, this is expressed directly in fragments of a sentence ("I feel that my arousal increases all the time - but I feel no arousal anyway") What sounds like linguistic paucity or a semantic paradox is a reliable perceptual equivalent.

As a result of having been violated seductively, yet another modality of Bourdieu's "symbolic violence," Pia's arousal is non-arousal. Sexual arousal has been unmade, not by pain and danger, but by a seductive abuser's skilled hands and tongue. Arousal equals abuse. Consequently, arousal cannot be enjoyed, as she says in the following passage. And, finally, since her arousal equals her own participation in her abuse, she becomes the abuser, she herself is what makes her say, "I sicken myself so much I throw up afterwards" which, in the Norwegian original, "jeg spyr av meg selv etterpå, jeg er ekkel..." is virtually identical to Kristeva's description of the rejection of the abject. An English translation does not render visible that Pia perceives of herself as the abject she tries to get rid of. The reliability of this interpretation, however, is evident in Pia's repeated suicide attempts and her self-destructiveness which is different from Mary's self-assaults, and Tanja's self-neglect and risk-taking.

2.8 HANNA AND ANNABELLA

The next exploration of judgment confusion brought about by silenced abuse is connected to Tanja's formulation of the impossibility of knowing "with certainty." Hanna addresses perceptions and sensations "in search" of their explanation and meaning, in several passages. One of these is the following excerpt:

> **H:** ... But I believe something has happened - I believe something has happened - real sexual assaults of me as a baby. I confronted my uncle with that, I tried to make contact with this when I made this video. Because I have to tell about one more thing. There has most probably been something more, since I - Sunday night, my friend stayed with me, we planned to go to bed early since both of us had to get up early the next morning and we were fairly tired. All of a sudden, I wake up with a horrible pain in my anus, which is a kind of pain I've been awakened by and have had nightmares about as long as I can remember. Well? Usually I go to the toilet and think I might ... but I know there is nothing, but I go anyhow, and there is nothing ... and I (laughs slightly and as if a little bothered) it's good to be allowed to say such things which are so strange, because there is something I want to quit, so I try to pull it out, but there is nothing. Well, it's a little disturbing to say, to pull something out of one's own anus. But there is such a pain. Then I thought that it was so painful this time, I would go and tell my friend, because I can tell him whatever is there. I said: 'I'm miserable. I feel pain in my stomach. No, I really have a pain in my anus, horrible pain.' He's very good at calming me down. He gave me a massage on my lower back. That released the pain, and then my tears came, and then the pictures began to emerge. The first picture was one of my uncle washing himself in the fjord, which he used to do in the summertime. Then there comes - exactly as if a strange film which speeds up - then I was in the bathroom, and there I look down into a dark hole with a strong smell of urine. These were the pictures which came that night. And a whole lot of crying. I linked this to the pain in my anus. And before I fell asleep that night, I saw a picture I can't grasp, of a lamp, a glass globe with green stripes, and it was broken so there was a hole underneath and I can see the bulb, it whirls round and round, and whirls and whirls ...

What does Hanna tell about? What is at stake here? A pain at night, interrupting her sleep, known to Hanna for as long as she can remember and representing in her imagination something she has to get out, leading her to do an improper thing which she, before naming it, feels obliged to "announce," and which she excuses with a particular laughter covering the awkward confession. When receiving help to calm down by the caressing touch of loving hands, the pain disappears and gives way to moving pictures, as of film fragments, presenting her abuser first, and then, fragments linked to a particular smell which cannot be placed in a context wherein a proper meaning could emerge. She is directly receptive to the interviewer's "translation" of the lamp as seen from beneath and the whirling bulb as a visual distortion at the moment of losing consciousness. She has thought so herself; she believes this to be an adequate association - the question is, to what. She cannot provide an answer; the sequence has never emerged before; she only sees the film fragment; there has not been anyone, as yet, who could provide the place where it fits. But Hanna had an irresistible impulse to draw the next day. Everything she shaped on the paper was like exploding, red phalluses. She had never painted anything like that before. There is another passage about something waiting for a proper place to fit in:

H: I know after all a bit more, and I asked my uncle about that, because I remembered that he had tempted me to come to his room to get some chocolate. This is classic, isn't it? He tempted me time and again with chocolate. I remember him locking the door, that I'm inside. There are a lot of things I remember which don't need such a lot of logical sense before one thinks: 'why did he do that, why did he lock it?' Why was the chocolate in his room. There is a strong logic.

The above quoted excerpt (page ten) leads, however, to another topic. The topic of what happened in her uncle's bedroom reemerges (page fourteen) after a passage which will be explored in the next chapter. She has spoken extensively about a strange sensation in the lower part of her body. The excerpt is as follows:

H: There is something else I've directed my attention to now, which I also have felt all my life, shall I show you what it is? There are two areas here (she points to two places on her back besides both shoulder-blades),[179] and two areas here in front (points to both shoulders),[180] if you can imagine two points pressing here and two points pressing on the back side, this way ...[181]

A: ... who has held you around your shoulders this way?

H: Yes, isn't it like that? How many times have I mentioned this, have told it to therapists that I had two painful areas on both sides, that there is something, *and then and then and then* I've thought of this in relationship to my uncle, that he has gotten me into this disgusting bed of his, several times I guess, *and so and so and so* I will get up all my strength, all my strength I will get up, and there is this pressure which is, that is what I say, this pain is there, the body has no option to get up, and there is something which presses down, and then it's just as I said to doctor S. that it feels just as if you step on the accelerator of a car and simultaneously hit the breaks as hard as you can and the car stands still and I feel this kind of exhaustion, I feel it physically, but I feel also that it's like all the problems I get into - I have so much energy, I have so much imagination about what I could do, and then there is a counter-force[182] in the system that comes, that holds back - and then I get my headache - I guess this is what it means (her voice has been very clear and strong all the time, the last sentence is said with some insecurity - or perhaps as an apology).

Hanna reveals that she has an association to what might have happened in her uncle's bedroom when he locked the door and she was, as she can see, inside. She could not enter this associative field through the material by which it was touched upon and thematized in the previous part of the interview. She enters it from her own body, from an as yet unexplained and not comprehended aspect of her corporeality which she has experienced as a health problem for many years, and for which she has initiated many contacts with physiotherapists. There is a particular pain in front and behind her shoulders, which seeks to be understood. She "demonstrates" the pain by demonstrating its probable origin: a grip from hands around her shoulders from her front. The very moment she receives confirmation of this possible "reading," she can place the pain as a grip in a context which had not emerged previously, although she searched for an entrance to that psychic and physical room. Suddenly, an access to the room is provided. The pain and the grip lead to her uncle's bed, and the grip becomes the force holding her down against her will, while she struggles to get up, and is left exhausted even by the mere telling about the situation. Hanna also offers syntactical/phonetic expression of her physical effort in a struggle to understand the force which was directed against her. She stutters twice, as if pinned to the spot, representing a remarkable anomaly in this interview with a very self-aware and verbally skilled woman.

A last exploration of confused judgments resulting from sexual abuse takes its point of departure from the question of responsibility. The interview with Annabella provides a very particular demonstration of the problems and conflicts emerging from such a confusion. Annabella is one of the two women who reported to have been abused by a woman, in her case by her mother. The abuse started when Annabella was a little girl, and lasted until the mother died of cancer. Since the time of her death, Annabella felt some sort of trouble, a feeling of guilt or responsibility was linked to her mother; Annabella had never dared mention it before, but offered it to the interviewer, who, as a physician, she felt might conceivably be willing or able to answer some specific questions Annabella had thought about for years. The excerpt from this passage is as follows (A means Annabella):

A: But it happened as well when she was at the hospital, at the end, when I came to see her, then I brought our special equipment to the cancer clinic. It wasn't every time, and at the end there wasn't so much space in her bed, she had an ovarian cancer which had spread all around in her stomach, she was huge at the end, so I had to stand besides her (pause). The worst I did was really the last time, then I was called for from the

> hospital in the middle of the night, they said that she wouldn't live much longer. Whether it was by old habits or why it happened, I don't know, but I brought the equipment. That was almost her last experience. This is a little ... a little disturbing[183] ... it was a long time ... far into my own therapy ... *that I thought that if I hadn't done this she might have survived.* I thought for quite a long while that it really was me who had caused her death, or something like that (her voice totally flat). Yes, it was almost a given, I was completely sure of this. Because my uncle had told me that if I reported something they would die. Early in my therapy, I recognized that she might have developed cancer anyhow. But as she had cancer, after all, I mean, I performed an act with her near to her last breath, it took me quite a while to come to grips with that I had nothing to do with her death. It just shortened the process a little because her heart pumped a little faster that short period ...? Her heart couldn't stand that ...? She didn't even ask me to do it, I did it quite automatically. So it was an act of my own free will. This abuse situation seemed the most difficult of my adult life to me, although it was me who did it, which was to turn it upside down in a way. It was as it should be, according to my understanding. I couldn't do anything else...

There is much which could be commented upon concerning this excerpt, but the deep confusion of the guilt for having caused her mother's death through an act which led to arousal and orgasm will be the main focus here. In a long and extended story of abuse by her mother, the only occasion where Annabella did to her mother what was usually done to her, burdened Annabella for years after her mother's death. The core of her guilt was that she had performed this sexual act by her free will shortly before her mother died. The mother had not asked for it. It was an exception. She did something which was morally detestable. Had she provoked a situation creating the very kind of physical strain her mother's destroyed body could not tolerate any more? Although Annabella had struggled to sort this out, she still posed questions when telling about it, she was still in doubt, and she still felt responsible, not for the disease, but for the death which perhaps occurred earlier than it otherwise might have without this last orgasm, which Annabella's manipulation caused. It was also linked in her mind to a threat from a family member who had abused her intermittently, making her responsible for future deaths in case she revealed the ongoing abuse. Annabella had been certain: the guilt was hers; were it not for her, her mother might still be alive. The strain and harm of years of being abused by several persons in the family, had been turned into a burden of guilt for a death the one and only time Annabella had acted not on command and as an object, but without being asked and as a subject. She had felt guilty and accepted her responsibility without hesitation or resistance, as she had never been able to distinguish clearly what constitutes right and wrong. The "from" of moral judgment was twisted. She could neither know nor judge with certainty.

2.9 SUMMARY

I have explored interview excerpts thematizing a variety of confused judgments as a particular aspect of sexual abuse experience in childhood. The explorations allow an understanding of the long-term consequences of foundations unmade, resulting in blurred differences and confused distinctions. Theoretical frameworks developed by Scarry, Bourdieu and Kristeva, offer options for comprehending the intrinsic logic in apparently illogical reactions, responses, and processes.

3. MALADAPTIVE ADAPTATIONS

3.1 LINE

Line has been mentioned, concerning her resistance to naming the unnameable, and her pattern of reply-hesitation. Her adaptation to hidden abuse is thematized in the following excerpt:

L: I remember my father or mother, I don't remember precisely - made me see a psychologist or a psychiatrist once; possibly he was a psychologist. He made me very nervous, I didn't feel well. I felt misunderstood He had asked me something and then he said: 'Now you have to answer me.' He got angry at me. He made me quite nervous.

A: Why should you see him, what, in your opinion, may your parents have thought?

L: I thought there was something wrong with me, as it were. 'Is there something wrong with me?' was what I thought. I was afraid of the psychologist. But I saw him only once, only on that occasion. I was about 12 or 13 years old at that time, it might have occurred after the second abuse. Sure it was, because I was - I guess I was in high school at that time. But I wasn't sent by the school, but by my parents.

A: Do you remember why?

L: I've never gotten a proper answer to that question. I thought there was something wrong with me. I'd been teased a great deal at school, I remember, and I'm uncertain whether it was due to that. But I don't know for sure whether it was my mother or my father who sent me.

A: Perhaps your mother had registered that you were different?

L: That may be right; you may be right. I remember, now and then, her mentioning that I'd changed. I believe it was about her observation that I'd become more withdrawn, I can imagine. That I didn't talk as much as before, that I was very silent. And, of course, she was quite right. I couldn't tell her why.

A: What did your father do to you?

L: He fondled me, he touched my breasts and my genitals, and he forced me ... to ... forced me to ... forced me that I had to... (long pauses between every attempt)

A: This is difficult to say?

L: He forced me to suck him. I had to suck him. Yes. (voice fades) But I didn't want to, but he forced me.

A: Was it then you stopped talking - afterwards?

L: Yes. I felt all the time that I had such a lump inside me, for several years, I swallowed and swallowed ... I don't have it anymore, but I felt it for many years and also frequently later - before, I mean. And the next time he abused me I felt a lump deep in my vagina, I remember, because I'd started bleeding. The social worker I talk to now, regularly, when I started therapy with him, I started with drawing, I made drawings, and I painted my father in different colors of red. So we agreed that it was this I wanted to talk about, about the time when I was bleeding. But he didn't try to have intercourse with me then. That occurred the first time when I was twenty-five, when he raped me.

Line has been abused three times by her father. The first time, when she was almost nine years, he fondled her genitals and penetrated her orally. The next time, when she was almost twelve, he penetrated her vaginally without performing intercourse, and the last time, when she was twenty-five, he completed a vaginal rape.

After the first abuse Line stopped talking. Her conduct and her sudden silence must have been quite remarkable because she recalls she was teased for it at school and that her parents consulted a psychologist. Line reports uncertainty about who took that initiative. She recalls her mother's repeated comments regarding Line's silence. Line admits her mother's observation was relevant, but she "couldn't say why" she was so withdrawn. From this remark, one might conclude the silence to be both a direct effect of the abuse and the lump in her throat, and a strategy not to talk in order to avoid accidentally bursting out with what was impossible to express aloud. She neither can nor will talk. Her mouth is concretely and symbolically closed. A physical experience and a social necessity converge. She is effectively silenced. This double closure of her mouth provokes an intervention. No mention is made of whether the child, Line, is aware of what psychologists usually do. Perhaps she is afraid he shall understand. She does not answer his questions at all. He gets angry, and she becomes afraid. Her habit of keeping silent is accentuated. Does this strategy have any consequences later? Can she recall situations where keeping silent created conflicts? The topic reemerges when we talk about her son and the child-care authorities who have placed him in a foster home. The excerpt from that passage is as follows:

L: They hadn't warned me in advance. I didn't know that a person who was a psychologist intended to talk with me about myself. You should have seen me, I was really furious, and this psychologist ...[184] he... I despised his face (her voice is very intense, she strokes the tablecloth in front of her so intensely that it can be heard on the tape). That was no comfortable visit. He hurt my case. He delivered a report which contained only negative information about me. That I was so *apathetic* (her voice, very *indignant*). Have I become *apathetic*? Well, my lawyer had to engage another consultant. That was a judgment about me, it was meant to be used in my case, the authorities were convinced that I couldn't take care of my son. It was meant to make clear how capable I am of taking care of my son. I was so let down not to have been informed in advance. The authorities didn't tell me that such a person would appear, and neither did my son's foster parents.

A: But you are sure that the authorities had 'ordered' him?

L: Yes. Most definitely. He was so unpleasant as a person. I didn't trust him at all. I'm not the sort of person who can trust just anyone. I'm not. He was personally very unpleasant. He asked me uncomfortable questions, about myself.

A: What was their content?

L: What I did and what I occupy myself with and when I'd been in therapy, and things like that. He was unpleasant. He asked whether I lived alone, and much more. I regarded these as very private questions, I couldn't understand the point of them.

Prior to this passage Line has told about her flight from home and her relationship to a man who became the father of her son. This man was an alcoholic who gradually became violent. When Line understood that he might be capable not only of battering her but also the child, she escaped to a women's shelter. The authorities became involved and determined that Line was incapable of taking care of her child, who, consequently, was placed in a foster home. Some years later, Line tried to regain custodial rights, which meant another evaluation by a psychologist. Her visible and audible rage concerning this unannounced and unprepared visit culminated in her expression of intense indignation about how he had characterized her: as apathetic. She was most furious about that. For her, their dialogue was an uncomfortable interrogation. Obviously, she was unaware that her frequent and habitual hesitations in dialogues had caused the psychologist to perceive of her as either retarded or withdrawn. This observation was made the main argument against her in her claim to regain care of her son.

Had she met similar difficulties with other therapists, as for example with physicians? The following excerpt provides a consistent overview of the life-long impact of a childhood strategy:

A: You aren't skeptical towards physicians?

L: No, I'm not. In general, I've been lucky. But, of course, there are differences. There was a doctor who was rude to me. He insisted that I answer his questions too.

A: I wonder whether this is due to the fact that you often hesitate a long time before answering, I myself have, several times while we've been talking, thought you hadn't heard what I'd said or that I hadn't made myself clear enough. Do you know that you hesitate unusually long when answering or talking?

L: I've experienced this. That's correct. It doesn't mean that I don't comprehend what's said. *But I have to take my time. I have to think it over now and then.* But then I experience that people react, they've gotten furious and abrupt. (very long pause).

A: Have you always waited like this, been so hesitant?

L: (hesitates a long time) I don't know. It may be so. *Perhaps that's the reason why they're abrupt and angry. It makes me nervous. Then I get insecure. Perhaps now and then that also I begin to stutter.* That I stop myself.[185] (long pause)

A: I understand this. Several times while we've been talking and I become somewhat more intense, - for example, when you told about your visit to a doctor's office after the rape -, then I could see that you withdrew even more ...

L: *I'm unaware of doing so.* That's quite possible.

A: Has anybody told you how you were as a child, before the abuse?

L: I'm not certain, but I know that people have mentioned that *I was open, much more open* than after I started school. They thought it was possibly connected to the school. I've talked to several people who knew me before. They told me they'd wondered *why I'd become so silent*, but nothing else. I don't know what they thought.

A: Did your father threaten you with something to get you to keep silent?

L: Yes, he said *I would be taken to a children's home* if I told anybody. I got terribly scared. And the next time he threatened me by saying *I would be imprisoned.* (long silence).

This passage of the interview ought to be commented upon at two levels. The first concerns content and comprehension in a phenomenological frame. The second concerns methodology.

The key to the interview with Line, the access to understanding the impact of three separate abusive acts performed by her father, is an observation of a pattern representing a strategy. Not telling about abuse because of a threat has been transformed into not talking at all. Silence as an adaptation to secret violations has affected all communication. In the beginning, the abuse of her mouth, the lump in her chest, and the threat of her being sent away, silence the child totally. Consulting a professional helper causes damage as the silence is misinterpreted, rendering the intervention an anti-help. To feel interrogated is threatening, given the background of yet another threat, and complicates a conflict between "Tell!" and "Don't tell!"

This overwhelming conflict becomes a fundamental internal sound within Line's corporeality, influencing her meeting with another psychologist decades later. The expressions of rage and disgust in her recollection of that meeting are beyond what is appropriate as a reaction to the factual situation. They cannot be comprehended unless one applies the perspectives of temporality and relationality to make visible the constellation of "psychologist/ interrogation/ authorities/ and child-care institution." In retrospect, this is nearly identical to a former constellation as it was recognized and interpreted by Line, twelve years old. By adapting to childhood abuse with silence under the threat of being separated from her parents and institutionalized, Line, guided by forces beyond her will, prepares the ground for her own child to be separated from her, fifteen years later. So effectively has her mouth been alienated that she cannot "give voice" to her own and her child's interests. In an almost magical way, her father's threat is fulfilled, primarily, and ultimately, because she cannot and will not speak. *Adaptation turns maladaptive.*

The methodology applied within this passage of the interview might be read as representing two distinguishable phases of interpretation simultaneously. Here, an on-the-spot validation takes place regarding the presence of a pattern of speech which reveals a strategy, and represents the core of conflicting interests. The interviewee can immediately confirm that she usually hesitates. However, she rationalizes: 'It just takes some time. I must think it over.' When concurred with in this explanation, but also challenged to reflect upon a possible temporal relationship, she understands something which she has neither known nor understood before. She can suddenly link her hesitation and other people's impatience, their increasing insistence and her intensified withdrawal, to the conflict between *to tell* and *not to tell*, created by imposed silence. In the sense of Kvale, this process of understanding reflected in the speech turns of this last passage equals the sixth phase of interpretation: a new insight allows the interviewee to see her own actions in a new perspective and offers her options to act differently from how she has acted in the past; a potential for healing has been facilitated.

3.2 RUTH

Another variant of adaptation to sexual abuse by means of silence is exemplified in the interview with Ruth, the eldest of the interviewees, who also stopped talking. She views the impact of abuse and her silence as crucial to her intellectual development, her femaleness, and her life course. The thread of an adaptive silence after sexual abuse in early childhood must be followed through the complete interview. Only thus may its impact on a life-span, influenced by changing frameworks for understanding and acceptance, both in the social sphere and in the realm of the health professions, be rendered visible. The abuse is mentioned and elucidated in the first phase of the interview, recorded and transcribed verbatim. The excerpts concerning the silence will be explored successively:

> **A:** How old were you when the abuse started? It lasted for a while, didn't it?
>
> **R:** At any rate, it lasted one winter. That means for half a year. And I can't say precisely what my age was because I hadn't started school. I may have been five or six years old, approximately. It happened throughout half a year, until he attempted to have intercourse with me, then I got so scared that I didn't dare to be alone with him anymore; after that I never went alone to the places where he was.
>
> **A:** Who was he? Was he an adult?
>
> **R:** Yes. He was an adult. In his forties, I guess. He must be dead by now.
>
> **A:** He was an acquaintance?
>
> **R:** I lived on a farm; he was one of those responsible for the cattle; he was employed by my family.
>
> **A:** This means he was familiar but not a member of the immediate family.
>
> **R:** No, not a family member. And he was the only one.

A pre-school girl is offended in her own home by a member of the household, yet not a relative.[186] His offense has particular consequences which lead to kinds of estrangement and types of conflicts which render the girl lonely within a huge family, unprotected within a sheltered life, and ashamed within a frame of respectability. She changes immediately.

When talking about her family, she describes what happened as follows:

> **R:** Yes, I know that my brother was different, he'd been very sick as a child, my mother told me. But I wasn't like that. I dressed myself before I was two years old, I started talking early, I was alert, I had my stubborn period, my terrible-two's, at the right time, I was a joyful kid, I've been told. (pause) But then, after the abuse had happened, I didn't try to get close to anybody anymore.
>
> **A:** I can understand this, it seems a reasonable reaction.
>
> **R:** Yes, it seems so to *me* too (stress on the pronoun). But it stopped me in my development. In any case, emotionally. Perhaps also physically.

Ruth's parents, aware of a new strangeness in their child, consult a relative, a psychiatrist. This uncle gives them some advice which Ruth cannot remember, but he does not propose that the situation be explored or suggest a therapy. Ruth does not relate any further details about her childhood in a home where she had always to avoid an abuser. She does report becoming "anxious," however, not daring to go anywhere, afraid to be alone, afraid of darkness and dependent on the presence of others, although not able to be physically near them. Proximity means danger, even within the home. Adaptive withdrawal, silence and bodily distance, invoked as protection and maintained as a habit, become a problem in her marriage, as described in the following:

R: I really intended to not get married. In fact, I believe that would have been just right. But I longed for children, which also guided my choice of profession. And in those times, you couldn't have children without being married.[187] I have two very nice children.

A: But in a way, your relationship to men had been ...

R: ... it had been disturbed.

A: ... very early in your life.

R: I suffered from this fact. *I've always suffered because of this. I'll never get over it.* Because this horror and this trembling I experienced then (at the time of abuse), can come back whenever I feel a little lonely, somewhere where there are no other people.

A: It affected you profoundly?

R: Yes, it did so. They haven't been so good helping me, because when I got sick, they didn't know so much about it. I had to be treated because I had a nervous breakdown.

A: You were admitted to a psychiatric ward?

R: Yes, to doctor J. at the regional hospital. When it comes to the abuse, they were not very good at helping me, I would have to say.

As a young woman, Ruth had been admitted to a psychiatric ward due to a nervous breakdown a few years after she had gotten married. What did she tell about the background for this?

R: After I got married, I developed these nervous problems. I'm sure the reason was that I couldn't tolerate intimate contact with my husband.

At this point, a shift of level is required in order to allow a reflection upon a methodological challenge and a theoretical problem. The aim of this study is to explore the interface between abuse and health. One of its central aspects is the viewing of a transformation initiated by an act of violation and leading to impingements on health. In order to catch sight of the possible pathway by which this occurs, the steps must be made visible. These, however, are not provided as separate, in the form of clearly demarcated "states" presenting themselves as self-evident. Rather, they are embedded in narratives and statements about the translation process from the abuse experience to the sickness experience. The transformation is not present as a fact, neither in the narrative nor even in the awareness of the narrator. Thus, the pathways have to be searched for in their emotionally, maturationally or relationally coded form. They have to be traced, so to say. *The pathway cannot be seen unless it is looked for.*—

However, looking for a possible path of development comprises several factors: first, the translation must be addressed, which means, the interviewee must be asked about her health in different phases of life; second, whenever she offers statements concerning a relationship between her life and her health, these have to be recognized as encoded information; third, the encoded information may be offered in a disconnected way so that what belongs together in meaning may emerge in very different contexts; fourth, the code may be multiple, which means that encoded information may, due to the context in which it is given, also be the key for decoding another statement. *Consequently, to look for a path with the aim of understanding, "is ultimately always to construct, to constitute, and to bring it about,"* as Merleau-Ponty has stated. An insight into a development has to be elaborated. This must be done explicitly, providing potential readers with an insight and thus enabling them to judge the appropriateness of the mode and the reliability of the conclusion. In terms of method, this might be done by a presentation of a collection of statements about relations to a health problem which the interviewee has thematized herself. As to the point of departure for such a purposeful "search for the path," a statement will be repeated:

R: After I got married, I developed these nervous problems. I'm sure the reason was that I couldn't tolerate intimate contact with my husband.

The problems linked to or provoked by sexual intimacy are thematized several times during Ruth's interview. They shall be quoted again consecutively, so that their intrinsic inter-relationship and their consistency throughout the course of a life may be revealed. This technique is chosen as an attempt to grasp a process or a lifetime development, though in a textual rather than a chronological sequence. The first statement emerges in relationship to her discussion of therapy. The second relates to the interviewee's state of widowhood. The third introduces a health problem leading to a hysterectomy. The fourth concerns sexual pleasure in her marriage, and the fifth emerges when talking about Ruth as a child. These five excerpts from pages five, seven, twelve and thirteen of the interview text, are as follows:

R: The psychiatrists understand very little, in my opinion,[188] because when I told my therapist that I had problems in my having intimate relations with my husband, can you guess what he responded? 'And what if I tried?'[189] he said. (pause) 'No,' I replied.

A: How could he dare?[190]

R: Of course, this is beyond the limits. Isn't this ... well ... and after that *many years passed before I mentioned the abuse again, because I didn't dare. I became afraid of the helping system and such reactions.* (long pause) No, I've suffered a lot from this. In my opinion, they haven't shown much skill treating me.

A: Here is an important message for me to get: nobody would hear what you talked about, and those who heard what you said made it into something else.

R: Yes, they made it into something other than what it really was.

R: Now my husband is dead, he died some years ago. Now I have no intimate relations with anybody. And, in fact, I find my situation improved.

A: You don't miss it?

R: No, *I do better. I'm less anxious too.*

A: In a way this grip has never let go, this (connection) between *sexuality and anxiety...*

R: ... *these have been closely tied together.* I'm less anxious, but I'm still not courageous enough to go out in the evening when it's dark outside. Of course, there may be other reasons not to go out alone. But, after all, for me it was a decisive reason not to go out on my own.

R: Yes, I menstruated every two weeks during the last years, I almost couldn't participate in social life, sometimes the bleeding could be so heavy that I couldn't go out. Well, I can see it ... I remember that on our honeymoon, then our sexuality wasn't so bad. But I recall *I felt a feeling of release* whenever I menstruated, because *then I escaped.*

A: Yes. Then it can't also have been guided so much by desire?

R: No, it wasn't. And by and by I always started crying. And that couldn't have been so very pleasant for my husband, I can see now that he was very considerate with me.

R: Sometimes *I sensed that perhaps I felt a little pleasure.* But if my husband said something, that he'd recognized this, then, then *I closed up immediately.* I couldn't tolerate hearing that.

A: Was there something *shameful* in it?

R: Yes, yes. I suppose so.

A: To feel aroused wasn't appropriate?

R: No, that wasn't *appropriate.*

A: Where did this come from?

R: Well, I don't know. Really, I believe ... I believe ... I believe it results from *the total experience,* because it's been very bad for me, after all (several times, when she begins a sentence, she almost stutters, talks in fragments). And you can say that I changed character, I, really, after this, after that ... *me, who had been so involved in things,* I shouldn't do anything wrong. I had to be perfect to be able to accept myself in my own eyes. And of course, I couldn't be acceptable in anyone's eyes, *me who was like this,* because how much I ever tried, I turned out to be an impossible child.

R: ... After all, I can't remember all the experiences I've had in my lifetime, but I know for sure that after that *I didn't try to get close to* anybody anymore. And it was very early to stop having close contact, before you've started school.

A: And it continues through your whole life, and also in relationships where contact should be the only natural thing.

R: Yes, surely, it really should. In my mind I understand that it isn't wrong, but *in my emotions it feels different.*

A: Yes, both your body and something else rejected it. But when you felt arousal - and your husband said he sensed that, the arousal disappeared. Within this, there was something improper. You shouldn't feel this?

R: No, *I shouldn't.*

A: Might one think this could be connected to the little girl back then who felt both *arousal and anxiety?...*

R: *... and anxiety simultaneously. That may very well be so.* I've heard about, and also read, that kids have a sexual drive, but kids must be allowed to be children, one shouldn't awaken the sexual drive in children.

Through these statements, the interface between intimacy and anxiety in Ruth's life becomes visible. It is represented in Ruth's initial comment about the reason for which she was admitted to the psychiatric ward. The origin of the problems mentioned is not clear, however. When connecting the statements chronologically, phases of a development emerge, allowing a conclusion to be drawn as to the origin of her anxiety as she lived and experienced it. When Ruth dares mention her sexual problems for the first time, she is offended by the therapist's comment that he might try to teach her. Her wording allows the conclusion that she experiences the offer as quite similar to, and, in meaning, a repetition of, the childhood sexual offense. Her response is identical. She regards the offender as representing all people, withdraws from further proximity, and keeps silent.[191] Her sexual problems, most intensely present during sexual intercourse with her husband, were such that she was glad to menstruate because it enabled her to escape from sexual obligations. Some of what made intercourse difficult was her own arousal, or rather, her awareness of her husband's awareness of it. Whenever she recognized that he sensed her arousal, she broke off immediately. To be seen aroused brought shame and the feeling of it being inappropriate. Although she knew "in her mind" that her marital intimacy was not wrong, her emotions and her body felt "different." She should not be so shameless as to become aroused by sexual acts. *To be seen in arousal triggered anxiety.*

Why this link, and what had engendered it? Could the little girl have felt arousal during abuse, causing the shame of impropriety and the anxiety caused by a conflict? Yes, it appeared simultaneously in acts of boundary violation on several levels. Pleasure and danger met, thus rendering sensual pleasure a specific source of anxiety. Could this linkage be loosened? Yes, it could. When Ruth was freed from all sexual obligations by her husband's death, she voluntarily chose celibacy. Did she miss her sexuality? "No, I do better. I'm less anxious too." This short answer *invalidates psychiatry's interpretation of her forty-year sickness-history,* and *validates the presented interpretation of her abuse history.* As was Pia's, Ruth's corporeality was informed by the conflict between abuse and arousal. It had been imposed on her at the age of five, had informed all aspects of her life, and had affected her most intimate relationship until she chose to have none. As sexual pleasure had been unmade and converted into abuse, shame, and impropriety, she had to "live in anxiety" as long as she "lived in sexuality."

3.3 NORA AND HEDVIG

Adaptation by silence and withdrawal represent two modalities in a wider Gestalt which I have named "avoidance." The strategy of avoidance as the paramount means to cope with states of conflict or situations of ambiguity was "invented" by many of the interviewees.

Nora had been abused by her father from the age of five until she was eleven, when she began to suffer from an intense pain in the area of her right hip. The pain became continually more intense, and Nora had to consult the family doctor who immediately referred her to a specialist in orthopedics. From there she came to a tertiary clinic for further diagnosis and treatment. The diagnostic process must have been very difficult because Nora remembers there being so many x-ray examinations that she felt she had been "on that table for hours all together." Finally, after almost two years of frequent examinations interrupted by periods when Nora was at home, a conclusion was drawn. The diagnostic tools could identify neither a certain nor even a probable cause of Nora's pain. Thus, the physicians decided, in Nora's own words, "to open me up and look at my hip to find the cause of the pain." This operation was performed when she was thirteen years old. Her rehabilitation was unusually long; she stayed in the tertiary clinic for one year because she would not start to walk and was unable to move alone. The excerpt from this point of the interview is as follows:

A: When you returned from the hospital, was the abuse terminated?

N: Yes, *it ended right before.* But I couldn't cope with coming home after all. I wonder whether it was because of my feeling that all the people around me were so different from what I'd expected. I'd become accustomed to the atmosphere at the hospital during that year. *I longed to be back at the clinic for several years. And it had affected me so deeply that when I returned to the hospital as an adult, I didn't want to go home.*

A: You mean, after you received your first prosthetic hip?

N: Yes, then I *once again became a difficult patient.*

A: First because you broke down, as you mentioned when we filled in the sickness list, and then because you didn't want to go home?

N: Yes, everything collapsed for me. I didn't manage anything. I told about a lot, I don't know what, it just ran out of me. This was after the operation, *when I was supposed to start walking again on my legs and get out* of bed. I don't know what I said but I said a whole lot. I was miserable, which everyone could see.

A: But you were prepared for this operation; it was no acute intervention?

N: I had to wait half a year. I should have been prepared. Then I asked to see a psychologist. They sent a man whom my husband had consulted. He suggested I was burned out by an overload of work. I thought he didn't understand a thing. He came once more. But he only sat there. I thought he wasn't a person to go to. Then I was sent from the hospital to a place in L. for rehabilitation. I couldn't cope with a thing at that time. I was totally desperate, because I wanted to be in the hospital. So I had this breakdown. The physician responsible for my treatment recognized what was going on and he asked if he and I couldn't think through together how to go ahead. That made it easier. He wanted me to talk to a social worker. He understood that a lot was lying under the surface of what I was saying, I mean the social worker. We met frequently during those three weeks I stayed there. But I didn't say a thing. And I didn't refuse to go home after those weeks. But when I had to be operated on again, the year after, *the same thing happened. I collapsed that time too,* but by then I had my own psychologist, he came and saw me several times in the hospital. Once more I was sent to L. for rehabilitation. *I refused to leave there. I dug my heels in totally.*[192]

A: Have you thought about why?

N: No, I don't understand it completely.

A: Was it a safe place, really?

N: Yes, I felt that *the worst thing a person could do to me was to send me home.*

A: Because there it was so horrible?

N: Yes. I couldn't cope at home. Of course, I was at home on short visits in between, but then I only slept. I looked forward most to going back to L. on Mondays. *I slept through the whole weekend.*[193] But then the doctor in L. planned to refer me to MB.[194] He felt I needed it.

A: Why?

N: Well, *he couldn't make me go home.* So, he had to do whatever he could to get me out?

A: But, no matter what, it was not because you had told something or they had understood something?

N: No, I said very little. I was visited by the doctor several times, but I didn't say much. *They* were sitting there and *tried to make me say something,* I understood (*laughs slightly*).

A: That didn't work.

N: *No, (laughs) it didn't.* They started to ask a lot of questions about my husband and the like, I guess they understood that there was something...

A: ...but that was true, after all.

N: Yes, that was true after all. But I couldn't cope with it. I don't recall what I said, but I didn't say much. The doctor called the psychologist, they had agreed to refer me to MB, but by the next time, the psychologist had changed his mind, he didn't want to send me. He called me at the hospital and we talked together. We agreed that I shouldn't be referred. I didn't want to, by then. I wasn't bad that way, *I just didn't want to go home.* (long pause). But since I didn't want to be referred to MB, I had to go home, it was useless to beg.

A: And there everything went on as before?

N: Yes, *everything was as before.* (hesitating, long pause)

A: When did your husband stop putting pressure on you sexually?

N: *I worked overtime to get some days off,* in a way, after many years of marriage, then I got free one or two days every week somehow.[195] By and by it increased, I got free more and more. Then I used a special method for doing it - he didn't give up, in a way. I used a special method. *I ... let it happen ... in a way,* I was so tired of all his insistence, because otherwise he wouldn't let me sleep, so I could let it happen as well. At least I got my sleep afterwards. *But he was totally insatiable.* He pressured me all the time. *Then I wouldn't get my sleep.* So I began to develop problems falling asleep. Finally, I didn't sleep day nor night for several years. Then *I went to a doctor with my legs,* then I said that I couldn't sleep. I was totally desperate. I felt as if I had sand-paper in my eyes. I didn't sleep at all. I got sleeping pills.

To quote so extensively from Nora's interview, covering a time span of more than three decades, demands a justification. It is simply this: this very passage represents at least seven different texts, all of which are relevant to an understanding of the impact of sexual abuse in general, and for the comprehension of a person's corporeality in particular. The first and most obvious text is the sickness-history. Inscribed in this, and made visible in the italics, is the second text, the avoidance-history, which, in turn, holds the encoded third text, the abuse history. Hidden behind the sickness-history is the fourth text, yet another abuse history, completely invisible apart from in one clue, "his insatiability", and concerning Nora's adopted daughter, Hedvig. The fifth text is the inside of the first, so to speak, and is a longing-history. The sixth text emerges from the meta-perspective and concerns the connection-history. The seventh text must be read from the micro-level and reveals the withholding-history.

Withholding and avoidance, however, are two modes of Nora's adaptation to sexual violation during three decades. They are the driving forces for what appears on the surface to be a chronic disease and in its core to be a realization of longing; it is the guiding force behind what must be seen as the maintenance of abuse; and it is the absent force which renders a second person unprotected against abuse. To interpret these seven texts adequately, goes far beyond the frame and aim of the present study, and it would go even further beyond the capacity and skills of the present researcher. However, what should be mentioned is that the first text calls upon medical skills, the second phenomenological, the third sociological and victimological, the fourth victimological and criminological, the fifth psychological, the sixth philological, and the seventh linguistic. This complexity, calling for an inter-disciplinary approach, or, rather, demonstrating the inadequacy of the traditional distinction of disciplines in the face of life-world experiences, might be illustrated as follows:

Connection-history
about how the narrative of Nora's experiences makes meaning
Literature

Abuse-history I
about how Nora was abused by her father
Sociology/Victimology

Avoidance-history
about how Nora acquired a strategy of adaptation by withholding
Phenomenology

Sickness-history	Longing-history
about how Nora escaped via pain	about how Nora released her longing
Medicine	Psychology

Abuse-history II
about how Hedvig, Nora's adopted daughter, was abused by her stepfather, Nora's husband and abuser
Victimology/Criminology

Withholding-history
about how Nora consistently, even in speech and language, expresses withholding
Linguistics

Nora tells about more than three decades of her sickness-history, comprising the time up until the interview and including four operations on her hips, the fifth, at that time was still to be scheduled; two Cesarean sections when giving birth to her children due to medical advice concerning possible complications during vaginal delivery after a previous operation on her hip; and a tubal ligation after her second delivery due to pregnancy complications, which were assessed to present a potential risk to her overall health in case of another pregnancy. The first operation was due to a chronic and intractable pain. The pain was maintained by sexual and physical abuse by her husband. She had escaped from her first abuser through a pain which led to an operation, a year's absence from home and total safety provided by fatherly people, the physicians.

This "solution" laid the groundwork for a strategy, resulting in a pattern she repeated in adult life. From the onset of marital sexual violence during her second pregnancy, creating serious complications for her and putting the fetus' life at risk, yet unacknowledged by the health care workers, her condition became increasingly impaired. The pain in her hip, continually present since the first operation, increased. Two operations and a strong resistance to recovering offered escape from abuse, peace at the hospital, and care from "good fathers," personalized by male health care workers. During the months of hospitalization she felt safe, though the pain brought with it painful surgical interventions, rendering her helpless and dependent on medical staff, medications, and psychological assistance. Efforts were made to understand and to help her to open up. The first psychologist was unacceptable; he happened to be the one her husband had consulted. He presented an explanation Nora knew was irrelevant. Consequently, she did not trust him. The social worker, the physicians, and the next psychologist made attempts in various ways to approach Nora's incomprehensible non-improvement. According to medical criteria, she should have been able to leave the hospital and go home. When confronted with a choice between going to a psychiatric institution and her home, she refused the referral. Not that she chose to go home. She just had to.

So far, the sickness-history has opened perspectives on abuse histories I (Nora) and II (Hedvig), the longing history and the avoidance history. The last shall be explored more extensively through the remarks in italics in the previous excerpt. Once more, an attempt will be made to find a "path" which, in a phenomenological sense, can represent the abuse embodiment, visible as a process leading from sexual assault to impaired health. Simultaneously, I shall explore the interface between the one assaulted and the health care system, expressed in the dynamics of interventions.

After six years of childhood sexual abuse, Nora's health becomes consistently worse. She suffers from increasing pain, which occasions two years of frequent diagnostic efforts which yield no concrete conclusion. Between her short stays at the hospital, which are difficult as she has to allow others access to her body, she continues to be abused at home. The very moment she is admitted for surgical intervention, albeit based on a rather dubious indication, she escapes from abuse.

it (the abuse) ended right before

I longed to be back at the clinic for several years

When back home a year later, she feels estranged. Despite the painful operation and the demanding rehabilitation, the hospital has come to represent

a place for care, safety, and dignity.

Nora knows about this effect and can herself draw the direct line from the first admittance to the next, and the evolving circumstances.

It had affected me so deeply that, when I returned ... as an adult...

I didn't want to go home

I once again became a difficult patient

Nora says this without any paraverbal comment, as a statement derived from the perspective of her caretakers. One may wonder whether she "chooses" this perspective because, when speaking from her own position, she must say: I withheld my own recovery. This is what happened, after all. The remarkable delay is not due to complications caused by a surgical intervention, which, just as the first, is not mentioned with a word, almost as if it were completely negligible. Her breakdown is provoked by the demand to stand up and walk, which inaugurates the demand to walk out and go home.

I collapsed when I was supposed to start walking again on my legs and get out

Nora responds to this demand by literally withholding her legs, refusing to walk at all as she cannot bring herself to walk back to the abuse. By collapsing, she once more secures herself a stay at

the place for care, safety, and dignity.

Her home situation becomes ever more intolerable, because her violent husband has begun to abuse their two children physically, in the name of "punishment," and he abuses Nora physically and sexually. The only person he apparently does not harm is Hedvig, their adopted daughter. Nora's pain increases, again she is admitted to the same hospital.

the same thing happened. I collapsed that time too

I refused to leave there

the worst thing a person could do to me was to send me home

This time, and even more decisively, she withholds her recovery for months, causing further referrals and the suggestion she might need a long-term therapy in a psychiatric institution. Perhaps in hopes of making her "walk over," Nora is sent home for weekends. However, her caretakers do not know that Nora refuses to be consciously at home. She practices absence while obliged to be present where she cannot be.

I slept through the whole weekend

It seems as if only absence in sleep could make her physical presence tolerable. It provided a little of what only the hospital could give fully:

a place for safety and dignity - she can renounce caring if necessary.

On the background of knowledge about the daughter Hedvig's abuse by her stepfather in her stepmother's absence, one may wonder whether Nora chooses literally to keep her eyes closed when at home those few days. On the other hand, Nora is convinced that her husband, who harms others deliberately, would never harm Hedvig, his beloved favorite. Of course, Nora also has reasons for withholding her conscious presence when she is close to him which relate specifically to herself. Therefore, she continues withholding her recovery

he couldn't make me go home
they ... tried to make me say something (laughs slightly)
that didn't work (laughs)
I just didn't want to go home

Withholding her recovery, Nora secures for herself the concerned and collected interest of her male caretakers. These men represent safety. Nora never feels endangered by their maleness. This will be explored in the longing-history. She is aware of her own strength, transferred to a strong, passive resistance. She is also a little amused by all the obvious, but futile, attempts to "move her." Finally, confronted with the choice between psychiatry and home, she knows why she cannot choose the first. Though aware of the reputation of the actual institution, it is only

a place for care and safety - but she cannot renounce dignity.

Finally, therefore, she has no option. She returns to the place from which she cannot escape by means of her primary strategy of withholding, but not because she chooses to, as it is too horrible. Therefore

everything was as before.

Knowing that, sooner or later, the situation will bring her once again into the state of intolerable pain, she tries to create

a place for dignity, although forced to renounce care and safety

which means one or two days a week without sexual abuse by virtue of a means consistent with her defensiveness and thus not really marking a boundary

I worked overtime to get some days off

This implies accepting more the other days. Gradually she invents or reactivates the most passive mode, mental absence

I ... let it happen ... in a way

thus leaving her body to him and shielding her mind so he can get what he wants without reaching all of her. Thus, she argues, he might leave her sooner so she can fall asleep, which is

a place for dignity.

Her consistent lack of explicit resistance, however, invites him to invade ever more of her sleep, because

he was totally insatiable
I wouldn't get my sleep

Even her method for practicing mental absence does not grant her a faster termination and a few hours of "space," as he never becomes satisfied. Ultimately, she cannot sleep anymore. Her legs become consistently more painful. She knows it does not help to ask for another operation. She has to ask for sleep, for drugs granting a sheltering mental absence.

then I went to the doctor with my legs

once again, she went to the doctor not "on" or "about" her legs, but "with" them. She does not view her legs as means to walk, stand firmly, stand up for herself, or even leave. "The legs which cause sleeplessness," are a new embodiment and yet another modality of the intense withholding she had invented very early in her childhood as a survival strategy, and which, however, in her adult life and as regards her relationship, became maladaptive.

A view from a meta-perspective now permits the following conclusion: by means of a phenomenological interpretation, the path from sexual abuse to impaired health has become visible, and the phases of a consistent process have been reconstructed. These are not fabrications that are constructed, nor are they inventions. The process allows the validation of the reliability of the present interpretation, and a comprehension of the appearance and form of Nora's latest health problem.[196]

The complex web of preconditions comprises two cases of childhood sexual abuse, an imprisonment, vast amounts of health care, and the breakdown of a family. To deepen insight into abuse history II, some information provided in the interview and in the comments to the sickness list will be introduced. When severely sick during her second pregnancy, Nora accepts the suggestion that she be sterilized as a means of birth control, in connection with her second elective Cesarean section. However, the physicians are unaware that it is not the state of being pregnant, but rather the fact of frequent marital rapes which she cannot avoid, that are the real threat to her health while pregnant. Consequently, this intervention is of no protective use, is meaningless, evokes in Nora immediate regret after the birth, and fuels her desire to adopt another child. After years of lobbying for her plan, her husband agrees. In fact, he comes to adore this daughter, "his girl," as the mother admits with an air of resignation. Can Nora, herself abused by her father, remain unaware that Hedvig is being abused during her absences? Nora answers:

N: That's what I mean, that's what's so crazy, and what I have such a bad conscience over, too ... or rather guilt for it. But it happened when I was in the hospital. He never entered her room when I was home. He only did it when I was in the hospital. Then she'd lie in my bed. She slept in my bed. Both of them say so. I believe this, because I'm very attentive, I would have heard if he had left the bedroom at night and had gone to her room. Hedvig confirms it, and I'm sure. (pause) But, after all, I suppose I didn't want to know it. (long pause) I had to acknowledge this later. He was so unnaturally fond of her. I used to say that he carried her on a golden tray.[197] He was very caring, there was only Hedvig, Hedvig. He kissed her every night, he would prefer to join her when she had a bath, and it was he who took over. In a way, I gave up fighting for her many years ago, he was the dominating one there too. All these things with defining limits and being a little strict, that was my job. That's why she can't really take it in that he isn't kind.

A: He was aiming for something.

N: Yes, he's worked for it. Once he took her on a camping trip, only the two of them, he was always initiating something. And I asked him many times: 'if I ever find out you're doing something with her, then you'd have to leave,' I told him. He only laughed, as crudely as only he can laugh, of course he wouldn't do something like that. And so he did it.

When Hedvig's abuse is discovered because she finally tells a teacher what happens to her at home, and while the court prepares the case, and while Nora tries to come to grips with the situation, her father, her abuser, falls sick. When Nora's husband is sentenced, her father dies. By a paradoxical coincidence she is suddenly freed from lifelong threats of humiliation, and with them the desperate need to fill her longing. Since then, she has not been hospitalized. The longing-history may be read from the following excerpt:

N: I absorbed all the attention I wasn't used to getting at home, and in the meantime I found it a bit troubling ...

A: ... because... ?

N: I thought it wasn't right ...

A: ... you didn't deserve it?

N: No. No, that's correct.

A: Were you also afraid they might find something out?[198]

N: No. No. I really thought of myself as not being exactly a human being.

A: What weren't you?

N: I wasn't a human being. In a way. That is so difficult to explain. That's why I also found it difficult to be at the hospital. Because there I was a human being. If I was anything, it was a room number. The doctor who was going to operate on me, coming in, standing by my bed, shaking his head. So young and such bad hips, he said. I couldn't cope with that. He talked to me, in a way.

A: Did he treat you like you were some kind of wreck?

N: No. No. It wasn't that. He was so worried, in a way.

A: Oh, well, there was something positive in this. He pitied you. That was what you couldn't take. [199]

N: No, I couldn't take that. He cared for me in a way. I wasn't used to others caring for me. My father had always shown me that I had no worth, I was just to be pushed away, after all. He didn't ever care for me. I guess I longed for a father. I saw fathers in all doctors. I gradually began to look forward to their visits. Then I got care (laughs sadly) which I never got otherwise . .

A: . . . from men . . .

N: Yes. I wanted more and more of that. (long pause) That's why I didn't want to go home. There I wouldn't get this care the doctors at the hospital showed to me. And the social worker . . .

A: . . . who was a man . . .

N: Uh-huh. I always asked for male psychologists.

A: Hello, that's a new one! (completely astonished) [200]

N: It ought to be the opposite, I suppose. But I prefer a man. Not a woman, no women. From men I got the care I hadn't been given. The psychologist gradually became like a father to me.

A: You really can't get enough of these good men, can you?

N: No. No. I can't get too much care. I've been starved for a man's care, I've recognized recently. I didn't realize this when I clung to exactly those guys. And apart from that, they were my age and younger. But it was in a way at the age my father was when I was young. He should have treated me that way from childhood to adolescence to adulthood, given me the help I never got. I wasn't worth it.

Care, safety and dignity was provided by the men of the health care system. That was what made them so important. Therefore Nora had her best times when she was in hospitals. She saw them as substitutes for the father she had never met, as they were even at the age her father was when she had needed his help and his appreciation so profoundly. Nora has understood how it could be that even men younger than she could be "father" and not "man." She never saw them as men, she never felt erotic attraction, and she was never approached indecently by any of them. Asked explicitly about her longing for a man, and not just a father, she responded that she had not been "there" until very lately. Now, she suggested, she might be ready to open for friendships with men, not for more. There was no longing for sexuality, which only meant indignity. She loved the good men in the hospitals, where she found nourishment. In a way, Nora, to an astounding degree, had gotten what she wanted due to, and not despite of her withholding. Freed from humiliation, could she now let go of her "bad old habit?" The withholding-history has been told already, in part. Is there a next chapter, so to speak, a breakthrough, a promise? Nora's answer, in connection with the male care takers, is as follows:

N: I saw my father. I wanted their care. I only saw my father. It seems a little crazy when I look at it now.

A: Compared to your 'crazy' past that can't be crazy? When I think about what you've gone through, the most 'crazy' was what your two men, your father and your husband, have done to you.

N: Yes, of course, that's what's made me what I am. But I didn't understand this, at that time, that it was like that. But I've thought so afterwards. I've gone through it, and I think and think and think and... I almost go crazy with that too. I never succeed in relaxing my thoughts.

A: But outside you've kept artificially 'quiet' since you were seven.

N: Yes, on the outside I'll be calm and balanced, which I always am.

A: What did it cost you?[201]

N: Well you may ask. It has had it's price. But I didn't see it then. I never cried. But I never laughed either. There were many years I couldn't laugh. That came clear to me this week. I saw a movie entitled 'The Man Who Couldn't Laugh.' It made a bang! in my head. Then I understood how I'd been. I couldn't laugh for years. And suddenly I laughed about something, I don't recall what. I almost felt shocked, that I could laugh.

A: We've laughed several times together during our talk.

N: Yes, now I can. But there were several years without laughter.

A: There wasn't anything to laugh at, actually.

N: I couldn't. I couldn't laugh. I couldn't even smile. I guess I was dead inside. Probably that was the case. But nobody told me that I was dead. So I was shocked when suddenly I was laughing. And then I was glad. There are so many crazy things about me (the voice is laughing and clear). I was glad to be able to laugh after all that's happened. Now I'm almost afraid that I may laugh too much...

A: If it's not one thing, it's another... (both laughing loudly and long).

Yes, she has now begun to let go, both tears and laughter, as is demonstrated right there. In a previous passage of the interview, she tells how she used both her crutches to molest a huge pillow filled with pieces of isopore. The pillow exploded, in a way, due to her insistent battering, the isopore pouring out and spreading all over, because she had been so nasty. It took her a day to "collect" them again, but she did not regret the extra work at all. Still there is a reservation in expression and speech. In the initial excerpt, even her syntax mirrors her withholding of emotions and experience while talking about how she obstructed her recoveries. There are long passages of question-answer turns and few narratives about an experience; she has to be "drawn" out by questions and comments. However, she seems nonetheless to be about the business of releasing herself, in some areas of emotional life, from the imperative of the calm, balance and quiet, so that the once engendered strategic adaptation which turned maladaptive is no longer her only option.

3.4 ANNABELLA

Annabella, like Nora, applied a "pain-full" strategy of avoidance, although in a different way and evoking a different medical response. Most of the details of her account are confirmed in extensive records from the hospital where she stayed most of the times she was hospitalized. She reports her abuse history as follows:

> **A:** The abuse occurred from when I was four until I was nineteen. The abuser was a woman,[202] mostly, but also an adult man (an uncle) ... and then several rapes. There were four boys my age, which means young men between sixteen and twenty-seven, and then two adults, one woman and one man. The latter raped me twice.

The sickness list contained: contact with school psychologist before the age of ten; many injuries, several of them due to self-assault such as burns from cigarettes she extinguished on her own legs, knife cuts, etc.; a prolapse in her early teens, causing several referrals and constituting the main health problem; anorexia in adolescence; abdominal pain for years in early adulthood; an abortion. The interview text from this point is as follows ('A' is Annabella, 'a' is the interviewer):

> **A:** The first time I had problems in my lower back, I'd recently turned fifteen, I was referred to KHS.[203] I'd had a low-grade pain for such a long time, they wanted to check it. I was placed there with all those old people, one of them was 32 years old, the rest were older. They did a spinal tap there, and I had to lie calmly for a long time. I guess they also did a contrast x-ray examination. *They diagnosed a prolapse between L4 and L5.* After a while, I was sent home again. And that's how it was ... at regular intervals I "broke my back"[204] and was examined and lifted by the legs and those different things they do, where you have to bend and so on. They took computer tomograms - that was a bit later - and otherwise I've taken all these tests. They never did anything else. Knock wood, since I

was twenty-five I haven't had any problems with my back. *I find this a little strange, there have only been some small pains, but never so much that I had to lie down for weeks and rest and calm my back.*

a: You needed that before?

A: Yes. I was lying in bed for a pretty long time, many times.

a: If you look back and reflect - did your lower back pain have any relationship, as to timing, to what happened in your life?

A: I really don't know. In adolescence, after the problems with my back, actually before and after, the one abusing me then was a so-called kind molester, *but of course, it might have been a strain, because it happened every day apart from Tuesdays and those times when I ran away.* It is a strain for your back having sex almost every day. On Tuesdays, mother attended the sermon, so I could go to bed before she came home, and she didn't come to my room. Sometimes I ran away, so nothing happened these days. In terms of body posture, it surely was a strain for my back when I was asked for special positions and the like, *I don't know whether this has something to do with the frequent recurrence.*

a: What do you think if you look back and imagine how you were handled?

A: Well, what I was taught at the hospital didn't match it, I took part in a 'back-school' whenever I was there. All this twisting I did and the like, and the strain I felt when I was ordered to stand or to get on my knees, *it was precisely what I learned I shouldn't do* . Surely, if what they said out there was true, it's possible that it was so ...

a: That the abuse inflicted a strain on you which made bad become worse?

A: Yes, that's possible, I haven't thought about it before now. Clearly, the way I was treated regarding abuse, it's struck me a few times, when I imagine how it happened, then it's clear, *this and the prolapse, they didn't go so well together.* I don't know. I hope not. In that case, it would have been too bad. I really hoped that this defect was a family defect, perhaps.

a: It would have been easier to accept?

A: Yes, it would have been easier to do something about it. That the defect was maintained, I can imagine. If I think back on some of the abuses in my mind, then, it clearly also may be these rapes, that they contributed. As it was, they happened at the most incredible places, on gravestones and benches and in a container - and by God everywhere, and you're getting hit at the same time - *but some of it must well be... I mean ... as I've understood it... it's caused by strain, and when something goes wrong, this disc can't stay in its place.* So it slides out, or whatever it does. But that this defect has been maintained I am sure of. ... The last time I had this prolapse I had to lie down, I had to leave school because I couldn't continue. It was when I was twenty-five. I went to bed at home and phoned my doctor to ask whether he had something which helped so I could go to the bathroom. Apart from this, I had to do all this paperwork, he had to search for a place, which resulted in me staying in bed too long. Had I phoned the hospital directly, they would have operated on me. I was really a little irritated that time, but here and now I'm glad

> I wasn't operated on, because I've never suffered from it anymore. *It's unnecessary to operate on something which passes.* ...I had to lie in bed for three weeks or more at least twice a year since I was fifteen. What was bad, when I think about it, was that also those times when I was in hospital she came and abused me. I mean my mother. When I had to stay in bed, I wasn't so fit and couldn't move as usual, but still she came. *Perhaps it isn't strange that it didn't improve faster.*

Fifteen years old, Annabella is referred to a tertiary orthopedic surgery clinic in order to have "checked some small pains" she has had for a while, though she leaves unmentioned for how long. According to her records, she visited the school physician several times for pain in her back and left leg, and she was examined by a physiotherapist who reports unusual stiffness along Annabella's back and a slight side deviation in the upper thorax which is not attributed any significance. The physician's examination concentrates upon verifying whether the young girl might suffer from Scheuermann's disease, which would explain "stiffness" in an anatomical or mechanical sense. However, this is not the case. Apart from a positive Lasègue's test at 50o in the left leg, nothing else is found. The "stiffness" remains unexplained; neither neurological nor radiological examinations can contribute conclusive findings as to the nature of the recurring pain. When examined at the hospital by means of extensive diagnostic instruments and procedures, the same conclusion is arrived at. The results of the tests deviate from the symptoms, which is noted by the physicians as being rather astonishing since the clinical signs so convincingly indicate a defect. What is initially called a young girl's "prolapse-suspect back" is not verified as representing a disease. The treatment, consequently, is "conservative," with drugs against inflammation and a longer period of physiotherapy, scheduled for the end of the semester. The medical findings during her next stay at the hospital are similar, but now, she is given advice on how to train the muscles of her back, and physiotherapy, which, however, "didn't have the effect one had hoped for," according to the records. Mention is made of the remarkable stiffness and her fluctuating pain, mainly concentrated in her lower back and left leg, and what in Norwegian is called "stramninger," denoting a palpable increase of muscular tension in wider muscular areas. The succeeding records from later admissions sound similar, though the diagnostic procedures applied become more advanced from one examination to the next. The clinical findings are described in almost the same wording every time, and so are the patient's symptoms. The conclusions are consistent: the pain can be explained by no verifiable defects. The advice is consistent also, though the importance of her own training activity and physical exercise according to the instructions given to her, are increasingly stressed. In later records, mention is made of several admissions to other hospitals and psychiatric wards due to intoxication and depression. The wording of these notes expresses the perception that these admissions are episodic, irrelevant to the current problem, which is consistently defined as somatic despite a persistent lack of any objective verification of its somatic nature. The concluding remark in the latest record concerning these other admissions is: "Lately, she has been relatively stabile with regard to her nerves."

These two "narratives," one, personally constructed based on sickness experience in relation to biographical time, and the other, medically constructed according to biomedical explanations in relation to linear time, stand side-by-side. They are characterized by different languages and perspectives, but they are not pure in the sense of consistently representing one frame of reference. Medical terms are braided into Annabella's history, and it is constructed in part according to medical concepts. Descriptive words are braided into medicine's history, and it is constructed in part according to non-specified perceptions. Annabella has integrated a biomedical gaze upon her body; she thinks with orthopedics and radiology, in the sense of Saris, when stating initially:

> *They diagnosed a prolapse between L4 and L5.*

Next, she draws a retrospective conclusion about the astonishing change in the amount of rest and calm she needed previously in order to improve, which she no longer needs.

> *I find this a little strange. There have only been some small pains, but never so much that I had to lie down for weeks and rest and calm my back.*

In her further reflection, the word 'strain' appears, which, both in the Norwegian and English language denotes the impact of either a physical, mechanical, mental or emotional force on something, as if it were concretely or symbolically pressed or stretched to the point that "it" might break, be torn, collapse or disintegrate. Annabella uses strain in the physical, or even more, in the mechanical sense by linking the pain in her back to the abusive activity in which she has had to partake.

> *but of course, it might have been a strain, because it happened every day apart from Tuesdays and those times when I ran away.*

She considers a connection between the frequent strain of abuse which she only rarely escapes, and the frequent recurrence of the back pain, though uncertain of its nature.

> *I don't know whether this has something to do with the frequent recurrence.*

In her attempt to explore the impact of strain, particularly the mechanical impact of her body postures during sexual acts, she visualizes the similarity of positions she has been advised to avoid again and again in order to protect her back from being strained, and to prevent a recurrence of the pain. According to the records, these educational parts of the treatment are given more and more importance. Finally, they make up the whole of the treatment she receives. But as to the physical activity during abuse,

it was precisely what I learned I shouldn't do.

Was there "a prolapse" caused by an unknown strain unidentified by her numerous therapists, or an inherited and unambiguous "family defect," most easy to accept as not being anyone's fault or responsibility? Or was there any other physical or mechanical origin? If so, there most certainly was an aggravating abuse, though unknown to all therapists.

this and the prolapse, they didn't go so well together

In Annabella's mind, the meaning of "this," the sexual abuse positions, and the meaning of "the prolapse," although never verified, is "strain." Both strains are perceived as mechanical. Consequently, both work in the same way, or they affect "the back" in the same manner, although Annabella becomes slightly confused about which strain may be the cause; though it seems reasonable to think of "this" as adverse, the only explanatory model she has been offered is mechanical, as outer forces squeezing something

but some of it must well be... I mean ... as I've understood it ... it's caused by strain, and when something goes wrong, this disc can't stay in its place.

On the other hand, she herself has experienced something which "was there," named "the prolapse," explaining all her frequent periods of pain, legitimizing all the weeks in bed, and resulting in all the educational and practical advice directed towards her muscular strength and her bodily fitness, a prolapse which suddenly "wasn't there" anymore, fortunately, even before she could be operated on, as

it's unnecessary to operate on something which passes.

The disappearance of that which was there, caused by whichever strain, creates a situation in which she, by means of therapeutic education and advice, gradually becomes responsible for her own pain. Whenever it recurs, she, after having undergone increasingly sophisticated medical diagnostical procedures, receives the same yet always more emphatic advice, to "be careful in particular ways," and to "do exercise in certain ways." This, however, is not so easy to execute as she cannot just simply do or avoid doing what she ought. The double strain present in Annabella's life, yet only known to her, takes from her the command over her body to remain in the metaphorical field of strength, exercise and control. Somebody else has command over it, even when she is in the hospital. This means that not even there may she avoid the strain termed "this."

Perhaps it isn't strange that it didn't improve faster.

Annabella spells out two kinds of strain affecting her: the physical strain of an unknown force on her spine, and the mechanical strain of the postures during abuse. She is influenced by a third kind of strain, however, imposed on her in the name of health, and transferred to her from the therapeutic field, from the health care system itself. It represents the impact of a theory, and the effect of a practical inconsistency.

Emotional or mental strains are involved yet merely mentioned and never explored or named. These different kinds of influences, a strenuous field, so to speak, have to be explored in order to become visible in all their complexity. The pain in the primarily "prolapse-suspect" back of a young girl growing into a woman while a patient, was initially named, descriptively, as "stiffness" (stivhet) and "tension" (stramninger). Despite the use of continually improving technical measures for diagnosis, the pain remained unexplained, or rather, "stiffness" and "tension" remained the only consistent findings. These, however, are non-medical terms as they are merely descriptive and not explanatory. Apart from this, they belong rather to a vocabulary used by physiotherapists than physicians. Indeed, they are introduced into Annabella's records after a physiotherapeutic examination. They lead to the suggestion that physiotherapy may provide a relief, which is, however, not the case. Here is the entrance to a professional intersection filled with tension, creating a strain which is transferred onto Annabella. In the words of physiotherapist Eline Thornquist:

> Physiotherapy has grown mainly out of practical experience. A characteristic of the profession is also that it has evolved under strong influence from the medical profession. Thus it has two main roots: knowledge derived from experience, and conventional medical knowledge. The experience-based knowledge is only partly rendered in words and systematized, while the medical knowledge is conveyed through extensive theory-building and formal knowledge generation. (Thornquist 1994:701)

Concerning the two types of knowledge addressed, she continues by presenting their difference:

> The concept of clinical knowledge is usually used in reference to the ability of knowing-how or knowing-in-action; an ability to 'see' the whole situation, 'understand' what it requires and act accordingly. In short, an ability to contextualize and know what to do when and how. Clinical knowledge can thus be regarded as a continuation of everyday competence, in which context is the very resource of understanding. But a view of context as the enemy of understanding has been characteristic of the natural science model, on which medicine is based... Such decontextualization of phenomena and events - the

> 'scientific attitude'- which removes them from particular personal and social contexts is the formal and accepted view of medical knowledge. *Tension and conflicts are thus embedded in the relationship between clinical knowledge and this 'scientific attitude'.* (Thornquist 1994:702-3).

This field of "tension and conflict" is present in Annabella's records; she is almost literally placed into it. The clinical knowledge, that which is "only partly rendered in words," as Thornquist says, but present in the hands and perception of a physiotherapist, is named "stiffness" and "tension." As the "stiffness" cannot be verified by means of the "medical gaze," in the sense of Foucault, the observation of a stiffened body is rendered irrelevant. What remains is a muscular "tension" in various areas of the body, a word denoting something very tight which must be loosened or relaxed. Although the word tension as a name for a clinical perception is derived from a context-informed knowledge, that which is "named" is now decontextualized and regarded as a medical finding, since no other explanation, requiring a more specific treatment, is found.

In the records, no mention is made of an attempt to understand: why this young girl suffers from such painful tensions and what she is responding to in this way. As they are conceptualized, one attempts to "relax" them. Annabella's body becomes the place to practice an inconsistency in medicine. Consequently, the treatment does not have the effect "one had hoped for." Since these tensions are the only consistent finding, and since they remain decontextualized, the therapists conclude that they can be prevented by avoidance of muscular strain, and relaxed by muscular exercise. What is to be avoided, and what has to be trained, is taught. Annabella attends the "back-school" whenever in the hospital, as she mentions.

She usually improves after some weeks of "rest and calm," and the pain decreases or disappears. However, it returns. For each referral or every acute period of pain where she stays at home, her therapists increasingly emphasize the paramount importance of the avoidance of strain, and the value of training. Ultimately, this becomes the only therapy. This means that, whenever Annabella has pain, she receives an implicit message about her failure to avoid and act adequately. How does she perceive this situation? The next excerpt, comprising the interview text directly following the first excerpt above, is as follows:

a: I can see the distance between what was done in the hospital so you would learn to use your back in the appropriate way, and what others did with it . . .

A: . . . it was rather contradictory, really. Because I couldn't tell them in the hospital. *Whenever they asked I had to lie.* They asked frequently how I did with the exercise and the training and all this, how I lifted and how I walked the stairs and, well you know, the whole package, and I couldn't do anything other than to answer that everything was fine, that I did as they had told me to, but of course, they asked me: *'But why do you have pain in your back now? What's been going on now?'* 'Well, it just happens like this, I don't know', I said. That was okay, they accepted it.

a: No, what should you say? Other things happened than that you walked up stairs and cleaned floors.

A: Yes, of course, *it was a bit mean,* at least towards the one, I never remember names, he was such a warm guy. *I had to sit there and lie.* He had such kind eyes and was so all right, *that felt incredibly upsetting.*

a: Might he have been someone you could have talked to?

A: No, I don't think so. *But I felt it was bad that I had to lie to him. I never would have told what really happened.* We talked together, and he told about how his family was and his children and things like this ... at that time I called him a cool doctor. But I wasn't admitted every time to that hospital when I had these attacks in my back, I may have been there four times, I guess. The first two times, I was so young compared to all the others that I became a kind of a mascot. In one way that felt all right, *I had a little freedom and could relax a little. I experienced how it felt to be cared for a little. That felt all right.*

A lack of a medically consistent explanation of her pain, compensated with an educational activity on theoretical premises, though obviously insufficient in this case, is transformed into Annabella's responsibility, as a compliance in the sense of Trostle, imposing on her the strain of keeping herself free from pain. Also, whenever her helpers ask her whether she does as they have taught her to, she has to lie. She is explicit about how she experiences this conflict imposed on her in the guise of care. The reality of what ought to, but cannot, be achieved, is a strain and evokes guilt and the shame of dishonesty. But she needs sometimes to be somewhere where she can have "a little freedom and relax a little," being the "mascot," the youngest, having the right to extra attention and care. She likes some of her caretakers so much that lying feels mean, upsetting and burdening. The burden is in the conflict between being there and lying, or renouncing what she only receives there. Just as the hospital is a place of care, safety and dignity despite the pain of the surgical interventions for Nora, the hospital means care and near safety to Annabella, despite this conflict.

Such an interpretation, and the obvious maintaining force of what "hospitals provide or mean" in the corporeality of sexually assaulted persons, may open a very particular perspective into a field of human behavior as yet scarcely understood, the Munchausen syndrome. This name is reminiscent the German Baron of Münchhausen, a man of unlimited fantasy and ability to recount his own incredible deeds and experiences, which, to everyone's awareness, were so incredible that they had to be "inventions," lies. The Munchausen syndrome is ascribed to people who go to unthinkable ends to be admitted to hospitals and get treated, in whatever way possible, preferably surgically. They usually arrive in "emergencies," entering the hospitals late at night or on weekends or collective holidays. They insist on operations and, if such an indication is lacking according to medical judgment, they often ask immediately to be released. This behavior, judged according to its outer appearance as exotic, strange, and irrational, thus qualifies for a psychiatric diagnosis. It has been viewed as expressing sadomasochistic traits which, of course, seems logical if one presumes the person is searching for a pain. It has been conceptualized as seductive purposefulness which, of course, seems logical if one presumes the person is exploiting an inexperienced staff to fulfill their own desires, since particular clustering of their arrivals coincides with times of reduced and less skilled staffing. As a consequence of this, and as they consult a variety of physicians, which blocks an exchange among doctors and hospitals of reports of previous interventions, they often "succeed" in negotiating the system and they "get what they want," which is, apparently, yet another painful invasion of their body.[205] These "exotic" behaviors may reveal their implicit logical essence as an embodiment of "hospitals" being the only means for receiving care, safety and dignity, even at the price of painful and incapacitating interventions - or rather, because of them.

The incapacity of medicine to understand chronic pain due to its own theoretical framework, is addressed by Isabelle Baszanger who writes:

> Chronic pain is a problematic reality for at least two reasons. First, pain is a person's private experience, to which no one else has direct access. Second, chronic pain is *lasting proof of a failure that questions the validity and explanations, both past and future, of all involved.* Because pain is a private sensation that cannot be reduced by objectification, it cannot, ultimately, be stabilized as an unquestionable fact that can serve as the basis of medical practice and thus organize relationships between professional and lay persons. This fragile factuality increases the work a physician has to do to decipher a patient's pain... Because of these characteristics of pain, *physicians are forced to work on the elusive information provided by patients* so as to bring into being something called chronic pain. When doing this *they tap various, nearly incompatible, resources.* (italics mine) (Baszanger 1992)

The international literature provides only a few studies having explored a possible connection between pain and sexual assault.[206] The first publication of a significant relationship between chronic pain and sexual abuse history appeared in 1985. Since then, chronic pain and sexual abuse have been shown, with an ever increasing degree of probability, to be connected. The relevant literature until now is dominated by studies among patient populations in pain clinics, in gynecology - that setting providing the first study on this topic in 1981 - and in gastroenterology.

As a very alarming and rather troubling "accidental" finding, some of these studies discovered an above average number of life-time surgical interventions having been performed on the indication of pain, either in the same region or in different regions of the body. Numerous critics of these findings have tended to attribute them to selection bias and interview bias among "very special" populations. Quite recently, studies of non-clinical populations have confirmed a statistically positive correlation between abuse experience and pain history.[207] The occurrence of lifetime surgery in abused persons, however, has as yet not been studied in non-clinical populations.[208]

3.5 SUMMARY

I have explored interview excerpts thematizing a variety of maladaptive adaptations, as a particular aspect of sexual abuse experience. The explorations provide possibilities for seeing and following the individual process of abuse embodiments as linked to strategies which are "invented" primarily as adaptations to unbearable tension or incomprehensible discrepancies. Theoretical frameworks developed by Ricoeur, Saris, Trostle, Foucault, Scarry, Baszanger, and Thornquist offer options for an understanding of both how adaptation to silenced assault is achieved, and how adaptations turn maladaptive in a life course influenced by a hidden agenda.

4. STRAINED RELATIONSHIPS

4.1 TANJA, BJARNE AND ODA

A strained relationship within a family is addressed by Tanja, who, during her investigations into and reconstruction of own past, has become aware of an "abusive net" in her family. Several of its members, she has discovered, have suffered from hidden sexual abuse. In the following excerpt, this is thematized as follows:

* She describes how she used to arrange to avoid meeting her uncle on the occasion of family celebrations and visits to her grandmother. In her family, there exist a great number of alliances designed to prevent confrontations and trouble, and it seems completely in vain to hope that her mother, father or grandmother would take any action were they aware of what her uncle has done, which they have ignored.

* Mother believes that Grandmother has been abused by her own father, the mother's grandfather. Tanja has not asked her mother why she believes this, but is confident that her mother has good reasons for her suggestion.

* During the years, she has taught herself to register certain reactions, comments or avoidances linked to particular topics, or when the names of particular persons are mentioned. In this way, and guided by her own ob-

servations, she herself has come to the conclusion that her grandmother was abused.

T: I can see it. I can also hear it. The way they talk, the way they 'laugh things away,' the way they defend the idiot. I do know for sure that my grandmother's father drank and battered his wife and children. That's how it is. The way she speaks of her family, the way she speaks of other women, all this ...

A: Then you recognize something and think 'look at that'?

T: Yes. I even believe that my uncle, the one abusing me, as a boy was abused by their grandfather, exactly like my mother. He behaved so strangely, apart from abusing me. Grandmother told me that as a little boy, after having gone to the bathroom, he always washed himself. Norwegian children don't learn this, it's not usual for boys, mainly not for boys, to wash your genital area after having gone to the bathroom. He washed and scrubbed himself every time. Then I wonder why he did that, because I know that from myself, I've washed and scrubbed all the time. So this is my theory about my uncle.

Tanja, having recognized a web of abusive relationships through at least three generations in her family, addresses the different kinds of strains this inflicts on the various current family relationships. She can register patterns of avoiding, making light of, and keeping silent. She describes how what shall not be said or known is reacted to; it seems that the unnameable is "in" the reactions, in one way or an other. To "laugh away" comments or statements is to make them no longer dangerous, to key them as funny, nothing to take serious, in the sense of Goffman. *In the avoiding, unnameable things are encoded.* Also the making light of things, laughing them away and ignoring them. As Tanja says, it is not mainly what is done, but how it is done, which carries the mark of the silenced. That which is never a topic, or which can only be talked about in a certain way is what is dangerous, that which shall not be known and which no one wants to be knowledgeable about or aware of. In order to maintain this consensual, as it were, avoidance or ignorance, alliances are established. One knows what is to be no topic ever within the various constellations of familial encounters. Nor is the implicit strain burdening the members addressed, though it is very heavy, as its force might be destructive, expressed in the following statement by Tanja:

T: I feel cheated by my mother. I've told her that. I've told her that I hate her, which I regret because now we can't talk to each other anymore without touching on this. Still I feel so irritated with her that I might beat her up. My father, he's a very persistent person. After all, it's my mother who's supported me in a way during the last years.

Obviously, Tanja has a lot to balance as her mother does not cope with her daughter's accusations; she neither protects her nor supports her by avoiding any contact with Tanja's abuser. There is, at least, a dialogue and a kind of openness. The mother has said that she too has been abused, which does not make things easier for Tanja. Her father is as if absent. He seems to be unsure about the abuse. His daughter has accused a family member of sexual abuse twice, publicly, and twice the court has found the accused not guilty. No matter what his daughter says, the voice of the court and the public embarrassment of having a shameless, promiscuous daughter, is enough to keep him persistently unapproachable. He is the one who does not believe what cannot be proven, regardless what his only child may express. He represents society's opinion concerning what cannot be, what cannot be done, and what cannot be trusted.

A clear parallel to Tanja's relation to her father exists in Bjarne's interview. He describes his father expressing a persistent lack of confidence in a multi-voiced report of the sexual abuse committed by one of the father's friends, who, as a leader of a religious boarding school, abused several pupils. Bjarne's father found it unbelievable that a respected colleague holding a trusted position in a respectable congregation of high moral standard could be guilty of having assaulted children. He could not even believe his son, at that time more than fifty years old.

Perhaps, as happened to Tanja, Bjarne's own conduct during his life was used to discredit him, as it deviated in all respects from the moral standard of his parents and their societal and professional frame. Bjarne abused alcohol in periods of his adult life; he was ambivalent about his sexual orientation; he was divorced; he was admitted to psychiatric institutions and had attempted to commit suicide twice. All this flawed Bjarne's trustworthiness in the moral universe of the Norwegian Lutheran Mission Congregation. His father doubted the very existence of child sexual abuse generally, and insistently refused to believe it might be practiced in an institution of responsibility, by a friend and colleague, and during his personal presence in the very school to which he had entrusted the youngest of his own children. He would not believe it until his friend himself admitted it. In the same way, Tanja's father, in order to keep his world view intact, was persistent, even if the price was an increasing estrangement from his only child. She had to be wrong about the uncle. Everyone knew he was kind. And everyone had witnessed that Tanja was "easy." Her morally contemptible conduct during adolescence was not only proof of her having constructed the allegation, in the court's judgment; it was also the means by which her father chose to remain blind. Bjarne and Tanja, branded by molesters who violated them at "their weakest spot" so to speak, at the place of love, devotion and unlimited trust, wherein no human being ought to be threatened, became outcasts. They were thrown into uncertainty, estranging them from their known interiority, their familiar "from." The confused children, no longer at home in their own families, began their confused lives which, however, were thereafter attributed to their own responsibility.

The kindness which created danger, the "symbolic violence" in the sense of Bourdieu, in Bjarne's case initially linked to a caring touch and in Tanja's case expressed in gifts and attention, can be present in professional kindness, and in the pretense of help. Professional kindness can be expressed by health care staff members' giving extra attention to a patient, on the pretense of offering help, and may provide a false promise of improvement to a sick child. The relational strain this imposes on a child, who is dependent on help and has no options to escape, is destructive. Once again, Bourdieu's theory concerning the threat and violence in acts of "kindness" will be explored. Oda's abuse history is to such an extent and degree the other side of her sickness history, that her interview alone could demonstrate convincingly the futility of comprehending the one unless one knows the other.[209] The following excerpt is from the very beginning of the interview, after having filled in the sickness list:

O: To start with the experiences from the hospital in my early childhood. I was born with a defective bone construction, 'osteogenesis imperfecta.' That resulted in fractures and defects in my bones. So I was admitted to the hospital frequently. There were two male nurses especially who were very, very eager to wash me down there. They said all the time that it was important for me to be clean there. I accepted this, of course. I've thought about it many times since then. It wasn't necessary to almost scrub me down there. I remember this as a repulsive experience.[210] Of course, I've been admitted to the hospital many times later also, but then it was all right. No other situations came up which could be similar to what I mentioned now. Since then, I've been raped twice. I mean, I couldn't defend myself because I was so afraid it would cause a fracture. I was in high school the first time it happened, the second time happened after I'd moved to this place.

A: And that was in addition to those three...

O: Yes, in addition to those three who abused me for a long time. The one who raped me was in the class above me. We, all of us, were on a tour. At that time I still could walk. It happened in the woods far away from the school. I was deathly afraid. Of course, I couldn't do a thing. Primarily because where we were was so remote, the others had gone ahead. And he was so much stronger. I knew how little could cause a fracture. After that I was afraid for a while I might have gotten pregnant. That was painful to experience. And he was a schoolmate.

A: And the second?

O: The other one was an adult. He took advantage of a situation where I couldn't do anything. I tried to hit him, but that didn't help. I knew him, but never imagined he would exploit my helplessness. That was really painful. And, of course, the same thing happened, and I was so afraid to have gotten pregnant. Luckily it passed that time too. All this, in addition to having such heavy bleeding and pelvic pain, I asked to have myself sterilized. I couldn't stand to imagine it (the fear of pregnancy) happening again and giving me this psychological strain in addition.

A: You were simply prepared to be raped again because you had no means of defending yourself without risking fractures and a lot of pain?

O: Yes, I was. I was.

A: How can one live with that?

O: It's very, very painful. You're in danger all the time. You don't dare to initiate a relationship for fear of being misunderstood, that the other person might think there's more than it is, although, I wish with all my heart I could make contact with other people, with somebody, to receive the love all people long for. I regard myself as ordinary, but I can never tell about my wish.

(long, long pause)

I had somebody I in a way ... my contact to my mother and father wasn't so good. It wasn't. And sometimes, when I was a child, when we were out as scouts, I joined the scouts for several years, I adopted one of the leaders on my own, inside myself I called them mother or father. But they left, some of them they died eventually. Then I was back where I started. Then I had a very good girlfriend who lived with me for seventeen years, I'm a lesbian. I'm a lesbian because I experienced all this. She lived with me for seventeen years. Then she died. Again I was alone. I feel very lonely, really. (Her voice becomes so soft that the next sentences disappear).

It took Oda a little more than one page to tell her sixty year sickness and abuse history, characterized by layer upon layer of the vulnerability of a girl with an inherited disease which renders her periodically helpless and permanently defenseless. Her life was virtually shaped by a radical incompatibility between her inherited condition of fragility in the most literal of senses, and a paramount omnipresence of male violence and total male disregard for female integrity and boundaries. This incompatibility resulted in life-long, defensive and regressive struggle for maximized adaptation through minimized femaleness and reduced degrees of freedom. At home, her father, highly respected in the local community, humiliated her mother with his frequent marital infidelities, and battered her whenever he felt that she disapproved. Due to these problems at home, Oda often stayed with her aunt and uncle. The uncle abused her from when she was six years old until she began to menstruate, at sixteen. He also gave access to her, to two of his best friends who then abused her. He had never used force. He had told the little girl that *what he did would heal her.* And, of course, she wanted to be healed. At the hospital she was abused by caretakers. They had told the little girl that *they helped her to be clean.* And, of course, she wanted to be clean. It was as if she was born and destined to be a victim, and she was doomed to accept it. Her life was a permanent exercise in letting others use her, not because she chose or desired to be used, but because of the paradoxical non-choice between being abused and suffering excruciating pain. Her sickness, progressively more incapacitating, caused such extremely painful, often multiple, fractures, that already during her frequent stays at the hospital in very early childhood, she had been given morphine for months at a time. She always was sent away to a University clinic due to her very special condition, which meant long separations from her family and a total dependency on the kindness of strangers. The following excerpt concerns the subject of the hospital admissions during childhood:

A: These periods in the hospital which you said you had at that age, were they safe, did you feel safe?

O: It felt very quiet, because I was away from my father and the others. But there were these nurses. I returned again and again to the same hospital all those years. All together, I've been in the hospital for five years of my childhood.

A: Oh, were they the same nurses, you met them again and again every time you were admitted, oh my dear. And no one recognized what they did? There was no one who noticed what they did?

O: Oh no. They felt such pity for me. It in a way was so violent. All the others must have thought these two were so very nice, that they wanted to help me and care for me. I was lying there in the children's room, there were only children, no other adults. So they just continued. That was hard. (long pause) As a child, I was always convinced that what others did to me was my fault. I've thought so a long time, until fairly recently. Then I understood that it's not the child's fault when adults offend it. Earlier, I used to wonder whether I'd done something which might have encouraged them to do what they did, also when I was raped those two times.

In the first excerpt, Oda mentions the male nurses at the hospital. In the next sentence, however, she denies any indecent approach during later periods of hospitalization. The interviewer, consequently, regards the report of the washing ritual as concerning one solitary episode. But in the next excerpt, when touching upon later childhood hospitalizations, this assumption proves false, mirrored in the probing wording of the third speech turn. Oda meets those two men again and again, during a time span totaling five years, all before she is ten. This means, the once explicitly rationalized, and as such legitimized, abuse lasts for years in periods of pain and dependency. Initially, when describing the act, Oda chooses the adjective "ekkel" to describe her perception of what was done with her, although she has explicitly agreed to the "explanation" given. The acts which are legitimized as "washing" have a nauseating effect. Oda's sensory perception categorizes that which is done to her as abject, repulsive. This discrepancy between a verbal acceptance and a sensory repulsion demands theoretical reflection, as it thematizes the interface between the legitimate and the violent, linked to acts of kindness, and to cognitively acceptable rationalization. The Norwegian anthropologist, Christian Krohn-Hansen, has explored the *legitimate which is violent*, by analyzing different kinds of interactions from a basic assumption that, "statements about violence are statements about legitimacy."[211] Krohn-Hansen refers to the English anthropologist David Riches who "raises questions about violence by considering actions through the concept pair of strategy and meaning," explored in the context of the socio-political relationships between perpetrator and victim. Krohn-Hansen explains this as follows:

> Violence is a way to a social advancement characterized by the fact that the others - the victims and witnesses - are by definition *unwilling* others. But at the same time others *must* to a minimal extent be convinced that the action is acceptable, because the perpetrator wishes to minimize the possibilities of retaliation. The purpose associated with violence, and which best handles this contradiction is what Riches sums up as the notion of "tactical preemption": securing a practical advantage over one's opponents in the short term, by forestalling their moves. His assumption is that this notion is of vital importance to the perpetrator. By making it explicit he can, when pressed, attempt to defend himself. ...
>
> To put it another way: according to Riches, the significance of the core purpose is that it is the means which, ultimately, is used to justify violence; it represents, so to say, the substance of the legitimacy of violence. (Krohn-Hansen 1993)

Oda was exposed to a manipulation, a tactical preemption in the sense of Riches. Her perpetrators knew or suggested that she would otherwise react negatively. This means that they were aware of their acts' being illegal. The making explicit of their presumed intention made their victim cooperate by not protesting and not telling. Oda, silenced by arguments she could neither resist nor invalidate, was, so to speak, violated legitimately. This, of course, rendered her very confused concerning the question of responsibility. Not resisting their attentions, she thought it possible she had contributed to them or even encouraged them, although she was uncertain as to how. The confusion of judgment resulting from exposure to legitimated violation influenced Oda's interpretations of many situations later in life, mainly related to questions of moral obligation, duty and responsibility. Thus, Oda experienced a deep conflict when her uncle committed suicide while she was still a young woman. She felt immediately, and almost reflexively, guilty for his death, supposing to have caused him such unresolvable conflicts that he could not live. She was convinced that the responsibility for his death was hers.

Her permanent or "latent" readiness to take responsibility was addressed when we talked about an encounter with a psychiatrist:

O: I didn't know how to behave or what to say. He didn't help me. There we sat, watching each other. No one said a word. I felt so empty when I left. I was in need of help, really. That's why I'd gone to see him. But I didn't know how it ought to be, and what he could do. I left as helpless as I came. Of course, I thought I was the one who'd been stupid.

A: Is it a habit of yours to try to understand everyone else and excuse them when they react negatively or don't help you according to your needs?

O: Yes, that's right. I always try to find the positive sides in everyone.

A: But does that mean that the guilt very often seems to be on your side?

O: Yes, that's correct. I tell myself that because I've carried so much, I can take a little more.

A: Perhaps you can, but ought you to?

O: That's another matter. Very often I have no choice. If something offends me, when I feel pain, I can't manage to tell the one it concerns. It's been like that for a long time, that I've known not to show it when somebody had offended me, or to complain about somebody. I'd already learned this as a child in the hospital. There, they told me that, when I was crying I wouldn't be allowed to have visitors, that my mother wouldn't be allowed to come and see me. So I didn't cry because I wanted to have visitors. I've continued that way, to appear hard and a little tough. But I'm not. I've learned to keep my reactions back. I've learned to keep closed no matter what.

A: This means you allow other people to take a lot of space, and very many liberties. And you don't make any demands on them in return.

O: Yes, other people confide a lot of things to me, I just listen, I hear about a lot of pain this way. They can be very exhausting. I can get quite exhausted from that. I get to know a lot of things and I keep my mouth shut. But it's not so easy to carry all these painful things for other people too. I've been working in a youth prison for a while. Once, I'd hoped to become a social worker. I call myself a quack-social worker. I've been so for years. I've been in this prison. I was meant to be a trainee. And of course, I carried all the problems home with me. Little by little, I learned to sort out what to do and whom I could say something to. Since then, I've been a personal contact. Only, I myself never have anyone to talk to about my problems. I don't know who it would be. I have no one. Not any more. So it might be a lot, what I carry around and think about. (long pause) After all, I suppose I have an enormous amount of life experience. But, well, that may be a little one-sided. I've done many strange things and been many strange places.

A: Wandering a bit...?

O: Yes, a lot of wandering. Very restless. My relatives have been worried about that. I've been much too restless. Perhaps that isn't so strange. I've often wondered whether there's a connection between everything I've experienced from people I've been dependent on. I guess I'll never be able to find rest and peace anywhere. I may want to meet people. And at the same time, I'm afraid they might get too close. You don't exactly cling to places where you've experienced painful things. (long pause)

Oda describes the process of estrangement from people, places, and oneself, by virtue of violence. As a destructive force, violence, in all obvious or disguised forms, results in alienation. In one way or another, the offended person becomes estranged from human relationships; and the world of the known becomes undone through the offending act represented in forms of help, kindness, gifts, promises, threats, pain, objects, places, danger, and so on.

4.2 SYNNØVE

The impact of abuse by means of symbolic violence, impossible to estimate by objectifying methods, is also addressed in the interview with Synnøve. As a result of hidden sexual abuse, she experiences deep conflicts and encounters severe obstacles in her relationships to relevant others person. The abusive acts she has been exposed to may, in terms of the current research categorization, be characterized as follows: three or four occasions of genital fondling by a non-relative in early adolescence without threat or violence. Given these characteristics combining the criteria "physical contact," "a few single episodes," "not forcible," "extra-familial," and "no penetration," Synnøve's experience is attributed the lowest degree of impact severity in the category "physical contact abuse." How does it feel to Synnøve? In her interview, recorded from the beginning, Synnøve describes her abuse history as follows:

S: I'm not completely sure how old I was, but I guess I was twelve years old, and it lasted until I was sixteen. He was my mother's partner.

A: No one later got too close?

S: Yes, but nothing besides what was natural, so to speak.

A: What is natural?

S: Well, one of my friends, he once tried in a car,[212] and I didn't manage to say no, and it went too far. But, of course, he understood that I didn't want to, so he stopped. This is a situation I find very repulsive[213] and which has clung to ... but I think of it myself as not his fault, because it was I who was the one who didn't stop it, in a way. But he wouldn't really listen to what I wanted.

A: Was this the only situation you can remember which was like that?

S: I think *there have been many such situations* with boys. Those years during adolescence, they often took liberties with me[214] right away, and that *I didn't say no but instead I pretended like I wasn't there anyway,* in a way ...

A: And you hoped they would recognize that you weren't there?

S: Yes, or get finished. The first time I really talked about this was when I met the man I married. Because he sensed that something was wrong. But that was in a way different.

This passage shall be examined in order to show the variety of culturally, socially and personally constituted presuppositions which conflow into a Gestalt of the passive vulnerability and the sexual non-integrity of an abused girl. The key to all three levels is the word natural. What is considered as natural in a culture, in a society and by a person with regard to sexually appropriate, and as such acceptable, or inappropriate and as such unacceptable, approaches? Synnøve's answer to the question about experiences of unwanted approaches after the termination of the abusive relationship is: "nothing besides what was natural." Which norm makes up the invisible background when the "natural" is the foreground? What has informed Synnøve's perception of what she has to accept and what she may be allowed to refuse? Thus, by necessity, the next question must address what *she* perceives as 'natural,' in contrast to some not formulated 'unnatural' or 'unacceptable.' The answer emerges in a description of a male challenge of female boundaries, expressed in the wording of her first sentence: her inability to "speak her boundary," his proceeding, her inability to stop him, his perception of her passivity, her responsibility that "it went too far," and her feeling of the futility of voicing her own interest. What Synnøve calls "natural" is the composition of that which she has learned to perceive of as a normal development in a sexual encounter between a young woman and a young man in her culture. The man takes the initiative, even without the woman expressly asking for or wanting intimacy, and he proceeds unless the woman says or indicates her 'no' - so far Synnøve is in the situation of all women; but then she "leaves," which in this situation is sensed by the young man; *however, in the similar and frequently occurring situations also mentioned, it is not sensed.* Synnøve mentions the sensation of repulsion, however, in the very moment when she leaves herself, when she does not speak but rather practices 'no.' To practice a 'no' feels abject. Mental absence in order to shorten the sexual event is linked to repulsion and nausea. When and how was this link established, and what was it made of? She answers indirectly when reflecting about her problems giving guidance to her little son with regard to his intimate hygiene, being uncertain about the distinction between help and abuse:

S: When he was little I was very afraid to wash him and so on, because I was afraid he would feel it was abusive. I was very aware of that. I found it very repulsive ("ekkelt"). I really had to push myself and used to say: 'and now I'm going to clean your penis and there's nothing repulsive ("ekkel") about that,' but I don't believe that he felt it abusive.

A: Does it seem so to you?

S: No. It doesn't.

A: But this means it has influenced you significantly, both in the relation to your child and to other people.

S: Yes, more than I really believed ... or, well, when you hear about others who've been exposed to incest I think the focus is on the violent, harsh things all the time. And that's not what I've been exposed to. I've never been raped, he's never been inside me. And he's never threatened me directly, that he would murder me, or ... what he did to me was to come into my room and when I lay in bed he would begin to touch me and so on... clearly, he kept me, his hand held me down, pretty hard, he was so heavy-handed with me, but it was only his hands, and a lot of repulsive (ekkel) talk about all of this. And I felt very anxious, because we lived on a remote farm, I knew it was futile to shout. What should I do? Should I dare to tell him he should stop, or would he get even more heavy-handed then? I never dared to say anything, I just lay there and let it happen. It didn't happen so often, it happened three, four times during those four years. And I think it was worst in addition to this, that he was in the house all the time. *Because he grew to have such an immense power over me.* He tried to be as kind as can be. But for me he was just repulsive ("ekkel"), right from the first time I saw him. I guess I already understood this when I was ten, from the first time I saw him. That there was something wrong. He tried every way possible to ingratiate himself to me, but I just kept rejecting him, all the time. This really wasn't like me, because I was really a 'nice' girl. But to be in the house and all the time think of where he was, that seemed to me - and I've thought of this later, *that was what exhausted me most.*

When addressing her doubt concerning how to help her son properly, Synnøve thematized her confusion about the right or wrong touch as a result of a sexual offense (yet another modality of a confused judgment). Touch is mentioned in many of the interviews, and the wrongness and rightness of touch as to how it is perceived is explicitly described by several of the women in the preliminary study. There, mention is made of distinct situations in, for example, a ballet school, where the hands of some of the teachers "meant" something other than they should when correcting the posture of the girl's legs. The girls made each other aware of this by warning and sharing, and they found a collective protection in ridiculing these teachers in their absence. The women reported how their abusers "keyed" the wrongness of touch in the presence of other people by a very particular laughter. This laughter told all eyewitnesses that "this is not a touching of breasts as it seems to be, but only a joke," and, in the same laughter there was a proof of the abuser being aware of the obvious improperness of the touch. The reframing prevented a reaction from the other people present. In the descriptions of keying, in Goffman's sense, an abusive situation in public by means of laughter, implicit rules became visible. Men who touched a girl's breasts or thighs improperly could do so, even in the presence of others, provided they reframed the situation from what it was, a disrespectful insult to physical and personal integrity, to "a game," an as-if situation. By virtue of a particular laughter, all those present were suddenly in doubt as to what they saw, or thought they saw something other than they in fact did. This could work, of course, only because everyone took part in the "game," witnessing everyone else's tacit knowledge about how this combination of laughter and touch ought to be perceived. The sudden agreement that this was done in play evidenced the existence of a tacit and shared knowing lying latent in wait of that which activates it. The key was a particular coinciding of a male laughter and a man's hands at intimate places on a girl's body. Societal agreement regarding which kind of improper acts men may allow themselves to perform in public provided they are reframed, is evidenced in the lack of protest on behalf of an offended girl. In the preliminary study, it was the third variant of examples, those concerning wrong touch, which seemed to create the most confusion as to judgment of perception. This confusion seemed to interfere greatly in later personal relationships. Disturbing doubts were generated by hands whose touch was at once gentle and exploitative. These ambiguous messages from the same hands, for example a father kissing and hugging his daughter frequently, but touching her differently in other's absence, engendered a problematic ambiguity concerning sexual touch in adulthood.

Synnøve's doubt and the repulsion she felt when touching her son's genitals, could only be overcome by her telling her son aloud that this had to be done and was not repulsive at all. She herself seemed to be uncertain whether the child would find her hands exploitative. She did not consider that the boy might feel, and therefore knew, that her hands "meant" care and not "offense," which he certainly did.[215] The very fact of her uncertainty on behalf of her son's ability to discriminate, however, can be read as testimony not only to her own lost confidence in her capacity to perceive the clear difference between good and bad touch, but also in her lost memory of having had this ability as a child. At the same time, she did recall her immediate feeling of discomfort at the first sight of the man who later molested her, expressing no doubt whatever as to the correctness of this "sensual" perception and its discriminative power.

When returning for a moment to the initial notion of what is "natural" with regard to sexual situations, the reflections about touch have to be integrated into the background of tacit presuppositions which make the "natural" emerge as a foreground. What is considered natural indirectly defines what is unnatural. "The natural" as a cultural construction is not reflected upon.[216] Synnøve speaks about herself and her experience from a perception of a triple-layered construction of "the natural" in sexual relationships, that is: grounded culturally, socially, and personally. Consequently, her view upon herself from her zero-point of moral judgment concerning sexuality, that which is taken for granted, is *a man's right to take liberties*: to initiate sexual approaches; to ignore a woman's unwillingness to be approached; to challenge her boundaries; to remain ignorant of any kind of resistance if not explicit; to touch a woman how, where and when he wants; to legitimize this in different ways accepted by the society; to disregard her feelings concerning touch unless she explicitly protests; to be the arbiter of defining what form of protest shall be deemed explicit and thus perceived and thus heeded. On the other hand, it is *a woman's obligation*: to resist a man's approaches; to defend her boundaries as long as possible; to accept his disregard of her objections; to tolerate his disrespectfulness; to feel responsible for his offensiveness; to bear the responsibility for having protested inadequately if her protests aren't heeded. These common, culturally and socially informed and maintained presuppositions, valid for males and females of all ages, interwoven with personal experiences, add some crucial dimensions to the complex question of right versus obligation. By seeing her mother's lover batter her mother to the point that the police have to be called, and having herself been threatened in this connection, Synnøve knows of her abuser as a violent and dangerous man. Therefore, she adds to the list above: to be prepared for threat and physical violence. Yet another "fact" is integrated which will be addressed more deeply in the next excerpt: to accept that it is futile to hope for help or assistance. This is Synnøve's "from," the place from which to judge, to wonder or to decide.

Against this background for judgment of adolescence sexuality, Synnøve measures her sexual experiences. In the second excerpt, she states explicitly that she has never been raped. This remark demonstrates what Synnøve has learned to think of as a rape. She herself speaks of sexual intercourse, unwanted and not initiated by her, which she only can tolerate by means of mental absence. Her perception of "the natural" allows her to remain convinced she has never been raped, reflecting an internalized cultural view that only forcible penetration is rape.[217] Consequently, no sexual intercourse which she has not actively refused is called a rape, even if she finds the act so intolerable that she needs to absent herself mentally.[218] Against the same background, Synnøve wonders what is and what is not abuse severe enough to cause harm. She herself registers that her life and her personal relationships have been much more deeply influenced than expected.

When talking about the impact of the abuse she experienced, she expresses both her astonishment about her case not fitting the model determining abuse severity, and her indignation about the common underestimation of the impact of non-violent abuse. This apparent contradiction is engendered by two facts. The first is her acceptance of an established norm. According to "the opinion" Synnøve refers to, reflecting society and scientific research, her experience is not to be regarded as serious abuse. This, however, does not reflect with how she feels. "The opinion," however, the voice of society and authority, contradicts her individual and lay voice. This conflict of the public and outer with the personal and inner voice may be read in the syntax and the wording of the second passage of the excerpt. Synnøve refers to the public reference by telling what has not been done to her, all these "violent and harsh" things which usually are focused upon. Gradually, she voices her experience, speaking from her world of emotions, feelings, thoughts and perceptions. She depicts a situation of powerlessness and loneliness which no one can see, and its implicit danger. The heavy-handed touch of this man can easily turn much more "heavy-handed," as she very well knows from his treatment of her mother, which she witnessed, and from the chair he threw through a closed window in order to hit her as she stood outside. She knows his violence. She dare not evoke it. Not on those "three, four" occasions, and not otherwise. And all the while he is around, and in her thoughts and awareness. She is always attentive to him who occupies her room, time, thoughts and activities, as expressed in the following:

> **S:** I could never just sit down somewhere in a chair and relax or read a magazine or something, in a way, I had to be on my guard all the time, and whenever he came into the room, I left and went to another room. I couldn't stand being with him. (long pause, deep, heavy breathing) I've thought a lot about this when I read what's written about sexual abuse and incest, they focus all the time on these serious things, knives and other objects, threat and ... but there are similar damaging and long-term effects of the other things too.

Synnøve has to collect all her strength to contradict the voice of "the opinion" concerning the severity of impact. This public voice speaks from a perspective which does not have the potential to imagine and visualize the power that may be exerted even in absentia, the danger that may be embedded in a relationship, and consequently, the strain of coexistence. These can feel so overpowering that they lead to both *perpetual vigilance and perpetual exhaustion.* To be left alone in such a state felt increasingly burdensome to Synnøve who describes how both her mother and her father sacrificed her with "open eyes," as it might seem and surely was felt by Synnøve. This is thematized in the following excerpt:

> **S:** Once, I told my mother, I lay and wept one night, he had done it again, then she asked whether I wanted to be in her bed together with her. I didn't say anything, but she asked me, I just said 'yes' and 'no.' I've always wondered, 'what did she ask me, what did she ask me,' and I've tried to remember. I recall her asking whether he'd touched me and where - I answered yes and no and then I remember we had a scene in the kitchen where he apologized and promised never to do it again. He held my head and pressed it against his stomach, while he was sitting in a chair. I said 'okay,' that was all.
>
> **A:** And your mother was satisfied by this?
>
> **S:** Yes, I guess so. But he was like that, she regularly threw him out and he came back. Whenever he had his drinking periods, she didn't want to see him, then he was sorry and he phoned, and then she took him back But *first she asked me whether it was all right,* you see, that was a way to express how difficult she felt it was without him. *I got the feeling I ruined her life* if he couldn't come back. It was ... well, well... so he came back. I knew he was thrown out ... mother knew about it, but he returned.

And concerning an episode with her father, who visited her now and then after her parents had been divorced for years:

> **S:** Now, after having told everything to my father I discovered that mother knew he was a sexual criminal. Father had told her about the prior case where he was found not guilty. Father asked her to be aware that she should never leave me alone with this man. And I told my mother that I remember a walk with my father, and that he tried to tell me, that is, I don't remember how, but he tried to warn me against him. My father says now that he did that and that he also told me that I should come to him immediately whenever I got in trouble. Then he would help me. I remember this walk, I remember that he tried to tell me something, but I couldn't remember what it was. I've repressed it, *because it had happened already,* but I couldn't tell him. I knew he most probably would be sent to prison. Then mother would be alone and I would be sent to a children's institution. I didn't dare.

This violent man who only touched her a few times estranged her from her home, from her mother and her father, and she was burdened with a responsibility for the adults' welfare which was far beyond her capacity. He heavy-handedly interfered on all levels of her existence. His hands even entered her marriage and broke it. This was addressed in the following excerpt:

A: When you think of how this has influenced you, and what you said: that you had understood later that your pelvic pain and your pain during intercourse, the reason you went to doctors and searched for help, were connected to your experiences; could you please think aloud about this?

S: Yes, those are connected because whenever I had intercourse with the man I married, I almost never - or very often - felt no desire. But I'm uncertain whether I recognized that then and there. I wanted to feel desire. I remember whenever he touched me I thought: 'I don't feel it, I don't feel it...' because this is how I thought during the abuse. *Those hands of his were irrelevant to me in a way. And that got mixed up with the hands of the man I was married to.* Then I tried to feel, I tried to convince myself that I could feel. I really worked hard with myself in this intercourse performance. Because either I had to try to join him and feel that it could be good, *or to disappear totally and let it go.* I very seldom felt desire.

A: Did it happen that you took the initiative?

S: No, I don't do it now either. But I don't feel it as objectionable anymore, now I feel it's good, that it's right. I've never felt repulsed by the man I love now. But I remember when I met my former husband, I don't know how I reacted, but I was stiff like a stick, so he understood that something was seriously wrong. Finally he just took me and shook me and said: 'if you don't tell me what has happened I'll walk out.' Then I had to tell. But then that was that, in a way. Then it was nothing to talk about anymore. I could just as well talk about it but I believe he didn't understand the significance.

4.3 ANNIKA

The modalities of strained relationships can be even more complex, as shall be explored in the interview with Annika. She was abused from the age of eleven to seventeen by an elder brother. The abuse terminated when she "fled" into a relationship with a man who began to abuse her sexually during her second pregnancy. As a consequence, she rejected the child which she perceived of as misshapen. Her partner became continually more abusive, confronting her with the ultimatum that she either obey him or leave. The following excerpt contains Annika's presentation of the further development ('A' means Annika, 'a' means the interviewer):

A: Yes. It became more and more intolerable. But then he only told me that either I make myself available or I get out. Which I did. She was just a baby. I brought her with me. But the boy remained with him. He was three years old then. I had no place to go, no place to get rid of the baby, although I could put her into a shoe box. But what are you going to do with a three-year-old? I had no choice. I didn't know what else to do than to leave him with his father. I ran away on the spot, I couldn't stand anything else, but I've thought upon this a lot afterwards...

a: What could you have done, in retrospect?

A: I could have told him that he had to move so we could have lived there, the children and I. I shared a house with my grandparents. It was their house. It really was more my home than his. Had I been as I am today I wouldn't have moved (laughs resignedly, helpless), but I believed I had no rights and just had to leave.

a: He remained there with the help you could have used?

A: Yes. In the beginning he in a way was meant to have the boy. He had caretakers in the house. The boy was really safe with those two. But little by little, I found out that he more and more often left the boy to himself. I wouldn't believe what the others told me, I had to shield myself, I didn't manage to carry more.

* She tells about the turbulence and insecurity during the first year after she fled, finally finding a job and a room and a baby-sitter and having the boy with her on her off-duty weekends - then the father had to participate in a military reserve duty for some weeks. The boy was to be with her. But the father did not come afterwards to take him back, even after she knew he had already been home for weeks.

* She describes the boy's reaction to this, making it clear that he had understood that his mother did not want him initially, and now his father did not want him either.

A: I began to ask for help that year, the boy was a little more than three years old, but I could see him already becoming hard as stone. He reacted violently when I took the bus to my job ...

a: He knew you could disappear?

A: Yes. Sure he knew. He had to try to keep me as tight as he could. I saw this so clearly. I've talked about this for years now, but only recently the school and the psychological counselors have recognized that he has great problems.

For the second time Annika is estranged from her home by unacceptable sexual demands. For the second time she flees. She cannot accept the permanent humiliation, so she has no choice. What seems strange is the fact that she herself emphasizes, when talking about her options, that she did not recognize having them before she left. She has the greater "right" to stay. The housing for her, her partner and the children is a part of her grandparents' house. But she is not even aware of these rights. Perhaps this is due to her previous experience of having been "left" by her entire family, having been made available to her abusive brother even after her grandmother addresses the fact of abuse. Annika has no reason to think of herself as a family member who can ask for support or take for granted being protected by her family. She does not even consider her family as concerned about her safety and rights in case of being violated. She knows she has to escape from the abuse, but she "knows" as well it is futile to ask for support to maintain her ability to take care of her two children. She has to leave the older child, which causes her conflicts, as is revealed by the wording in the three speech turns concerning her ambivalence and her feeling of being overwhelmed. What she does only feels right for herself, but wrong for her children. Therefore, when confronted with the fact of her partner's total lack of responsibility, forcing upon her the strain of being the sole caretaker for both children while working and living in one room, she asks for help. She is not heard, however. Indirectly, this constellation validates her assumption of the futility of asking her family for support. Not even when both children actually live with her is she invited back to her home in her grandparents' house. Her partner, though alone, stays on, but no one in her family objects.

Annika's problems with her daughter, emerging gradually as she grows older, are addressed as follows:

A: Then I started seeing a psychologist, I talked and tried to sort out my feelings and my relationships, mainly those with my children. I experienced that *they were harmed by my way of behaving.* This was what I wanted least, that they in turn have to go get treatment for a childhood I impose on them. Hopefully I won't abuse them, neither sexually nor any other way, there's still enough ...

a: You feel there's a long-term effect by which they are affected?

A: Yes. I see it very clearly. Mainly the eldest. He's so hard. He's able to say whatever and use words - he's only eight years old but seems very hard, as if he knew that feelings are dangerous to have. And the girl, she has tics (long pause). And I know that ... *I feel guilty for everything with the children.* (nervous laughter) (pause)

a: How does that show?

A: She sits and twists ... and she's in the kindergarten, and the children recognize ... it shows in her jerking her head, like this, and then she's better, for periods, when I feel that I calm down ...

a: You experience that you're the driving force in this?

A: Yes. I feel like that. Perhaps that's just something I imagine, *but when I'm feeling harmonious, and when it's peaceful around, it shows less in her.* But when we're under stress - it returns. Right now she's in a period with a lot of it... and then I get angry and say 'ohh, you mustn't do that,' 'stop it,' though I know she can't control it, that it's something ... (laughs nervously, a little helpless, resignedly) ... she's been examined at the children's clinic, *they haven't found a cause, not one that's medical* ... (her voice lowers, hesitating)...

a: It was a neurological examination?

A: Yes, with these brain connections and so on. It was quite okay for her. She wasn't afraid. It didn't hurt. But otherwise ... she cries a lot and is a very sad kid, *which I feel is painful, because I can't give her enough* ... My mother lives in G., and I've lost contact with the rest of the family, it turned out like that after I'd talked about what had happened. This is all the social network I have, my husband, my children, and my mother in the south. When we visit her, my daughter sits on her lap and clings to her from when we come until we go. When we were there the last time she cried all the way home, for several hours, she doesn't want to go home, she doesn't want to go home.

* Annika tells how incompetent she feels towards her daughter and this unlimited sadness of hers, and her son who is so aggressive and emotionally cold. She expresses how intolerable it feels to her to be close to her children, take them close. When her daughter cries and says that no one loves her she in a way is right, and Annika feels it is painful not to have control over the revulsion which is provoked by body contact with her two eldest children.

Annika has, so it seems, the same tendency to feel guilty, and the same readiness to make herself responsible, as has Oda, though the relationships concerned are different. Annika does not here place more of the blame on the father of her two elder children for his part in these difficulties the children express or represent than with those few words in the first excerpt. She herself and she alone is the reason for the boy's cold, as if "frozen," aggression, and the girl's sad, as if "misshapen" twisting. Annika demonstrates how her daughter jerks her head while talking, saying "like this" while describing what happens. What she mimics, factually, is a strange and distorting movement of the head moving to one side combined with a change in facial expression as if it were asymmetrical. In her answer, when visualizing her daughter "sitting and twisting," she immediately introduces others' eyes (the other children's perspective), probably expressing a concern about what is so very visible that everyone must see it. But it may as well reflect her own view upon her nervously contorting daughter, since *the girl's twistings are the mother's mirror.* The mother is the driving force of the daughter's tics; they increase whenever there is no harmony and no peace. In other words, given Annika's explicit and perpetual responsibility, she is the reason, she is the strain. The more she rejects her daughter, the more the girl becomes visibly misshapen. Annika's very expressive, and possibly exaggerated, demonstration of a misshapen child becomes immediately reactivated in the interviewer's mind when Annika, at a point much later in the interview, mentions her initial rejection of her newborn daughter, which she perceived of as a freak, mentioned as an example of "Confused Judgments." There Annika makes the following remark at the end:

A: ...Half a year went by before I could look at her. She was born pre-term, she looked like a monster, I got chills up my back every time I nursed her. That was awful. Since then, I've thought upon that. Of course, I have photos of her when she was tiny. The child was quite normal, there was actually nothing wrong with her.

a: But you saw an ugly kid you didn't want.

A: It just became worse after the delivery. I'd had problems with sex all the time, it increased gradually, but after the delivery I couldn't stand the thought of it, after all those rapes. It was as if he was fixated. Very often, when leaving in the morning, he said that we would have intercourse that night. Then I went around filled with dread all day. And when he returned in the evening, I knew what he expected from me.

a: He ordered intercourse in the morning during breakfast?

A: Yes. And it became more and more intolerable...

This last statement was quoted above as the opening of Annika's narrative about her moving out. Here and now, the point of these speech turns is the interviewer's validation of what she has heard Annika say about what she saw when looking at her baby, and *Annika's non-answer.* In the interview text, her description and her demonstration of her daughter's tics appears on page three. The reflection about her view of the baby is mentioned on page six. Both presentations are so similar in content, that the interviewer "sees" what Annika tells, and asks for confirmation of the connection between the view of the ugly baby which she doesn't want and the view of the misshapen daughter she, as the driving force behind the contortions, has to face herself in. Annika, however, does not confront this statement. She jumps, as it were, into another topic, ignoring or avoiding the implicit question. A mother's distorted view of her virtually normal baby is, years later, reflected in a visible distortion of a normally shaped girl, the distortion becoming accentuated whenever the mother's view of her is rejecting. As if by magic, Annika's initially symbolic perceptual distortion has become a reality to which she herself is the contributing force. As Annika avoids what is "present-between" her and the interviewer by abruptly changing the topic to her dramatic flight from sexual pressure, this interpretation cannot be validated here. However, Annika's overall guilt for the children's problems, and her painful awareness of her own incompetence as a mother is thematized once more when Annika describes her continual distress about her lack of ability to get sufficient help for her children. She has a suspicion about her own contribution to this which is expressed in the following excerpt:

* She tells that she is very critical of herself and her reactions and often feels stupid. She is aware that she must smile as soon as somebody looks at her directly, or, at any rate, when she tries to say something which she feels is difficult to express. Therefore, she considers the possibility that it may be impossible for others to understand how serious things are, that she can't express herself adequately enough for others to be able see it.[219]

* We talk about such "automatic" reactions. Perhaps she has to train herself to tell those she talks to: please don't look at me, because then I have no chance of expressing the pain I feel and the severity of the situation.

A: I've thought that myself when I'm talking to my mother on the phone, when I don't need to think about her seeing me, then I can say things differently, then I guess I'm more serious than I can be when I'm looking at her. *This distance helps.* (long pause) Then I can say other things. Then I can also say that she shouldn't phone me so frequently. But then I have a bad conscience and phone her again, *she helps me so much*, after all.

a: It's struck me that she uses you, I mean that she uses your feelings to hold onto you when telling you that you're so difficult that she feels pain? Does it do you good to talk to her so often, do you need that for any reason?[220]

A: No. I must say it's most often she who phones me. She always wonders about the children and how they are and whether I'm coping. Especially the girl, then.

a: Does this mean she does a lot to maintain your conscience feeling guilty?

A: She worries so much about the children. And she recognizes my faults. And she points them out. She's very explicit about how I have to control myself and ... but I know that myself. I can't tell her that I know it myself. *I can't shield myself from her. But, really, she's the only one I have.*

a: Yes, but there's not an either/or. It could be she accepted that you've grown up, that you train yourself to keep a little distance, it seems obvious that she wouldn't let you ...

A: Uh huh. But she was right in all her worries concerning the children, I shouldn't have had them because I shouldn't have moved in with this man when I was seventeen ... so every conversation is about what a bad mother I am. (a long sigh) But I can't tell her that I'm aware of not being able to give my children what they may deserve, *but that she herself wasn't able to either*, that she didn't manage to protect me from my brother although she was so anxious to protect me.[221]

a: Perhaps she's aware of it now. The best way to comfort oneself for what one didn't succeed at is to point out that others do it worse. I wonder whether you aren't good to have for her.

A: Uh huh. I've thought so. We never talk about what has happened. That's forgotten. We only talk about my children.

a: Then she can blame you?

A: (long pause) Uh huh. (long pause) (Starts to cry, is searching for a handkerchief by steering her left arm with her right.)[222]

Annika expresses a high degree of ambivalence toward her mother who uses every possible chance to tell her daughter, in the guise of care and concern for her grandchildren, and especially the girl, that she deems Annika an incompetent mother. This, however, is what Annika is more than aware of herself. She strives to limit the amount of negative feedback she receives from her mother. It seems she feels she has no legitimate boundary to mark. In her word, at least in the beginning of the passage, she concedes to her mother an almost unlimited right to criticize and judge her. She admits to needing shielding, but she cannot be shielded because of her obligations and gratitude. The conflict between a wish for distance and an obligation to remain close is here again thematized in relation to children. Annika needs an explicit provocation by the interviewer before she allows herself to criticize her mother. She herself addresses the astounding parallels between her mother's story and her own without directly expressing her accusation against her mother. Her daughter's repetition of a teenager-pregnancy was the mother's greatest concern, probably due to own painful experience of the difficulty of being in such a situation. She did whatever she could to prevent whatever she thought was a risk to her daughter's future. The direct danger of sexual abuse by a family member must not have been in her mind. When she came to know about it, she did not regard its true nature carefully enough to take it seriously into account. Consequently, Annika was driven into both a destructive relationship and premature motherhood. This development was then, in an attempt of attribution, made out to be Annika's own responsibility. Thus, the mother could declare it as doomed to fail, Annika's ever lasting fault and the proof of her inability to judge and to heed her mother's warnings. She is right, Annika is wrong. The fault is Annika's, and so is the obligation of gratitude for all the "help" she receives. Her mother has succeeded in "turning the mirror." Instead of recognizing her own defensiveness and its contribution to Annika's flight from home, the true "reflection" of her failure to protect her daughter within the family, and instead of involving Annika's brother and father as co-bearers of the responsibility for Annika's need to flee her home, she blames the daughter. Annika sees herself doubly accused and twice blamed: in her mother's view, focused on her, and in her daughter's sight, reflected upon her. Whichever way she turns, she faces a sentence.

Annika's interview may allow some theoretical considerations. Research on the mediating forces or "factors" of victimization with regard to the impact of assaultive events has considered many possible constellations. Finkelhor and Browne have proposed a model called the, "Traumagenic Dynamics Model of Child Sexual Abuse." As they define it, a traumagenic dynamic,

> ...alters children's cognitive and emotional orientation to the world, and creates trauma by distorting children's self-concept, world view, and affective capacities. (Finkelhor & Browne 1986:181)

They hypothesize that the impact of abuse can be accounted for by four traumagenic dynamics: stigmatization, betrayal, powerlessness, and traumatic sexualization. These terms are defined as follows:

> Stigmatization refers to the negative connotations - for example, badness, shame, and guilt - that are communicated to the child around the experiences and that then become incorporated into the child's self-image... Betrayal refers to the dynamic by which children discover that someone upon whom they were vitally dependent had caused them harm. ... powerlessness refers to the process in which the child's will, desires and sense of efficacy are continually contravened... Traumatic sexualization refers to a process in which a child's sexuality ... is shaped in a developmentally inappropriate and interpersonally dysfunctional fashion as a result of sexual abuse. (Finkelhor & Browne 1985:531-2)

After this model had been presented, further research has emphasized the importance of characteristics of the abusive event itself, the degree of support an abused person receives from parents or relevant others following the disclosure, and the strategies of coping the abused person develops. The amount of research concerning these aspects of abuse and its possible impact is contrasted by the very limited interest given the underlying or mediating psychological processes in general, and the possible influence of cognitive processes in particular, specifically, the manner of understanding and attributing meaning. The model of Finkelhor and Browne, however, has not until recently been empirically validated as to the relationship between the four factors. A study, designed to explore the possible pathways from violation to impact, was presented in 1996. In the original model, due to methodological problems with its exploration in an adult population, traumatic sexualization is replaced by self-blame. The conclusion of the study emphasizes the feelings of stigma and self-blame as mediators of an adverse long-term impact, in the words of the authors:

> These results suggest that the survivor's current perceptions of stigma and self-blame associated with the experience of childhood sexual abuse mediate the relationship between the experience of being sexually abused and the long-term adjustment problems often present for survivors of child sexual abuse. Because the path analysis tests a particular mediation model it is also fair to say that these results support the hypothesis that stigma and self-blame may underlie the long-term negative impact of child sexual abuse experience. Thus, partial support is provided for Finkelhor and Browne's traumagenic model of child sexual abuse. (Coffey et al. 1996)

Another recent study has addressed the mediating role of the feeling of shame in the development of stigmatization linked to childhood sexual abuse. Again, the circumstances and outer characteristics of the abusive event(s) are stressed, and the importance of social and emotional support following the disclosure is emphasized. Even more explicit than in the previously mentioned study, however, the decisive role of the cognitive processes in the victim are underlined. The process of victimization is presented as not fully and consistently comprehensible without regarding the "subjective" side, which means that of the experiencing person. A consistent understanding is, however, stressed as a prerequisite for identifying those victims most in need for intervention, and for an appropriate choice of therapeutic approaches to victims of abuse. The authors conclude as follows:

> Unless future research elucidates the process and circumstances whereby the experience of sexual abuse leads to poor adjustment, little progress will be made toward developing more effective treatment. (Feiring, Taska & Lewis 1996)

The mental and emotional atmosphere in the childhood home has been explored with regard to "outer circumstances," or the characteristics of abuse as defined from the perspective of an observer. A hypothesis was examined concerning a relationship between chronic illness in women, and parental alcoholism, and what was termed "parental rigidity," denoting a harsh and unsupportive home. The study, conducted among adult women in primary care settings, supported the suggestion of a strong relationship between chronic illness in women and the experience of parental alcoholism and physical and/or sexual violence. The chronic states comprised a variety of conditions, characterized predominantly by regression and pain, and sickness histories including surgical interventions far above average in number. The authors conclude as follows:

> Prior reports have suggested the importance of inquiring about abuse, particularly when chronic pain exists or over-utilization of health services is an issue. Primary care physicians by the nature of their specialty are oriented to prevention. They may wish to consider screening for abuse. Perhaps early detection may prevent establishment of some chronic pain problems and unnecessary surgeries. Research in primary care practices is needed to investigate these possibilities. (Radomsky 1992)

In order to elucidate a possible mutual influence - or a kind of shared vulnerability - between abused children and their mothers, a study was designed with its main focus on the relationship between child abuse and violence against mothers. A critical sociological analysis concerning the cultural and legal preconditions for domestic violence against children and women had been the study's background (Stark & Flitcraft 1988).

A study previous to this one had revealed a high occurrence of abuse experiences among the mothers of a group of children referred to a child-protection team. The study group in the actual setting were mother-child pairs where a protective report was filed *on behalf of the child* at an emergency room visit. Maternal victimization was found in far more than half of the assaulted children. The authors concluded as follows:

> The striking overlap in this study between the victimization of children and their mothers suggests a need for a serious redefinition of both problems, focusing on violence in the family. Such a family-level conceptualization may

be difficult to bring about in today's environment of specialized services and smaller human service budgets. But we believe that certain steps can be taken. Just as the staff of pediatric emergency rooms can take the time, pursue the training, and advocate for increased access to services to contend with the realities of the mothers' lives, so may the professionals who hear mothers' complaints, from physicians, nurses, to battered women's shelter workers, begin systematically to consider the safety and welfare of their offspring. (McKibben, De Vos & Newberger 1989)

The previously mentioned studies have all applied a classical reductionist approach, even those focusing on such subjective phenomena as shame and self-blame. They have all applied a methodology of measurement. They could all, in view of the methodology used, contribute to illucidating hypotheses about the impact of sexual abuse in general. However, all had to conclude that the relationship between the amount of abuse reported and the degree of adverse impact registered could not be accounted for. Nor did these studies permit the impact on the level of individual development to be grasped. Their common tenor was an agreement upon the impossibility to predict "outcome" on the background of the studies on abuse factors performed until now.

None of these studies provided an entrance into the life-world of the experience and, through this, access to an understanding of abuse embodiment. However, the word "experience" is used quite often in such studies, testifying to a lack of awareness about the difference between the researchers' position in reductionist research, and the life-world of the abused persons. This, by necessity, must lead to confusing results in studies which purport to explore experiences using a methodology of detachment, outer gaze and measurement.

An example shall be given. In a prospective study, children brought to a tertiary care unit after recent disclosure of sexual abuse were compared to non-abused children attending for other reasons. The aim of the study was to, "clarify the impact of sexual abuse on children's psychological well-being." The children in both groups were followed-up over the next six months. "Factors" defining the children's state were behavior, maternal psychiatric status, family function, and school performance. These were the means by which the researchers attempted to predict persistent disturbances due to sexual abuse. The authors commented on their results as follows:

Baseline characteristics of the abused children significantly or suggestively associated with persisting problematic behavior were older age, lower maternal educational attainment, poorer maternal psychiatric status, and lower family integration. These four factors accounted for 31% of the variance in the children's behavior. Unexpectedly, characteristics of *the children's sexual abuse experience* did not predict their later behavioral status. (italics mine) (Paradise et al. 1994)

From the description of the methods applied in this study, however, it is clear that the children themselves, between four and twelve years old, had not been given a voice. They had been reported on by their parents concerning their "home behavior," and by their teachers concerning their "classroom behavior." The data concerning the children were of an exclusively demographic nature. The classification of the abuse which had been inflicted upon them was descriptive. Still, the authors purported to have explored the children's experiences.[223] As they had not, they had to face an unexpected result. One may also ask why so much attention was given to the educational and psychiatric state of the mothers without any possibility for further exploration.[224] And, even more astounding, why was there no "instrument" to assess the quality of relationships within the family and, most important, to the abuser? At any rate, what was most definitely not explored was the children's own experience of what had been done to them. Therefore, the conclusion of the study has to be considered as correct as to the logic of its methods, but inconsistent, misleading or almost irrelevant to the postulated focus of interest. The conclusion was as follows:

> These findings suggest that preexisting, long-standing adverse psychosocial circumstances may contribute importantly to persistently problematic behavior and school performance among sexually abused children. The findings also suggest that *it is children's preexisting psychosocial circumstances, rather than the abuse,* that determine, at least in part, the nature of their functional outcomes. (italics mine) (Paradise et al. 1994)

The authors seem unaware that their study, regardless of how correctly it was performed in accordance with the chosen methodology, encompasses an error on the level of epistemology, which seriously flaws the results. In their discussion, they express no concern about basing conclusions regarding the impact of abuse on measures which, in no way access the perceptive, cognitive and emotional world of the assaulted children and their corporeality. Therefore, we shall return to the narrating subjects who hold the keys to such an understanding.

4.4 RUNA

In the interview with Synnøve, the effect of the threatening presence of an abuser in the home, provoking a state of continual vigilance, and resulting in a continual exhaustion, was demonstrated. Her situation was greatly complicated by the strain of feeling responsible for the welfare of her mother who, despite knowing about the ongoing abuse and the abuser's police record, believed the behavior to have ceased after a confrontation. This parallels Annika's experience, although the details of circumstances were different. Yet another way or constellation of the enduring vigilance, and the continual sense of being responsible for balancing the welfare of others emerged in Runa's interview. Her abuse history was as reported in the following excerpt:

R: I guess I was almost twelve years old when the abuse started. It lasted until I left home, then I was fifteen or sixteen years old. The first was a relative, and so - when one starts to think about it, but that's been more in the background, when you start to think and to talk about such things, more and more comes, there were several. There was a neighbor. And later there was a colleague of mine. All those three were adult men.

A: Do you recall other episodes when somebody came too close or when you didn't succeed in stopping what happened?

R: No. There were three of them.

Runa had thought of abuse only in connection with her main abuser, as became obvious when she considered the initial questions of the interview guide, which she was reading and filling out as she spoke to me. Her wording in the general, "when one starts to think..." may give the impression that she either supposes it to be usual to have been abused or approached indecently by several persons, or she felt it necessary to make it sound quite common in order not to stand out as exceptional. She continued with her recall of sickness and treatment, and, when asked which sickness episode or therapeutic relationship she wanted to reflect upon more extensively, she answered:

R: It becomes very clear to me that I already very early, as a teenager at least, was in a state of tension all the time. And I'd begun to bind myself up,[225] both at school and at home, and I always felt anxious. And it surely was due ... I got a headache from being tense.

A: Do you consider this as having been related to the abuse?

R: Yes, I really do. I absolutely do. I believe there was no one thinking along these lines at all.

A: Was it very well hidden? Was it your father?

R: Yes. It was him. I remember well that I felt more relaxed and functioned better at home when he wasn't there. But he was like this - he had this very strange power. He was the kind to be afraid of. He had this secret power no one else knew about except the two of us. I didn't understand ... I (stress on the word) didn't understand so much ... it wasn't until afterwards that I did understand.

A: He had a grip on you because he had made you cooperate with him on something?

R: Uh huh. It was unthinkable that one should tell anything at all, I guess I didn't even think about the possibility of it, that I should tell somebody to have it stopped. It just was like that. It was just like a knot here, hurting so much, it really disturbed any doing well at school and social contacts and everything.

A: Another woman recently said that she had understood that being 'on guard about where he was at any time' had been almost worse than the concrete situations where something happened.

R: I've also heard this and thought so myself, and found there's something right in it. I had my own room, as it were, but that was still no free-zone. I wasn't allowed to lock the door (pause). And to be anxious about what might happen, and to be as nice and kind as possible so nothing would happen, that meant that I had to put a strain on myself. That felt very exhausting.

A: Where in your body did it leave the deepest traces?

R: The area of my hips, down my thighs and the legs ...

A: ... to tip-toe with pelvic tension?

R: Uh-huh. I do so all the time. It's difficult. And it probably goes up to my back. I have a lot of pain in my back. But it may be rheumatic as well. Something in latency. Something about heredity which has emerged very early with me. The doctor said so the first time he diagnosed fibromyalgic pain that it was rather rare in someone so young. It was in connection with my delivery that it flared up fully. But it was diagnosed some years later.

A: What flared up fully?

R: The pelvic pain and the pain in my hips. I just walked around and had pain in my body without anyone finding out what it was. But it's also connected to stress, it gets worse. Always in the area of my hips.

A: May there be concrete reasons, I mean that you were forced to have intercourse - or more, this permanent tension, tip-toeing and pressing your legs together...

R: Yes, it's absolutely this tension. I guess it's only this tension. This tension, and that one is a nice girl all the time and in general, that one does everything so he's content with you, that he wouldn't do something you didn't like. But nevertheless, it happened, regardless how kind you were. And so you became a little too nice, and that is something following as well, which causes you trouble also in adulthood. One doesn't dare contradict and one says yes and yes - but one feels there's no reward for being nice. You say yes and mean no. (pause)

A: Precisely. And you have continued with that.

R: Yes, I guess so. It also connects a bit to the disease, that you can't tolerate so much, but still you feel you have to be available, both in your own and in others' families.

A: When you don't say no, your body must say no?

R: Yes, and so things just crash. (long silence) Because it feels like a hopeless situation. Something which has happened such a long time ago, it seems so hopeless to put it into order. (long silence) Because now it's not just a mental problem, it's gone over to becoming a physical problem. Ergo, you have both. And what is what, they may even be alternating, what is what, but it links up neatly, doesn't it, because do you have a lot of

> pain it isn't so easy to be strong, mentally I mean. (long pause) But I remember, now, to have been ill quite a lot, earlier, all the time. But I was strong in my body - at that time - but that I was very often ill earlier without anyone finding out what it was.

Runa's very consistent description of her pain makes visible how her corporeality is informed by spatiality, temporality and relationality. She has always been in a state of vigilance, straining to please, without a place to feel free, without the possibility of trusting herself, endangered in her own home, and caught in a secret and destructive conspiracy. The exhausting drain of such a state of permanent preparedness becomes obvious immediately. The development of a chronic pain on the background of a permanent threat of abuse becomes almost self evident. Runa describes the horror of the randomly occurring abuse. To inflict pain and humiliation on human beings at random is known to be one of the most powerful methods of torture. Elaine Scarry (Scarry 1985) has described its impact, and studies of political torture among female and male victims have made it clear. (Agger 1989, Agger 1992, Basoglu 1992) Abuse occurring at random, though attempts be made to prevent it through adaptation, is mentioned in several of the interviews, and the destructive potential of the feeling of powerlessness is paramount. Runa, as several others, tells about her repeated experience of the futility of "preventive" measures. She thought it would help to be nice and kind so "he would not do what you didn't like." This expectation of the effectiveness of pleasing has proved to be an illusion again and again. The child, Runa, experiences its futility - but still she tries to improve on the method which does not work. Others applied other methods, as for example Annabella, who ran away.[226] Laura adapted by non-crying and by "feeling strong" after breaking an unbreakable glass ball. As an overall pattern, attempts to avoid a contact with the abuser was, of course, doomed to fail when the abuser was a person of the nearest family. Any attempts not to attract the abuser's attention were likewise futile, as demonstrated earlier.

Runa's convincing description of this kind of no-win exchange, gives insight into its body-logical consequences. When asked where the strain feels most concrete, she, without hesitation, localizes it. The strain she puts on herself is transformed into a muscular tension in her hips and down her legs. She also accepts immediately the interviewer's "translation" of this localized tension into a Gestalt, the tip-toeing girl always ready to please. She not only confirms it on the spot, but even expands the imagery by adding that it is "from there going up my back." But, once having expressed this view on her pain, she shifts her perspective and suddenly "thinks with medicine and genetics." She received the diagnosis of fibromyalgic pain some years after her first delivery, which accentuated the pain. The doctor has commented on her early onset as being exceptional, so Runa presumes that a latent rheumatic disposition has emerged early. However, from the next speech turn on, she again has taken her inside position, so to speak, and continues to talk about her experience of the pain and its origin in her unceasing tension and lack of "free space," even in adulthood.

Runa also addresses the phenomenon of adaptation turned maladaptive, which has been explored in Line's, Annika's and Synnøve's interviews, to mention just a few. She knows about the futility of saying yes despite the wish to say no, but she sees no other option. This recognition leads her to a remarkable reflection about the ever more obvious impossibility of distinguishing between mental and bodily pain. Rather, "what is what" has become indistinguishable. The pain occurs, although it sometimes feels more mental and sometimes more physical. Runa, when considering that "it links up neatly," directly recognizes that it has always been so, because she has been "very often ill earlier without anyone finding out what it was."

The topic of chronic pain and its possible or probable origins has been addressed in innumerable medical studies. Its presentation in medical settings is one of the most important challenges to medicine, and one of the most frustrating tasks to respond to for physicians. One of its most frequent forms of presentation in Western countries during the last decade has been addressed in Runa's interview. It has been termed differently during the years, due to an always changing conceptualization of its etiology. As a consequence of this, no appropriate treatment has, as yet, been determined. The state of a more or less permanent yet varying pain in the body, in the medical terminology conceived of as being musculo-skeletal in location, has been considered as a "strain disease." The word strain has been dealt with in Annabella's interview. Its differing meanings and its various modalities make "strain disease" more a confusion than a term, but it is nevertheless used as if explanatory, although scarcely descriptive, at least not unambiguously or with clearly distinguishable designations. When listening to Runa's description of her embodiment of overwhelming powerlessness and random abuse, the pain acquires a Gestalt. This Gestalt is comprehensible. It represents a consistent corporeality. Does it need a name?

Studies on chronic pain and sexual abuse have been performed in the last decade, most of them mentioned earlier. The only study, to my knowledge, as yet to explore the possible relationship between sexual abuse and chronic pain in various musculo-skeletal areas, currently termed Fibromyalgia Syndrome, was presented in 1995. The authors have compared a group of women with this diagnosis to a group of women with a rheumatological diagnosis, with regard to the occurrence of sexual abuse. Once again, one confronts a piece of research built on an epistemological error. The authors do not take into account that these two medically distinguished conditions may not be two distinct conditions, but variants in a continuum of pain embodiment, and possibly pain generated from the same source or of similar origin. The study design, building on two diagnoses as appropriate measures for a selection of two groups, suffices to show that this selection might be flawed. However, the results may even emphasize my previous suggestion of both conditions being expressions of hidden assault experience. The authors write:

> Overall abuse was greater in FMS patients than in control patients (53% versus 42%; p not significant). Significant differences were observed for lifetime sexual abuse (17% versus 6%), physical abuse (18% versus 4%), combined physical and sexual abuse (17% versus 5%), and drug abuse (16% versus 3%). There was a trend toward a higher incidence of childhood sexual abuse (37% versus 22%) and of eating disorders (10% versus 3%) in the FMS group. (Boisset-Pioro, Esdaile & Fitzcharles 1995)

To the problem of the possibly indistinguishable "people in pain" as to groups and categories, adds another, not only methodological, problem. How can one know about abuse in persons who do not know about it themselves, who are, as yet, unaware? It may be permissible to suppose, on the background of the previously described narratives about childhood abuse, that the more the person has felt offended, and the more of her or his world has been unmade, the more the abuse is embodied without an access to it via "thought" or "memory." Furthermore, the current etiological model of rheumatic diseases as an immuno-auto-aggressive activity, might be reframed by a reading in another frame of references. Provided the indisputably verified occurrence of the self-neglectful, self-destructive and self-aggressive behavior of abused persons, could it be possible to think of an even more destructive, because more internalized and more deeply embodied, self-aggression that reaches the cellular level of function? In other words: persons in pain, expressed in a way which is currently termed rheumatic, which already is understood etiologically as being self-aggressive, and which is pathogenetically verified as self-destructive, could even more probably than many others, be victims of abuse. The proof of a difference in the study above could even be hypothesized as demonstrating the opposite: that even more, and more severely abused, victims are to be found among arthritics, only that being more traumatized, they are less able to recall it. Such possible pathways cannot, however, be found unless one looks for them.[227]

4.5 SUMMARY

I have explored interview excerpts thematizing a variety of strained relationships as a particular aspect of sexual abuse experiences. The explorations allow the comprehension of how the unmaking of trust by sexual abuse affects significant relationships, life-course, sexuality, and individual health. Theoretical frameworks developed by Goffman, Bourdieu, Riches, Haug, Flitcraft and Stark offer modes for reflecting upon both general and particular elements and their interactions, whereby it becomes evident that the core of the interpersonal strains is linked to the idiosyncratic abuse experience.

5. RECOGNIZED MEMORIES

5.1 BJARNE

In Bjarne's interview, a narrative about memories searching for meaning is provided, triggered by the hearing of a radio program which occasioned the rediscovery of abuse experience. This led to Bjarne's meeting with childhood friends and a period of memory work in a group. [228] The excerpt is as follows:

B: In this process afterwards, I went to see my mother, who was still alive then, and told her: 'I see two peculiar pictures which keep reappearing. The one is of me sitting in a narrow room, and I can see a person hurry down a hillside and into some bushes.' My mother replied: 'Can you see that?' And them she told me that I'd been deathly ill with dysentery before I was even two and a half years old, and my parents, in order to save me, had traveled through most of China down to Hong Kong to a children's hospital where they had quite recently begun using Sulfonamide preparations. And my parents had lost two children before me. On their way, there was suddenly an Air Force attack, and my mother could hardly walk because she had phlebitis so they had to stay outside, outside a building, and hide. And I had another picture. I was carried in someone's arms who was running among many screaming people. To that, my mother answered: 'But you had lost consciousness by then, and you awoke in my arms at the train station in Canton when there was suddenly an alarm because the planes were attacking, and everybody panicked. I was shown away in a hurry, you got cramps and lost consciousness again.' These are my first two memories.

A: Then you were two and a half years old.

B: Yes, I was no older than that. I was seriously ill. But apart from this, later, I haven't been sick.

A: I'm fascinated at how your pieces of memory fitted into an event you never would have been able to reconstruct on your own.

B: Yes. And if my mother hadn't still been alive. She died shortly after. I was happy to have these memories confirmed, because what happened afterwards, when I remembered the abuse, getting into contact with my feelings about it, my perception of my own body... has guided me to regain contact with the time when I'd been without a language. Early in my childhood, I had a Chinese nanny. And I found my way back to her. (his voice is almost strangled by weeping - whispering)

A: Does that mean she became a witness to a time which had disappeared for you?

B: Yes. That was marvelous. I don't have words for that, only lots of emotions. My mother had been sick at that time. She had lost two children before me. I was one year old, and from that time, it was my nanny who carried me. And she grew in my memory, she reemerged. It was completely fascinating to find my memory of her again. I've felt very lucky because of that.

In the sense of Pierre Janet, the excerpt reflects an active reconstruction of a fragmented past by organizing parts of memories into narratives. Bjarne searches actively for the right place for two "short movies," two sequences of visual memory which have no real place to belong to. He receives assistance from his mother who literally integrates *his sight* into hers, thus rendering Bjarne's memory, the authenticity of which Bjarne would never have been able to prove alone, meaningful, logical, and as something belonging to his early experiences. Without any hesitation, his mother tells the whole story, which he only knew in glimpses. She obviously sees the event from the same "point of view" as Bjarne does. His perception of something happening in his field of vision, obviously representing a view from a narrow place, is identical to hers. She has seen the same thing and recalls the context on the spot. Her outbreak 'Can you see that?' testifies to at least three layers of meaning, the third of which will be explored within the analysis of the next memory presentation. First, evidently, she too has seen "that," which demands the same perspective. Next, that which both saw must have been part of an extraordinary event since it immediately came into the foreground from the memory fund (background) of fifty years of turbulent life. His presentation makes it "present" for both. Thus it finds its place and meaning.

So does the next memory Bjarne presents: the sensations of being carried in the arms of someone who is running among many screaming people. A lot of questions emerge from this remark. The first memory presentation was much easier for another person to grasp or comprehend. To imagine a "view" on behalf of another person when having been on the same spot, thus having had the same perspective, the same position from which to see, can be conceptualized as representing a sharing of almost identical experiences as regards a particular mono-sensory perception, vision. The sharing of or partaking in a view from a place within a limited field of vision, is probably the simplest of common perceptions where one person can "assist" another to reconstruct experiences. The next presentation offered, however, is of quite a different nature and sensory complexity. Bjarne does not describe *a view* to share, but rather speaks from the knowledge of sensation, the feeling of arms around him while he is little and being carried, unaware of whose arms they are, who it is that is running among screaming people. Here, a man is speaking from an infant's corporeality, from an embodiment of a complex perception comprising time, place, motion, sound, and view. How can such an embodiment be shared by another person in such a way that it can be recognized, identified, explained and witnessed as the mother does? Most likely, this is possible only when having taken part, bodily, in the same flow of perceptions and thus having perceived "the same," although its complementary part. The one who carried, ran, and heard the same people "scream" *knows* it because what is described reactivates her own embodiment of the same situational experience. How does she respond? What he presents immediately becomes present for her. She is "there" too, replying: 'But you had lost consciousness then...' She speaks from the same past, shared corporeality of embodied temporal, spatial, visual, auditory and locomotive elements, rendering the experience a common memory. But how can he who was, according to her, unconscious, know this? Her exclamation does not express doubt concerning the authenticity of his memory, which she has already completely confirmed. She wonders at his ability to remember while being unconscious, which she has believed he was.

A final comment to this excerpt addresses the question of what children can remember from a very early age. Bjarne's mother gives a clue to this in her first response, as does the first remark of the interviewer's reaffirming of Bjarne's age. When regarding the context of the two episodes mentioned, the complete story, so to speak, they are evidently not two events of an everyday experience where a child and his mother are involved. The two memory fragments belong to a dramatic, threatening, exceptional and emotionally loaded time in the life of a family, and to a journey of several days duration in a foreign country during wartime. The situation is informed by the choice between a highly probable loss of yet another child, and the similarly highly probable possibility of getting killed on the way. If one wanted to construct a fictional situation of danger and drama, this would be an excellent choice.

The question of authenticity and accuracy of recollections from early childhood has been a highly controversial topic of debate and definition in the field of research on child sexual abuse during the last decade. The question has been linked to the methodological problems of surveying abuse experiences in various populations, and it has become a major topic with legal implications for lawsuits over accusations of abuse. There has emerged what might be called a movement of non-believers in early childhood memories, of believers in false childhood memories, and of believers in induced childhood memories. The memories one doubts, the dubious recalls, however, are predominantly those which represent recollections of physical and sexual abuse. Other kinds of memories have not been debated in this connection, or at least not been nearly so attacked, ridiculed and declared as dubious or flawed. Children who tell about earlier abusive experiences which they either have attempted to express nonverbally at a younger age or which they had "forgotten" for a time and remember due to different circumstances, have hardly been believed. Likewise, adults who have told about abuse experiences during early childhood have not been believed, regardless of whether they had remembered the experiences all the time, or whether they had forgotten them for shorter or longer periods of time and then recalled them at certain triggers. In the preceding texts, several examples of disbelief have been mentioned already.

Disbelief in the authenticity of abuse memories in general, and of childhood sexual abuse memories in particular, expresses primarily the societal resistance to acknowledge that violation of certain rules for interaction and relationship among people actually occurs. Next, it represents the partaking of larger groups of the society in the collective silencing of the voices which speak about such violations. Third, it contributes to a collective focusing on those who report having been abused, thereby steering a more critical or closer gaze away from those who are accused of having performed the violations. Fourth, it addresses the abused as individuals who may be strange in a way that renders them prone to becoming victims, thus effectively veiling the underlying societal structures which engender these violations. Fifth, it thematizes doubt in the integrity and competence of those who seek to give a voice to sexually assaulted persons, either through advocacy or therapy, with reference to traditional laws or theories. And sixth, it demands scientific proof of an appropriate memory function in early childhood, thus legitimizing *disbelief as the normative position* and burdening the abused and their supporters with the task of producing evidence.

The traditional site for research into memory in this connection was its biological side, within the fields of neurophysiology and neuroendocrinology. Studies of traumatic memories in children have, however, challenged these fields of research by demanding the development of quite new approaches. One has been forced to acknowledge that so-called "salient" experiences may hardly be simulated in a research laboratory. Constructed settings, regardless of the efforts made to mimic reality as closely as possible, do not provide the qualities which make the real situations what they are. An as-if test, provided it is carried out within an ethically acceptable frame, cannot "recreate" real experience. Jessica Wolfe, in her overview of the state of the art in trauma memory research and her reflections upon further challenges, writes:

> In terms of substantive content, it is evident that studies evaluating traumatic events from the distant past are faced with particular problems typically linked to retrospective reporting. These include issues of reporting bias, variations of forgetting, effects of subjective appraisal, and the probable impact of multiple intervening life events on event appraisal and recollection. Furthermore, as research in this series demonstrates, *an impressive array of person-based characteristics are influential in stress-related, autobiographical memory including, notably, age at the time of exposure, affective response, age at the time of reporting, and certain personal characteristics* ... To date, some research ... has suggested that *traumatic stress and related serious clinical conditions may be unique in their effect on the perception and recall of highly stressful events.* ... To improve methodologies further, research in this issue suggests that both experimental hypotheses and study constructs require improved definition and operationalization. (italics mine) (Wolfe 1995:724)

Wolfe, in her considerations on appropriate methodology, points to the necessity of a prospective design in studies concerning trauma experience. Such a design has been integrated into a study on children's memories of medical examinations in order to explore the accuracy of recall and the possible effect of time and other defined circumstances. The abilities of children between three and seven years of age, to recall details of medical interventions, a routine physical examination and an invasive radiological procedure, have been studied. The author writes:

> Medical 'stimulus events' have been selected for study because in some respects they are similar to the types of abusive experiences about which children are often asked to testify. Indeed, a visit to the doctor for a check-up provides a reasonable, although admittedly not perfect, analog for sexual abuse, which obviously cannot be studied in an experimental fashion. For example, during the check-up, children are undressed and are handled physically by an adult, often an opposite-sex, unknown adult. Moreover, certain aspects of the experience are stressful, at least for some children. Another medical event, a voiding cystourethrogram, provides an even more striking analog of sexual abuse, one that is both less familiar and more stressful than the routine office visit. (Ornstein 1995:582)

The author, referring to four different studies in a program designed to elaborate a theory on "children's long-term retention of salient personal experiences," at the University of North Carolina, presents four general themes about memory performance of the included children aged three, five, and seven years, formulated as follows:

> (1) Not everything gets into memory; ... this means that some incoming information must be selected for attention and further processing, while other

> information is necessarily excluded; ... this indicates, that the components of the physical examination varied markedly in terms of how salient they were to the children. (2) What gets into memory may vary in strength; ... this means that at least two factors have well known influences on the strength of representations in memory: the amount of exposure to a specific event and the age of the individual; ... in addition, obviously, even strong memory traces cannot be retrieved in certain contexts, i.e., without the appropriate set of retrieval clues. (3) The status of information in memory changes; ... this means that the passage of time, as well as a variety of intervening experiences can have an impact on the strength of organization of stored information, and this influence may vary as a function of age. (4) Retrieval is not perfect; ... this means that not everything in memory can be retrieved all of the time, and not everything that is retrieved is reported. A number of cognitive and social factors can influence the subject's ability to gain access to previously acquired information, or even to attempt to do so. (Ornstein 1995:589-97)

The author concludes as follows:

> The problems of accurate cognitive diagnosis stem from the high degree of *context specificity* that characterizes children's performance in a wide range of tasks. Because performance can be shown to vary markedly from setting to setting, the resulting estimates of children's skills are thus *dependent upon the setting in which observations are made.* ... This type of context specificity is *a salient and pervasive aspect of children's cognition, and the younger the child the greater the variability across settings.* One implication of these findings is that the conditions established by interviewers - be they researchers, police officials, attorneys, or clinicians - can *have a profound influence* on the types of reports that are produced and hence the conclusions vis-a-vis competency that are reached. (all italics mine) (Ornstein 1995:602)

The paramount message in this condensed report, which simultaneously provides a comprehensive overview of the research field on children's memory, is the following: What children can bring to consciousness of what they remember can be influenced, more, apparently, the younger the children are. Children may retrieve less: the more abstract the questions; the more they are alienated from narration; the more fragmented they are in nature; the more decontextualized they are; the more they resemble interrogation. As the authors of the studies reviewed in Ornstein's article agree, what may be retrieved is not identical with what children remember. In other words, children seem to be highly dependent on facilitating surroundings *to be able to know what they know.* Actually, this is not a unique conclusion valid only in this particular circumstance, as relevant research on education and the impact of pedagogical methods has shown. (van Manen 1990) However, there may be few if any "contexts" less facilitative to retrieval of abusive experiences in childhood than are male dominated societies, preoccupied as they are with, and structured for, the silencing the voices which speak about the amount and impact of male violence, especially in the private sphere.

This fact, and its consequences, has only quite recently begun to be addressed by feminist researchers in a variety of fields. Of great importance is the work being done in the helping professions, comprising medicine, psychology, psychiatry, and social work. The astounding delay of these professions in acknowledging and acting constructively against health problems among women and children caused by domestic violence is addressed in the literature.[229]

5.2 ELISABETH

Sandra K. Hewitt reports two cases of recall by abused girls who were abused at a preverbal age. These girls entered therapy at the age of four and six years. The author writes:

> These two girls were able to recall, with startling clarity, events that had occurred to them at ages 2 years 1 month and 2 years 7 months of age, even though at the time they did not have the verbal capacity to fully express their experiences. Smaller details have varied over time as the children have discussed their experiences, but the central details have always remained consistent and clear. The ability to recall very early abuse documented in these cases raises several issues and questions: how young can children remember? What is the quality of their memory? Is early memory affected by the nature of the abuse? What is needed to surface early memories? What do we know about the long term effects of early sexual abuse? And finally, what is the impact of this discussion on practice? (Hewitt 1994)

The conclusion of Ornstein's study about the crucial importance of circumstances, many details of which may still be unknown to experts doing scientific research or legal interrogation, may provide a partial answer to these questions. Another part of the answer may be provided in the previous exploration of Bjarne's active reconstruction, his search for the meaning of memories which had not been affected by the abuse. And yet another part may to be found in Elisabeth's interview. She talks about a memory recall linked to a referral to a psychiatric ward:

E: ... What I had was probably partly hallucinations. I was so afraid. I was afraid of everything. Possibly I was seeing things.[230]

A: Were they things you'd seen before?[231]

E: Yes. Indeed. Until then I'd been able to think back to when I was young, but not when I was little, I hadn't remembered how little I was - and this with my father. But all the time I'd had a feeling that there was something else, something before this with my stepfather. And while I was psychotic, I saw that somebody came and shielded me, that was in my childhood home, shielded me, took me from my father.

A: You saw that?

E: Yes. They came and got me out. And I don't know, I didn't know that he had been abusive, because it wasn't till two years after my psychosis that my mother told me that ...

A: (very soft voice) She told you what you had seen ... and then it became true ... that must have been very intense.

E: Yes, indeed. But I'd all the time, from when I started to open up, I felt that there was something, but I didn't remember, I didn't rely very much on myself. But that's why it was very good to receive this confirmation from her, yes, well, I'd been right, at least that time.

A: When you were in this locked ward, did you tell anything about what you saw?

E: No. I didn't talk. I didn't dare talk to anybody. I didn't trust them. They scared me to death. I didn't trust a person there.

A: You said something about that time: you were wordless, and in the mean-time you saw pictures from a time before you could talk.

E: Uh-huh.

A: It was there where you were, in the time before you had words?

E: Yes. I've understood this afterwards. (weeps in a tight, soft voice)

Because she was perceived of as being psychotic, Elisabeth had been forcibly committed, although she had never been a "danger to herself or others," as it is formulated in the law concerning compulsory referral to psychiatric institutions. She initially formulates this in a rather tentative way; however, when the interviewer accepts "the things" Elisabeth has offered, albeit reservedly, Elisabeth drops her previous reservation. Yes, she did see things she had seen before. She opened her eyes to a forgotten past. Prior to getting these images from earliest childhood, she had begun to recognize that she had been abused by her stepfather. This abuse had started when she was five years old. Her mother, divorced from her father before Elisabeth was three years old, had married again. Elisabeth had been unaware that the reason for the divorce was her father's abuse of her since she was a baby, which her mother only gradually and unwillingly had acknowledged. Thus, the child had been taken from the father by the authorities. Elisabeth had known that "there was something more, something before the stepfather," but she had not known what she knew. She had been unaware that her mother had gotten a divorce to save the daughter. And the mother had been unaware that once again she married a child molester who abused Elisabeth continuously until she was seventeen. Two years after Elisabeth, now an adult, was referred to the psychiatric ward, her mother dared to tell her about the earliest abuse by the father, which Elisabeth, in fact, had had visions of in her "psychotic" state, without understanding their origin.

Elisabeth receives confirmation of memories from the age of three and earlier; she views the time before the words. She is there, her past is present, she has no language, she only sees. What she sees is without a clear meaning. She cannot know that it is true. *She does not rely on herself*, as her life has been informed by so much which "cannot be true" or cannot be believed. Her mother's eventual report provides a place, a meaning and a confirmation for her vision, a testimony to its authenticity. That which Elisabeth has known in a very particular way, is now recognized or known anew. In the sense of Janet, a disintegrated part of her life is integrated into a narrative, whereby her memory, identity and selfhood are rendered more whole than before.

5.3 BJARNE AND ELISABETH

The healing process of knowing anew is described by Bjarne. A pain and a shriek, evidence from an experience without words, but with another significance than in Elisabeth's case, cannot be recognized until he dares to narrate their "origin," the experience of being in pain and shrieking. Statements about this process are quoted in the same sequence as presented in the interview:

B: The abuse took place when I was around ten years old. I've concluded after all, after having explored it, that more than one abuse must have occurred. But it was this final explosion, you might say, when I was raped. (page 1)

B: What I've experienced, bothering me a great deal, perhaps most in my twenties but later as well, were hemorrhoids. Every time I was in a tense situation - and I had this bleeding after the rape - I've only linked this theoretically, because I don't know whether there's a connection, after all - what is strange is that I'm not at all bothered by it anymore. It may reappear very slightly, along with constipation.

A: But one might think of these as connected, that the rape and the pain made this part of your body a difficult area for you ever since?

B: Yes. But before it was there constantly. And so painful. *I'm free of it now. I don't have that sensation any more.* (page 3)

B: ... because after the rape (hesitates) which was rather ... what should I say ... rather full of loneliness ... because I was taken down to a room in the basement by the man who had cared for me ... and he undressed me (almost stuttering) ... and then I was bent over a wooden case ... *I've reexperienced all this in the seminary period I'll come back to* ... so I regained consciousness alone down in the basement - (page 7)

B: But I'll never forget the explosion during the group session. And all the weeks after that. (pause) *I was awakened by my own scream, I was lying on this wooden case (his voice thickens) and I felt the wood in my palms* (his voice turns monotone, he weeps) ... (page 9)

B: And from that moment on, everything was wordless. But then I sensed him penetrating me - *and this pain* - it felt like I was split apart. I started to shriek. *It was a shriek.* I didn't dared to scream, because immediately everything felt dangerous. What had been safe suddenly felt dangerous. *And this shriek I've reexperienced* several times later in my life, it reoccurs.

Mainly at night, when I wake up, these nightmares, they returned. *But before I had no names for them, I didn't know what it was.* (page 13)

B: ... when we missionary children came together, when this explosion took place, *it was as if I myself had to reexperience what I'd experienced before.* And it was my body which talked to me and made me experience it. It happened in my dreams, mostly in what I remembered when waking up in the morning. Sometimes also I was awakened during the night by my shrieking ... *And the experience was very clearly the same as the one I'd had in China when I was almost eleven.* It was very strange, I had to get up and *walk and walk in my room for some hours so I could make my body return to me again.* From then on, it was in a way as if the pain was acknowledged, it wasn't me who'd inflicted pain on myself, but somebody else. I succeeded in transferring the responsibility back to where it belonged. But *I'd never gotten in touch with this before,* you see? (answer no.1 in the second reinterview) (All italics mine, and shall be referred to below)

These excerpts present a multilayered process, several kinds of pathways, so to speak, each of which shall be explored because they are of theoretical and methodological interest. The first is the path of telling about a rape and a particular pain in an interview to a stranger. The rape is mentioned immediately, just as a fact, embedded into considerations about other possible abusive situations which are mentioned in a summary later. Anal pain as "constant" lifelong trouble in "tense situations" is mentioned next, and a reflection upon the possibility of a connection between the rape and the "bleeding after the rape." Bjarne himself takes this connection into consideration, in a tentative formulation, however. The next step is an attempt to go into the rape situation and speak about the pain as it was inflicted. Yet Bjarne's narration stops right before it occurs, and he shifts narrative level to another time and another perspective. When returning to the initial level, he has evidently "jumped over the pain." The next time, he approaches the rape situation reporting what happened during a group session, which is what he had been speaking of when the aforementioned level-shift occurred, he dares go into the situational memory. Now he mentions a shriek, the wooden case, and the imprint of its texture in his palms. The retrieval and the recognition of the sensation, the perceptual components of the situation, are present - and overwhelming. Even some time later, he "goes down into the basement" again. And this time he speaks the pain, the penetration, and its shriek, the suppressing of a reaction which make shriek and pain one. This speaking the pain of rape has taken the space and time from the first to the thirteenth of the sixteen pages.

Simultaneously, Bjarne describes a path in the opposite direction, about how he has been freed of this pain. In the second statement, he expresses clearly that a sensation which has been in his body continually since it was imprinted there, is now gone. He himself addresses a connection, linked to the repeated experience of the pain's increasing during "tense situations," and by mentioning the sensory association to blood after the rape. This is expressed, however, with a reservation which may have a double foundation. The one is that Bjarne, educated in Western thought about the body, must "think with medicine." From an anatomical or a physiological perspective, he cannot be certain of a connection between increased anal pain and mental strain throughout the many decades since an anal penetration caused bleeding. The other is that Bjarne, aware of talking to a doctor, suggests that she "hears with medicine." This reservation in presenting one's own body experiences echoes Elisabeth's way of speaking about her visions. Most probably, these two somewhat reserved formulations address *the interface* between the official and culturally learned *truth about the human body,* the objective way of describing an object and its functions, and the personal and internally perceived *truth about one's own body,* the subjective way of knowing as a lived body with all its perceptions.

To explore this interface - and its impact on the interpretation of the interviews about bodily experiences, and before continuing on Bjarne's pathway of recognition, a shift of level is necessary. The point of departure is the culturally learned truth about the human body. Mary Douglas has dealt with bodily symbols for social constructions in most of her work. Regarding her notion of purity, this has been reviewed by scholars of religion, Sheldon R. Isenberg and Dennis E. Owen. They write:

> Through pollution rules a social value or classification system is socialized into the individual's body. The individual's body is presented to him, taught to him by society, usually in the manifestations of parents, and then by peers, perhaps also by schools. *Our attitudes about our bodies arise from society's image of itself.* So if we can learn how a person understands the workings of that complex system called the body, its organization, its spatial arrangement, and its priorities of needs, then we can guess much about the total pattern of self-understanding of the society, such as its perception of its own workings, its organization, its power structure and its cosmology. The human body, then, is a universal symbol system: every society attempts in some way to socialize its members, to educate its bodies. (Isenberg & Owen 1977)

According to Mary Douglas, humans "think with their body" in the way they have been taught to. The predominant powers of society define the body, and how it must be conceived of. A culturally constituted truth of the human body becomes the norm. Its normative power engenders the "objective" interpretation of what is felt, sensed and perceived of. It also provides the description of the "objectively" normal functions. Thereby, the subjective is rendered unimportant as a frame of reference.

What ever is felt, sensed and perceived by way of subjective experiences, has to be translated into the explanatory system provided by society. Whatever might be non-translatable according to the objective criteria, is, by definition and logic, deviant. If a culture has grounded its explanatory system of the body in biology and the natural sciences, these become the frame of references, the normative system. Thus, the scientifically produced knowledge defines the objective truth about the human body. Whatever might be inexplicable according to this truth, is, by definition and necessity, abnormal. The objective truth, culturally constructed, overrules and invalidates the subjective. Therefore Bjarne and Elisabeth do not present what they themselves have seen, felt, experienced and thought. They hesitate to trust their own perceptions, because these are deviant and abnormal, and they hesitate to tell about them, expecting adverse reactions. Only when another person confirms what they know, may they be allowed to believe what they know. Elisabeth's vision is confirmed. It is true. Nonetheless, she still talks about her "psychosis" and "hallucinations." When challenged on this, however, the following dialogue occurs:

A: Can you please tell more about the time when you were psychotic - although, I guess you ought to stop calling it that ...

E: Yes, I have stopped. At least with myself and within my family. But now and then it recurs. It does, indeed. Because it's been so intense. (pause) I was very afraid of people because I felt evil myself. I was evil, at least, as I experienced myself. I felt that other people condemned me. And that they had good reasons. I merged, very much, with this evilness, which I felt I was. The evil was in me.

A: In you. Not in the world or in others?

E: In me. It was me who was the origin of all the ugliness.

The culturally defined norm shows its (normative) power. Morals are at stake, not science. Elisabeth, though finally confirmed in the rightness of her vision and her refusal to label herself according to the societal norm, continues, nonetheless, to speak of herself in those terms, from time to time. The deviance she had imposed upon her, was deeply internalized, too long and too intensely to "quit" easily the very moment she "knows better." The deviant filled all of her. What she experienced, what she knew, was so abnormal, that there was only one possible conclusion. The evil was in her. She embodied it.[232] Therefore, she was both full of the evil and the origin of it, which invites, in a shift of level, a reflection on the concreteness of her apparently metaphorical language. Elisabeth had been abused orally and had been invaded by the abject, the ugly. She also spoke from a position of her self unmade to contain the ugly. Primarily, however, she spoke from the position of the one who deviates from the norm to such a degree that nothing in or about herself could be good. She had, still, not regained her mouth, which has been thematized in connection with her self-invasive training in opening her mouth.

Bjarne has regained his body in the sense of having left behind the pain of the rape. Although he pays tribute to the societal explanatory system by saying it "is strange," it is a fact. He does not have "that sensation any more." Through the steps of presenting the pain as it occurred and felt, he reports how he got to know what it meant. The perceptive synonyma, shriek and anal pain, have to be unlinked by returning to the experience of their simultaneous occurrence. As Bjarne's memories are confirmed and find their proper places, he can narrate, understand and regain. To be "in the shriek and pain of the rape" literally means to feel past presence. From there, Bjarne has to walk for hours to make his body return to him in the present. He has a way to go, a distance to overcome, because the pain cannot disappear before its real meaning is integrated and the true responsibility for it is placed "where it belonged." To regain one's body by walking back and making it return from the time when the disruption took place, means "re-cognition," literally, "knowing anew," reuniting as embodying the whole life, and healing as experiencing the whole body.

5.4 FREDRIC

As Bjarne and Elisabeth received confirmations of early memories from their mothers, making them confident of not being insane, so the other interviewees searched actively for pieces with which to reconstruct their past. Veronika collected hospital records and talked to those friends who were attentive. Tanja filled the blanks of her childhood by asking her mother, other adults in the family, and by watching and discussing a video of the child, Tanja. Line asked several people who could still remember her as a once vivid child who suddenly grew silent. Ruth asked her mother and others who remembered a child with a strong sense of self, a joyful little girl, who, suddenly and incomprehensibly, had turned strange, anxious, dependent and silent. Hanna received the confirmation of many of her memories by confronting her abuser, and documented this in the expectation of continued disbelief on the part of her family and the society. Annika tried to reconstruct what her mother and her father knew or did not know. And Fredric, when facing his own childhood vulnerability in the face of a ten year old boy, was overwhelmed by a conflict about asking his father what he had known. Did he ask others? How did he begin to understand? This topic is addressed in the following excerpt:

A: When did you start to tell about the abuse, that you understood you had to do something about it, that you needed help?

F: It became clear when I got divorced, then it was evident. It became in a way ... then I started to trace back to what had happened. *Perhaps, in a way, I had cut out a part of my life.* It was my wife who made me tell that I'd been abused. While she was together with me, she'd registered something, that something had happened to me. She had understood something.

A: How do you suppose she registered something?

F: What she had understood is difficult to say, it's not quite ... once she told me what it was that made her understand what had happened. *There were a few typical signs* she'd found.

A: I'm astonished, I guess that's well done considered all those people who don't register something like that.

F: Well, perhaps this was because of her knowledge about it, she knew about it because she worked in the health care system and she's been abused herself.

A: Well. She's met children and has been abused herself. She had antennae.

F: Yes, she did.

A: When she spoke it out, were you released or did you resist?

F: *I resisted a little at first,* in a way, when she made me say it I resisted, and then I began to think back to what really had happened, and then I began to recognize that *something* had happened to me *which had a great impact.* But then it took me quite a while before I began *to do* something about it.

A: This threshold is high, isn't it, it was so in all the cases I know of.

F: Yes. What it's resulted in, what feels worst, is... self-confidence, of course it's low, and to know what I want - and don't want. Well, and where I stand in relationship to myself and others. It's mostly that's characteristic.

A: Do others easily make you do what you don't really want to?

F: *Yes, well, but more so before,* not anymore nowadays. I've done something about it, *I've changed quite a lot of things.* I've had talks. I've become more aware. I saw a person they mentioned for me at the center. It's her therapy method which has done a lot. She practices Gestalt therapy.[233] It works very well for me.

A: Did you have an idea that it had to be a woman you saw to work with the abuse?

F: Yes, maybe a little. I make contact with women easily. It feels easy to talk to them. Really, in this sense it therefore didn't mean so much in terms of me getting started.

A: Would it have been possible to begin with a man?

F: It wouldn't have been impossible, but it would have taken more time. But it wouldn't have been impossible. I suppose it made my start easier that she was a woman.

A: I have an idea concerning your abusers, being men, and that men more easily talk about sexual abuse with women especially because it's difficult to admit vis-a-vis men. What do you suppose?

F: Yes, it's really very important and it meant something that she was a woman. It's difficult to talk about sexual matters - sexual abuse at least - it's difficult to tell it to a man.

* He talks about two years in Gestalt therapy, and how he had addressed, symbolically, a lot of people, one at a time, from his childhood and adolescence, family and neighbors, and tried to place himself in their situation. First, he talked to them. Then he reflected upon what they might have said. *In this manner he "talked it out" with all of them,* until there did not remain anything unclear between him and the others. He addressed one or two persons in each therapy session. In this way, he put a great deal behind him. After every therapy session, he saw his sister and told her about what had happened during the session. *In this way, they got very close to each other.*

This description comprises a remarkable three-year process from, slight resistance, to becoming increasingly aware that something salient may be "forgotten" which ought to be remembered, to two years of therapy and a symbolic "turning to" relevant persons from his childhood and home-town. After exploring a rather wide field of memories, persons and experiences, Fredric leaves quite a variety of thoughts and problems behind. In a way, he succeeds in *sorting out relevant relationships.* During this process, although settled too late to save his marriage, he builds a warm relationship to a younger half-sister. Then he, in fact, makes contact with several relevant people from his past. This is told at the end of the interview and, in a condensed form, was noted as follows:

* He encountered most of the people who had influenced his childhood. He talked to them and *asked them things he felt uncertain about,* and about events he could not make sense of. He received *confirmation* of a lot of connections he had *suspected* existed. An uncle in whom he felt great confidence could confirm that Fredric's mother and aunt had also been sexually abused as children. He *achieved clarity* about how particular traits of his family were *interrelated* with events which *had been partly hidden.* Events found their place, the pieces he remembered matched with pictures others could recall. *The pieces fit together.* The talks felt beneficial, everybody understood more than they had understood alone.

Fredric reverses what he has acknowledged to have done earlier, described as having "cut out a part" of his life. This lacking part is, most likely, what engendered the signs his wife interpreted as indications that "something had happened to him." He actively reconstructs what he has cut out, first, imaginatively, turning to relevant persons in a symbolic way by visualizing them, asking and telling them about himself. Then he encounters people he has known or still knows. In the interchange of memories he finds his own story, the past he had not carried in a conscious way. While filling in the blanks and receiving pictures from others, he recognizes the known anew.

"Forgetting" was a central topic for most interviewees. Although many of them were familiar with terms from psychology and psychoanalysis, only a few used the term "suppression." Berit said: "I suppressed it, the very moment it hit me what it was, I suppressed it," when describing the disproportionality she recognized. In this connection she used the word for a conscious act of "willing herself not to be aware." Consequently, it did not carry the meaning it is attributed in psychoanalytical theory.[234]

To forget, partly or totally, for a while or for decades - these constellations are mentioned. In the current debate about memory recall, induced memory and false memory, the phenomenon of forgetting is of central interest, as it is in the research on the nature of traumatic memory and the differences in memory organization under different circumstances. The core question is: Is complete forgetting of sexual abuse possible? Linda M. Williams performed a study among adult women with a previously documented history of child sexual abuse. These women were contacted seventeen years after their referral, as children, to a pediatric ward where they were registered by the authorities. Now as adults, they were asked about general health and life-time assault experience, though they were unaware the focus of the study was on childhood abuse. Of 129 women with documented abuse, 38% did not recall this abuse, although the interviewers addressed the topic with several questions. Women who were younger at the time of the abuse, and those who were molested by someone they knew were more likely to have no recall of the abuse. The author concludes:

> This suggests that large, community-based retrospective studies of child sexual abuse may misclassify as unabused a significant number of women who were abused in childhood. Furthermore, these findings suggest that retrospective studies miss information about a significant proportion of the abuse that the women have suffered. Therefore, understanding of the prevalence of abuse is affected, as is understanding of the nature of the abuse. For example, this study suggests that abuse of very young children and abuse perpetrated by individuals with a close relationship to the victim may be more likely to go undetected in retrospective studies. The problem of underreporting in retrospective studies may be even greater than these results suggest because this sample entirely comprised women whose sexual abuse was known to at least one family member and was reported to the authorities. Many victims of child sexual abuse never discuss their victimization with anyone, and they may even more likely have forgotten the abuse. (Williams 1994)

The study supported other findings of periodic or total "forgetting" of childhood abuse among adult women,[235] and it was followed by a report based on the same material focusing on periodic failure to recall.[236]

5.5 SYNNØVE

Hanna's approach was to reconstruct her own memory about the past by confronting the abuser. Such a confrontation may also be motivated on other grounds. In the interview with Synnøve, this topic is thematized in relation to what is currently her main difficulty. Her reflections are as follows:

S: Well, I feel it comes too close. It reemerges, everything which has been. I've discussed this with others who suggest that I haven't worked with it sufficiently, but I find I have to cope now. I have a partner and I have a son. I judge my functioning in everyday life as being adequate, so I want to keep it like this. And the abuser - I still think about him a lot. I go and read the obituaries in the newspaper and I have strange thoughts around that.

A: Might you wish he were dead?

S: Yes, indeed I do. But, of course, it wouldn't help me any if he did die, but I've heard he's in town, everybody else has seen him except me. But I've had the suspicion that he's tried to avoid me because, after all, he lives here. But I pulled myself together and thought I'd go downtown in order to meet him. Because every time I'm there I walk around and look for him in a way, I haven't been able to be in town without thinking about whether he might be there. But then I thought I shouldn't continue doing this but rather meet him. Now I feel so strong that, now, I can run him down. But then I saw him. And that wasn't quite what I'd expected. I walked over to him and sat down where he was sitting in a café, but it didn't turn out like that, I didn't run him over like I'd wished. I really feel I didn't win anything. But he hasn't understood what he's done. Or rather, he's understood but he ... he ... I guess he hasn't understood how bad it's been for me. That he's understood some things is certain, because when I - I had divorced and moved to a flat on my own, and then it wasn't long before he called me, I hadn't seen him for years, and I had nothing to do with him because he didn't live with my mother anymore, and then he asked whether my divorce, whether it had something to do with him ... whether he was the cause. Then I told him that yes, absolutely. Then he said this was a pity and that he wanted to talk to me.

Synnøve continued, telling that he retreated and never came back, although they had scheduled a meeting and she had prepared herself properly. Then she contacted his son and family, meeting the son's partner and their little boy. When seeing the child, she decided to tell her story and to warn the woman never to leave the boy alone with the grandfather. The woman told her that they had no contact with him because of his alcoholism. Synnøve stressed that the woman had been grateful for her warning. But she had to admit that the feeling of having taken revenge was not present, not even when facing him directly. He would not understand the significance of what he had done to her. Why did confronting him feel so futile as a way to mediate the impact of the abuse experience?

One of many possible answers may be derived from the excerpt quoted above. Synnøve presents herself convincingly as still under the influence of a person who disregarded her boundaries and disturbed her sense of self with his heavy-handedness, in the multilayered significance of the term explored previously. Just as she always knew where he was when he was living in her and her mother's home, she bears him in her mind constantly, even now. He still directs her attention. He still is present in her everyday life. She still needs to assure herself where he is. She still keeps her guard up continually. His presence in her thoughts is expressed in her everyday attention to the obituaries, her eyes searching for a particular name. His presence in her thoughts is documented by her awareness of all information from others who have seen him, her ears listening for particular clues. His presence in her precautions is mirrored in her being constantly on the look out for him whenever in town, her senses attentive to a particular Gestalt. He still occupies her, he still directs her, and *he still abuses her - as her time, attention, thoughts and strength are still his.* Synnøve feels so bound, occupied or invaded that she does what is almost as unnameable in her culture as what has been done to her: *she wishes him dead.* This wish, referred to by circumscribing it, is strong and concrete. Synnøve does nothing to avoid answering or to withhold a confirmation of the interviewer's interpretation. Yes, indeed, she wishes he was dead. This is a very courageous admittance, however, or from a different point of view, a very sinful confession. In a cultural frame stressing an intense appreciation of the individual's right to life, and of very elaborated rituals as how to speak of deceased people, Synnøve by will and with awareness, "commits" a contemptible disregard of the rules of piety: she hopes another person dies - and she dares to speak of it. Taking into consideration the degree of reflection, responsibility and regard for other people's welfare expressed in earlier excerpts from her interview, and her serious and concerned way of talking and responding, this statement carries powerful implications. No superficial or unresponsive woman is talking, but a desperate, subjugated human being. Only despair could explain such a forbidden wish being expressed by a person with such scrupulously held moral standards.

However, while maintaining her desperate hope that he might soon die, she immediately reflects upon her hope being in vain. It is not futile in the sense that she may never see her wish fulfilled, but that even its fulfillment would fail to have the effect for which she hopes. "Yes, indeed I do. But, of course, it wouldn't help me any if he did die...," expresses that she is aware that he would not leave *her* life even if he left *his*. She has fantasized how to confront him concerning his responsibility for destroying her life and her relationships. This demands strength, a physical power she imagines herself to need, and to possess now, at last. In her mind, he is still the strong and violent man, battering her mother and throwing chairs through closed windows. Now, she imagines her own strength exceeding his so she might "run him over." Synnøve uses a violent metaphor, again contrasting her consistently respectful rather than aggressive personality and speech. To wish to run over a human being is identical with a wish to kill. To run somebody down is war, either in warfare or in the streets. She knows what she is expressing, and that she, once again, commits what for her is a sin. But still she stands by her statement. However, to wish him dead will not work, and to kill him will not help. When she does see him, and both approaches and addresses him, her imagination pales, and her intention looses its force. From the interview text, it is impossible to know for sure whether Synnøve is confronted with another force within herself, of greater impact than her wish, representing the moral force of "Thou shall not," or whether she suddenly becomes aware of his impaired condition after years of alcoholism, rendering him somehow "down" so he cannot be run down any more, or whether she, once again, experiences his personality as overpowering hers. The result of the encounter, however, is expressed as resignation, not a relief, as might have been the result of the second of the constellations above. She wins nothing, because he understands nothing. Only his full comprehension of what he has done to her could be her release. She confronts a mismatch. Her moral and relational universe of personal responsibility is incomprehensible to a man who deliberately violates human beings and human relationships. The incompatibility of Synnøve's and her abuser's personalities are probably the core of her continuing despair and subjugation. She will not be released unless she can acknowledge her moral measures as fundamentally incompatible with his, and his power as grounded in her childhood perception of him.

Consequently, the answer to the initial question as to why it seems so difficult to mediate the lasting impact of abuse experience may possibly be embedded in every particular embodiment of abuse and the nature of every particular abusive relationship. A generally valid answer cannot be given; one must be elaborated for each and every narrative, the nature of which resides in its particulars.

5.6 BEATE

The particularity of abuse retrieval and the idiosyncratic nature of abuse embodiment are emphasized in excerpts from the interview with Beate. The report of her abuse history is interwoven with that of her sickness history, and the interview is presented from the beginning:

B: I was abused by an uncle from the time I was twelve, and later in adolescence I was used by several boys in the group I belonged to, guys who were rather a bit older than me or even young men. As a child, I often had throat infections. I had my adenoids removed for that when I was seven. That time I was admitted. Since I was ten years old, I've had problems with my back, lower-back pain. Apart from this, I had trouble with my stomach, almost constantly. But that didn't lead to hospital admissions. I often had urinary tract infections, but those didn't cause admissions either. The lower-back pains followed me into adulthood. They come and go, my lower back is totally stiff. *And I have a fungus infection. From my pelvis to my chin. Everybody says it's fungus. But the strange thing is that it flares up every time I'm nervous.* (she shows her exanthema by pulling down the neck of her sweater and pushing up her sleeves) Only one doctor once wondered and sent me to a specialist. But all examination stopped there, abruptly. No more tests were taken. Instead, I was given pills which damaged my liver. I didn't want to take them. Nobody wanted to put together my nerves and what I'd been through, because I got this eight years ago. *It was then that everything began. It came when I realized that my son had been abused. Suddenly all the other things flared up again. I didn't really remember my own abuse experience before two years ago.* First then I remembered that I'd been abused myself. This eczema is there, all the time. I've got creams, a treatment with tablets - but it will never go away.

A: And you register that it becomes worse every time you feel one kind of stress or another kind?

B: *Yes, psychologically and the like - then the eczema increases.* It doesn't make sense to me. After all, one should be able to stop a fungus infection. But one doesn't if it's connected with nerves. This is something that's kept me wondering for many years now. I connect it to a mild kind of psoriasis. It's just as intense. It itches horribly. *During these last years, I've been thinking about why it comes out every time I get close to something I can't grasp.*[237]

A: Could you think of writing about it and sending it to Dr. NN at the University clinic? He might be the only skin specialist I can think of curious enough to try to think innovatively together with you, if I may say so.

B: Yes, I'd like to go further with it. The only thing which helps is sunshine. It becomes less visible when I expose myself to the sun. But I haven't been in the sun for a while, it's like this, white around it and red in the middle. Then it itches. I should live in the South,[238] I've decided. And it's clearly demarcated. Look! (she shows the borders at her legs, her neck and her arms. All the skin is "painted" with eczema, the borders are quite distinct, the rest of her skin is unblemished)

A: It looks as if you're wearing a bright pink gym suit, one could draw a line along the borders, they're so distinct. Strange.

B: Yes. It's always like this. Nowhere else on my body. Once I met a doctor, I met him when I went to donate blood, but I wasn't allowed to since they didn't know what it was. He was the one to refer me. But I didn't get any further. He placed me in a room with a very strange light where you become almost fluorescent. He said no, it was only fungus, and I'd caught it in the South. But I'd never been in the South, I told him. Then I'd got it in the solarium. But I couldn't make any sense out of it. After I got this, ... I ... normally I perspire when I'm working, but even when I'm sitting very quietly, water pours down my body. It's precisely as if you open the faucet. It runs down me. That too is only in this part of my body. Then I'm completely wet. *There is a combination. That's beyond doubt. But to get a normal doctor to agree, that's not so easy. (laughs a little resignedly)*

A: No. And from there to find a solution - that's a way to go. But first one must understand.

B: *Yes. Because it started when our boy began to tell that he'd been abused.* Meaning, it started when it began to show that he had problems. Then came the next, and the next, but it took a while before we understood what it really was. Then this came, in the time before.

Beate thematizes a discrepancy between, on the one hand, a medical explanation of a health problem and the intervention measures taken to cure it, and, on the other, a subjective conviction of a connection or relatedness between this problem and mental strain. She has made some effort to get it examined, and to get a second opinion since none of the therapeutic measures has solved the problem as yet. The diagnosis, fungus, and its etiology, the "South" or the solarium, do not make sense to Beate. She mentions a particular time relationship for the onset of the itching, perspiring and burning of her skin. It started when she gradually understood that her son had been abused by her father, his grandfather, for years. The incompatible worlds within which to understand this chronic health problem, are biomedicine on the objective side of the experts on skin diseases, and embodiment on the subjective side of the affected person. From the excerpt above, a condensed view of the statements concerning the skin reaction is expressing this conflict.

And I have a fungus infection. From my pelvis to my chin. Everybody says it's fungus. But the strange thing is that it flares up every time I'm nervous.

Nobody wanted to put together my nerves and what I'd been through, because I got this eight years ago. It was then that everything began. It came when I realized that my son had been abused. Suddenly all the other things flared up again.

Yes, (when I am) psychologically (strained) and the like - then the eczema increases.

During these last years, I've been thinking about why it comes out every time I get close to something I can't grasp.

There is a combination. That's beyond doubt. But to get a normal doctor to agree, that's not so easy. (laughs a little resignedly)

Because it started when our boy began to tell that he'd been abused.

Within these statements about the start of a skin problem, there is also a statement about a recognition. When abuse of a son became a topic, a mother's own abuse experiences became recognizable:

Suddenly all the other things flared up again. I didn't really remember my own abuse experience until two years ago. First then I remembered that I'd been abused myself.

When a perception or constellation which is too similar seems to come too close to a dissociated experience, the embodiment of this experience becomes a topic as well. "Something" begins to show, the skin of a clearly demarcated part of the body reacts, as it is perceived of by the person whose skin is reacting. Beate herself expresses no doubt concerning there being at least a temporal relationship. But the immediacy of the statement about her own abuse recall, and its absence for so many years, linked to the disclosure of her son's abuse, is a valid testimony to Beate's thoughts of a connection. When understanding that something has been done to her son, she both reacts and recalls. In her presentation, however, a kind of delay is embedded into four connected sentences which represent the link between her abuse and her son's abuse:

It came when I realized that my son had been abused. Suddenly all the other things flared up again. I didn't really remember my own abuse experience before two years ago. First then I remembered that I'd been abused myself.

There seems to be a discrepancy in time between the described point of onset eight years ago, "when ... everything began," and Beate's knowing anew about having been abused herself. Or does she speak of layers of abuse, uncovered step-wise, so to speak? The next excerpt, addressing the gradual disclosure, opens up a new perspective on a relationship between her skin, her abuse experience, and her son's abuse:

A: Did you imagine why he had these problems, what hidden origin it might have?

B: No. He rationalized it using everything else. I really tried to find the thread in everything he said. He mentioned friends he fought with, it was the gang in our town which was bad, the teachers were crazy, it was everything. After we had found out of all these things he calmed down for a little while, but then it started again. And then we had to follow new threads. (long pause).

A: And then it proved to be his grandfather, your father, wasn't it? How long had this lasted?

B: My father had started terrorizing him a little when the boy was five years old. Now he is nineteen. The real abuse began from - I'm a little uncertain, he doesn't succeed in dating it precisely, but most probably at the shift from sixth to seventh grade, then it really started. But until then he had become so dependent on him (my father) that he couldn't say no. *It was carried through all the way.* And when we finally understood this, which now is five years ago, he was fourteen at that time, I got so bad that I was almost skinless from all the itching. *When he finally came out with the problem, I got a ring on my back. It felt like hellfire*[239] *but the doctor said it wasn't.* But I've seen photographs of hellfire, and it looked exactly like that. It burned so intensely, I had to sit like this (arches her back by bringing her arms behind her) in order not to have the clothes touch my skin. I still know exactly where I felt it. *It has remained,* in a way, I don't know how to put it ...

A: As a kind of an echo of the pain in your skin?

B: Yes, precisely like that. That's it. *And there it starts to prick me whenever I get worse, when the eczema begins to move because I react to something.*

A: Really, it's the place that answers first?

B: Yes. *It's the same, every time. And then my shoulder begins as well. I'm keeping something or other back...*

The onset of the skin disease is now precisely located: the very moment she understands that her father has abused her son, terrorizing him over several years in order to make him cooperate in sexual abuse as an adolescent boy, her skin begins to burn. *"It was carried through all the way,"* she says. What is carried through cannot be read immediately from her following statements, although the statement may relate to the careful preparation her father has done moving from psychic to physical to sexual abuse. However, she does react immediately, in her skin, a condition to which she herself gives a medical diagnosis, although her doctor does not agree. Now the situation is the opposite of the previous one. The burning pain, literally "arching her back," leaves an echo in her skin, a place of increased sensitivity or heightened attention, which reacts whenever Beate responds to something. Once again, a step is taken to link the skin, Beate and her son closer within the interactive field of abuse and its impact. Does Beate react so intensely on behalf of her son's experience as it is an echo of her own? Has her father abused her? The following excerpt is connected to a question concerning her youngest son:

A: He hasn't been abused?

B: (hesitates). Somewhat. By an older neighbor. But he was fortunately clever enough to tell. The neighbor was stopped after 25 years. He had abused boys of that age, between eight and twelve years. But he was a "kind" abuser. He'd never been evil with these kids.[240] As opposed to my father, he was evil in addition. Evil in all respects. *Ohh, he's been evil from when I was seven years old, it's the first thing I remember.*

A: Does this mean that you were physically abused by your father?

B: *He's never taken me, but I've seen everything.* Everything that happened at home. That means with my mother. *Chairs and knives flying around the walls.* It ended with the police the first time when I was seven (her voice is now remarkably soft and imprecise compared to earlier) Then he was thrown out. That was the first time. And since, there were many times after that. *All of this came up again when the boy began to tell. Then I suddenly was back. (pause) It also connects to my very stiff back which I never have managed to bend.* I've tried to keep my back straight and my shoulders down. That's something which has made many people wonder.

A: Does that mean you've received a lot of treatment?

B: It means that the doctors have sent me to physiotherapy and given me exercise, and apart from that I've seen alternative therapists. As a teenager, I went to a chiropractor, which I still do, now and then. Plus, I've gotten acupuncture, electric acupuncture, foot-zone therapy, and I've been to a person with warm hands ... (laughs a little)

From talking about her youngest son - but rather briefly and as if this were an unimportant topic in this connection, Beate immediately proceeds to her violent father and the first episode she can remember where the police were called. Chairs and knives were in the air. Her mother was battered. But, "He has never taken me," she says. When recalling these images, Beate makes a remark, all of a sudden, concerning her stiff back which then leads to a narrative about all the different treatments she has tried throughout the years. The stiffness and the futility in treating it in a way which solves the problem for any length of time parallel Annabella's history. Beate, however, presents this stiff back in direct connection to her recollection of violence in her home. Might this indicate her perception of a meaningful relationship between violence and stiffness? Two text pages ahead in her interview, she again addresses this topic:

B: About my back, the chiropractor has mentioned that I have 'too stiff stomach muscles,' that my muscles act like a tight belt, he calls it a kidney belt. I've tried to loosen my muscles in the stomach, but they're very tense. *Personally, I think of them as my readiness, just in case a punch hits my stomach.* Then I won't feel it hit. I understood this those times *when the boy attacked me with the knife,* then I became aware that I'd been ready for getting hit for a long time. He went right into me, but he didn't even knock the breath out of me. And I can see that I did the same when getting in between my father and mother, it was then I started to contract my stomach muscles. But I never got hit there.

A: You used your back and your stomach to make a wall? No wonder your back became stiff and tense.

B: No. It isn't strange. (long pause)

Beate has been on her guard for "a long time" by making a wall between her parents when her mother was battered. Her adaptation to random violence in her home, never hitting her directly but indirectly through Beate's determination to defend her mother, has turned maladaptive as it causes an unresolvable pain. And, all of a sudden, she has to repeat her defense strategy when her older son attacks her with a knife. She has, in one way or another, been prepared for a new "punch in the stomach," but why? What does she carry in her corporeal "preparedness?" When following the issue of the knife through the interview, the answer emerges, and the logic of Beate's embodiment is made evident. On page four, she mentions "knives flying around the walls," and on page five "the knife" becomes a topic in connection with a hospital admission:

B: I escaped every time they came for a blood sample. It has been a terror of mine for a very long time.

A: Needles and blood tests and things which are meant to go into the body? Does this link to the abuse, is that possible?

B: I don't know. I guess it more relates to my father. He used to sit with needles and knives and the like ... when he was in that mood ...

A: He sat with needles and knives - and looked as if he could imagine using them?

B: Yes. Sometimes he used the knife in such a way that the blood spurted.[241] (long pause)

A: And it always affected your mother?

B: Yes. My father has hit me only once, and I know why. I was fourteen then and completely drunk. (laughs) But otherwise, I always got in between him and my mother, since I was ten. He's never managed to move me. He hasn't taken me. Remarkably enough (hesitates). *I feel that this is simply revenge, that he took the boy because he didn't succeed in taking me.* He succeeded in subduing my sisters and my mother, but not me. Why, I don't know, he didn't manage. (long pause)

To abuse her son as a revenge for not having been able to subdue her, that is what Beate means when stating "it was carried through all the way." There is a will and a purpose, an intended development of escalating abuse to hit her "in proxy," to triumph over the only person he could not scare or batter into submission, the only one who threatened his demand for total control through violent suppression. Beate is sure that he aimed for hitting her at her only vulnerable point, her sons.

The impression of a delay between her reaction to the abuse of her son, and her recall of her own abuse experience is clarified by her answer to the question whether her terror concerning the "needles and things" is linked to the abuse. She located its origin at her father, thereby denoting abuse differences. He has battered and threatened her mother, but never her. Beate does not regard herself as abused by her father, never the direct object of his violence. This corresponds to the social conceptualization of violence and its degrees. However, it contradicts her description of the consequences of this non-violence as she herself still carries them in her corporeal reaction patterns. She does not doubt the impact on her son of her father's threatening conduct, his psychic violence. She regards the time from when her son was five, the onset of the "little terrorizing," until he was twelve or thirteen, the onset of the sexual abuse, to have been years of abuse. Therefore she sees something having been "carried through" when she is finally aware of the complete development.

Concerning the impact of having witnessed violence at home during childhood, some considerations have to be made clear. The topic has not as yet gained the professional interest it may require, as being a witness to an assault is still not defined as being a victim of an offense, or at any rate it is given lesser weight than being the one who is assaulted. Once again, this demonstrates the belief in visible, verifiable and objectifiable violation as being "real" assaults, according to the assumption that "proof" which is measurable and visible provides the only reliable basis for judgments about what is real and what is not. These presuppositions have been challenged during the last decades, as professional interest in the long-term impact of systematic violence in warfare has been fueled by the latest wars in Africa, Asia and, not at least, in Europe. Very little systematic work among children who are victims of warfare, though not, primarily, of bodily molestations, has as yet been organized in a systematic way. The long-term effect of the Gulf war on Kuwaiti children has been studied.[242] The overall impact of violence on children in general, of witnessing violence, and of witnessing violence against relevant others performed by persons of trust in particular, has only very lately been addressed as a topic holding relevance for the health care professions. This is expressed in a commentary, including twenty references on the topic, as follows:

> Most studies on the effects of traumatic exposure to violence against children have focused on elementary-school-aged children and adolescents. Preliminary evidence, however, suggests that preschoolers may be especially vulnerable to the effects of traumatic exposure to violence. Witnessing relatives and friends being hurt in fights or shoot-outs is especially stressful for young children who are already struggling with developmentally appropriate concerns about safety, competence, and bodily integrity... *Children who witness domestic violence may be particularly vulnerable to emotional and developmental problems.*

> Studies have not yet been conducted to compare the range, type, and severity of problems between children who witness violence in the home with those who witness it outside the home. However, based on our clinical experience, *it appears that witnessing violence between parents in the home results in more severe consequences.* Data are available showing that children who witness violence in the home *identify along gender lines* with their parents' relationship. Boys become more abusive as adults; girls become victims. (italics mine) (Groves et al. 1993)

Beate has witnessed her mother being battered and wounded with a knife. She has made herself a wall between her parents by stiffening her muscles, arching her back and contracting her abdomen in expectation of being hit by her father. She knows him to be capable of using knives and needles, which she mentions in an almost completely detached, non-concrete and vague way, attesting to the need of distance to even a slight description of the horror the knife commands. The vagueness of her statement, and the openness of the sentence, requires the interviewer to imagine how the events would look if what is left unsaid is to be comprehended. The affirmative response to a probing question comes immediately and in a firm, but short statement, given the impact of its meaning. There is a deadly threat in her home, which, apparently, only Beate resists. Her father does succeed in reaching her as well. He abuses her son, and again the topic of violence is relevant. Her son directs the knife at her. She has honed her preparedness and alertness throughout most of her life. He runs into the wall she made of herself earlier, though for quite different purposes. Thus he does not harm her. It seems she has embodied the threat of the knife, is prepared to meet it. Does she mention the knife once more? It emerges at the end of page five in a summarizing remark:

> * She tells how she understood by the clues her son gave her that her father had violated him. She saw how her son had begun to handle knives the same way her father did. She saw a pattern - and in this pattern she saw a message. That was what started the disclosure of a sexual, physical and mental abuse history.

A pattern, a particular way that hands hold a knife, inscribed in the vision of a girl by her violent father, is read as a message when this particular handling of knives reemerges, although the hands belong to her son. She knows by the sight of it that the boy is marked by abuse, and branded by a particular abuser. Simultaneously, when seeing this against the background of her tacit knowing about the message of the knife, she comprehends that also she, finally, has been hit by her father. That is what makes "it flare up" again and activates her own experiences. She is in the violence which she, only apparently, had escaped. All her body reacts, and mainly her skin, which is the primary organ of human contact. Her skin burns, beginning at a single point, almost as if a seismographic ground zero were inaugurating a spreading out of her general skin rash whenever she responds to something. But how do the sensation of burning, the rashes and the perspiring skin link to her own experience, why is it embodied in an apparently outer, or surface reaction of alarm and aversion? May it be interpreted as covert anger, expressed via a legitimate, albeit inadequate, mode of response in lieu of a more adequate, but taboo mode.[243] Can Beate's body be understood in this sense, that is, in the sense of Nancy Scheper-Hughes' "rebel body," expressing the subversive meanings of illness?[244] Before expanding on these theoretical perspectives, does Beate ever express any kind of anger during the interview? Perhaps there is an indication in a remark linked to her skin reactivity:

> *It's the same, every time. And then my shoulder begins as well. I'm keeping something or other back...*

But this remark is not expanded upon, most probably due to the interviewer's lack of awareness about its possible potential as a clue. However, on the last page of the interview text, in connection with a narrative about the relational problems all her siblings have, she addresses her own relationship to her mother. This is transcribed in a summary:

* She tells that she has given up talking to her mother about childhood and adolescence. Her mother remembers very little. She has broken all contacts with Beate's father. He was imprisoned for the abuse, but served only two years of a sentence of three. The mother has blamed her, and her father has attempted to make her responsible for what has happened. She tells that she once saw him in the town, *and that she felt very tempted to run him over. She hates him intensely.* He has married again. Apart from this he's written letters and has asked for meetings, he's had his punishment and wants everything to be forgotten. He doesn't understand that her son still has problems.

* She tells that, during the trial, it became known that her father had a former conviction, for sexually fondling boys and girls, and for several incidents of indecent exposure. This conviction had been unknown to her family, and her mother had also been ignorant of it. That case had concerned four children he had abused. The previous conviction resulted in his receiving a heavier sentence now.

Beate dares express her fantasy to kill him, to run him over, and she is even more explicit than was Synnøve. She dares, in words, to challenge the imperative of "Thou shall not kill," but only in the context of her expanded narrative about his overall destructiveness, his complete lack of empathy, his consistent disregard of others' integrity, and his total inability to comprehend the impact of his violent and assaultive acting out. Only in the perspective of his persuasive, repeated and unchanging abusive conduct through decades does she dare to mention that she has felt a strong impulse to kill him, which, of course, must be controlled the very moment it is felt. But how can this control be managed, and what does it demand of her in the form of substitutive or compensatory strategies? Which other options does she have?

Nancy Scheper-Hughes, in her theoretical reflections about embodiment as the soil of sickness, writes:

> While the body can be used, as symbolic anthropologists have indicated, to express a sense of belonging or affirmation, it can also be used to express negative, conflictual sentiments - feelings of distress, alienation, frustration, anger, resentment, sadness and loss. These two modes of bodily expression exist in a dialectical relationship, an expression of the tensions between belonging and alienation that characterize human social life everywhere. In social systems characterized by a great deal of institutionalized inequality and class, racial or gender exploitation, feelings of oppression, frustration and submerged rage are common, but disallowed, social sentiments. The most immediate way of dealing with feelings of anger and frustration is through direct physical and verbal aggression. For obvious reasons, however, the use of direct confrontation by the aggrieved and submerged sectors of society against their oppressors is only rarely resorted to and is more notable, historically, for its absence.

And she continues later in the text with regard to illness function:

> Likewise, some forms of physical illness and distress can also be viewed as acts of refusal or of mockery, as a form of protest (albeit not fully consciously) against oppressive social roles and ideologies. Of all the various options for expressing dissent and defiance, the use of illness is perhaps one of the most common, yet also one of the most problematic. ... If we start with a notion of the embodied person living out and reacting to his or her assigned place in the social order, then the social origins of many illnesses and much distress and the 'sickening' social order come into sharp focus. (Scheper-Hughes 1990)

Not yielding to imposed power and inflicted violation by means of sickness has been explored in connection with the avoidance by pain in Nora's and Annabella's interviews. In Beate's interview, the idiosyncratic refusal to be continually abused is expanded to include a refusal to accept either the abuse of others or the threat to one's own integrity, although suppressed by convention. As her resistance is overridden by an assault "in proxy," and as Beate is not allowed by social rules to act out, she "is sickened" in the sense of Scheper-Hughes. Her mindful body is affected and must be understood from this perspective, in accordance with her own perception of the onset, course, and aggravation of the disease. Once more Scheper-Hughes, in her reflection about the epistemology of sickness as seen from a critical perspective in medical anthropology:

> ... illness was rescued from the individualist, ethnocentric domain of doctor-patient relationships, and understood in more collectivist and social terms as

> personal or cultural narratives, as drama and performance, and as both individual ceremonies and collective rituals of bodily and social reform. (Scheper-Hughes 1990)

The medical history and explanatory system does not make any sense to Beate who has her own sickness history and explanation. Her perceptions of the relationships in time and origin, however, not only make sense, they rather allow a fuller view of her embodiment of abuse, violence, and suppression. Her reading is consistent, as is the story of the continual assaults against others by a man protected by social roles, rights, the notion of privacy and the traditional view of the impact of sexual assault on children, as reflected in the laws and their practice. Beate's burning, flushing skin, when read as a part of a protest against social injustice and personal violation, acquires meaning in a trans-generational family conflict. The medical view, and the medical measures, however, are inadequate both to explain and to treat. Beate does not experience herself as having received help. On the contrary, whenever she has tried to go further in order to understand the connections she lives, she is not met. Her personal sickness experience is incompatible with the medical disease explanation. This topic is expressed and thematized by the psychiatrist and anthropologist Arthur Kleinman in his latest book:

> The patient's and family's complaints are regarded as *subjective* self-reports, biased accounts of a too personal somewhere. The physician's task, wherever possible, is to replace these biased observations with *objective* data: the only valid sign of pathological processes, because they are based on verified and verifiable measurements. This is a view from a depersonalized nowhere.[245] Thus the doctor is expected to decode the untrustworthy story of *illness as experience* for the evidence of that which is considered authentic, *disease as biological pathology*. ... Yet by denying the patient's and family's experience, the practitioner of biomedicine is also led to discount the moral reality of suffering - the experience of bearing and enduring pain as a coming to terms with that which is most at stake, that which is of ultimate meaning, in living - while affirming objective bodily indices of morbidity. (Kleinman 1995:32)

A long sickness history, involving a considerable number of therapists and therapeutic models, did not reveal the origin of several chronic health problems. Nor did a considerable amount of treatment result in solving these problems. This may even be regarded as satisfying to every therapist, since it confirms the need for a quest for adequate interventions, in other words, that the treatment chosen must address the actual origin. Beate's interview statements regarding her health make evident why this quest was valid in her case as well. One might object that medical professionals cannot know what is not told. That is a reasonable objection, if the telling has been facilitated and possible, neither of which was the case in Beate's history. She was not invited to tell. Had she been, she would probably have been unable to "tell" the origin. She could not know it, since her way of knowing associatively, subjectively, bodily, tacitly, is, in general, not valued by medical professionals, as Kleinman emphasizes. Only provided a medical trust in the validity of subjective body perception, might she have been able to speak, which, however, is not identical with naming as medicine practices it. Persistent violation is still no diagnosis.[246]

5.7 SUMMARY

I have explored interview excerpts thematizing a variety of recognized memories as a particular aspect of sexual abuse experiences. The explorations render visible how impaired memory due to abuse affects identity and selfhood, and which kind of work has to be done to reconstruct parts of a painful past in order to reintegrate that which has been experienced as incompatible with the real and the true. Theoretical frameworks developed by Janet, Douglas, Haug, Munhall, Scheper-Hughes and Kleinman offer options for reading and interpreting narratives about cognitive and perceptive memories as part of a health biography, thus allowing an estimate to be made as to the impact of silenced and "forgotten" sexual abuse.

6. UNHEARD MESSAGES

6.1 CHRISTINE

Experiences from encounters with different representatives of the health care system shall be explored in the following. Christine tells of an encounter with a female physician, comprising an experience of both acceptance and denial, which results in a statement about deeply conflicted emotions. On the first page of the interview text, when recalling her previous diseases and her contacts with the health care system, she says the following:

> **C:** But this pain in my stomach has troubled me as long as I can remember, at least since I was seven years old. And because of everything I'd experienced, I'd refused to see a gynecologist until I was an adult. I didn't dare. I had to go to the emergency room because of heavy pain. There they told me that my ovaries were covered with scar tissue and they wondered whether I'd had many pelvic infections. But I didn't know. I just told them that I'd had a lot of pain. There was a female physician who examined me. And she must have understood something when she found out that this was my very first gynecological exam. She was very careful. And she talked to me all the time. I understood what she did, and it was painful and very uncomfortable, but she told me to warn her, and that made it so much easier. Apart from that she was a woman, she took me seriously.

Christine continues with a narrative about her next encounter with a gynecologist, again due to pelvic pain. That experience is quite the opposite of the prior one. She does not mention the female doctor any more through the complete interview. But in the last phase, when thematizing her experience of repeated betrayals on the part of adult society, the following dialogue develops:

A: We've been talking for quite a while now. I want to ask you about a few things, issues I'd like to return to. The one is the topic of betrayal where you mentioned parents, teachers, other adults, the child-care authorities. Do you also think of betrayal from the therapist's side?

C: If by therapists you mean the doctors I attended - then, yes. *They must have been thinking something, wondered about something.* They treated wounds - and nothing else happened. No questions, no follow-up, at least not as far as I can remember. When children give such clear signals as failure to thrive and repeated pain and wounds and bruises and the like. *Perhaps this has something to do with the whole medical profession.* In a way, they don't want to regard patients as people, that we have feelings, that things are interrelated. But all this is interrelating.

A: In the cases where you feel there was a betrayal, are you also thinking of those you understood had sensed something?

C: Yes. When I remember the one at the emergency room, for example, who performed the first gynecological examination I'd had, and who most definitely had sensed something. She was kind and correct and helped me with whatever I could ask for. *She had a very strange facial expression* as we talked together and as she examined me. *I still remember it, actually. I'll never forget it.* She was certainly kind and sensitive. *But she didn't help me to get going*[247] *although I was as close as I'd ever been before.* And when I think of the others, some of them general practitioners, *they should have seen that there was more, they should have recognized that there was something wrong.* That no help came during fourteen sessions of physiotherapy, or whatever it might have been. And such things as scolding you because you don't use your Cortisone spray - when I can't stand the taste of it and don't manage to spray something into the rear of my mouth. Or that I don't take my pills because I don't like to take drugs for all of this since I can't believe they'll help for what's actually wrong. *They ought to have understood something, these physicians. I showed them something, I cried. It's in a way a betrayal not to register things which are visible, or not to do more when really having understood something.* Finally, I've taken initiative myself. It takes you years to get in to see a psychologist. I sat down and wrote. I addressed it to the psychiatric out-patient clinic. But there are waiting lists everywhere. But it's a choice, after all. *To choose to be insistent a little longer, not to run away from it. It is serious. I wonder whether I can bear it. I might as well, after all.* (long pauses between every sentence) (starts to cry again) Now, I haven't anything more to say.

Christine is explicit in her judgment about the betrayal she has met in the health care professionals. She has not encountered one single physician who has been able or willing to help her so that she might tell about the neglect and abuse in her home, and the sexual abuse which was inflicted upon her from early school age. During her very long sickness history from early childhood on, causing numerous medical contacts due to wounds, bruises, infections, abdominal pain, asthma, allergies, eczema, and pelvic pain, and demanding frequent dental treatments, repeated surgery and physiotherapy, no therapist ever addressed the topic of physical or sexual abuse. She thematizes her wonder, or rather her indignation, at what she has been offered and what she thinks she ought to have received. She takes a perspective on the professionals as people skilled to see and to register, which means who might be expected to perceive and to react:

> *They must have been thinking something, wondered about something.*
> *Perhaps this has something to do with the whole medical profession.*
> *And such things as scolding you ...*
> *They should have recognized that there was something wrong.*
> *They ought to have understood something, these physicians.*
> *I showed them something, I cried.*

Her statements include reflections upon the physicians she has encountered, and upon how each of them added to her experience of physicians as either ignorant of or reluctant to respond to certain clues or particular presentations. They also include a view upon this reluctance as systematically or structurally inherent in the medical profession, despite the fact that health professionals are generally educated to be aware and attentive to patients' needs. She addresses the overall reluctance among physicians to react adequately to health impairment caused by domestic violence.

This phenomenon has very lately come into focus in several Western countries. The failure to react has, for the first time, been labeled publicly as unethical by the American Medical Association and the U.S. Surgeon General in 1992. This topic will be explored more extensively in chapter 3. Do, then, General Practitioners, for example, avoid addressing such violence and choose not to "see" it, as Christine asserts? A study among General Practitioners in a Primary Care Center has been performed, aimed at understanding the dynamics in patient-doctor relationships where domestic violence is the hidden background of the health problem. The design of the study is ethnographical, including in-depth interviews with 38 physicians. The authors present a summary of the results as follows:

> Repeatedly, the image of opening Pandora's box was used by physicians to describe their reaction to exploring domestic violence with patients. Eighteen percent of physicians used the phrase, 'Pandora's box.' (According to

> Greek mythology, Pandora was the first woman. Her creation was part of Zeus' revenge against Prometheus for providing mankind with fire. She was responsible for opening a box and thus unleashing the evils of the world - aging, labor, sickness, insanity, passion, and vice into the world.) Eight percent used analogous phrases such as 'opening a can of worms'. ... This metaphor suggests the fear of unleashing a myriad of evils. By examining the comments of the respondents, these 'evils' begin to take the shape of 'too close for comfort,' 'fear of offending,' 'powerlessness,' 'loss of control,' and 'tyranny of the time.' (Sugg & Inui 1992).

The physicians rationalized their reluctance in ways which expressed basic structures in their professional everyday life. The tyranny of time schedules, not leaving enough space for demanding topics, was mentioned by the majority. The underlying professional assumption of what is worth spending a physician's time on was, however, not addressed. Almost half of the doctors were explicit about a fear loosing control over the situation or the patient because of a lack of measures with which to intervene properly and effectively. The underlying professional ranking of 'acting' and 'controlling' above 'non-acting' but 'supporting' was, however, not addressed. Half of the doctors felt powerless and inadequate when confronted with the impact of domestic abuse and violence. The underlying professional disregard or underestimation of this topic as expressed in its absence from medical curricula was, however, not problematized. More than half of the doctors were fearful to intimidate the patients by expressing what they saw or thought. The underlying professional rationale of domestic violence as a private event and, as such, a non-topic in professional relationships was, however, not addressed. But finally, the evil the doctors felt they touched upon was both an evil in the sense that it felt uncomfortable to imagine violence among people in families, but also a familiar evil as a considerable number of them had experienced childhood or sexual violence themselves, the female physicians twice as often than the males.[248]

How does Christine reflect upon the professionalized reluctance she has seen and met everywhere since she was a child? The topic is addressed in connection with her recalling a tonsillectomy in childhood, and the excerpt from the interview is as follows:

> **C:** Merely to recall what I myself thought as a child and what happened and what the adults didn't comprehend. I only hope that, very soon, people become more skilled at recognizing the signs that something very concrete and very wrong is at issue in children who have a great deal of illness. (long pause) Of course, it's possible that there was someone who was aware of what happened. But, really, they must have hid it very well if they had seen or understood something. After all, there must have been someone who thought that there was something behind it all. Someone who wondered what was wrong. (long pause)

A: How do you imagine they could have indicated having seen, understood or observed something, suspected something behind it all?

C: It ought to be possible to say what one thinks. It ought to be possible to ask about what one suspects. It does no harm to be mistaken. Then only the therapist knows. One should at least not try to treat something which one doesn't understand. I understand that it feels uncomfortable for a physician. And it's difficult to answer. There are a few, I guess, who can talk about it without any problems. But most of those who've been abused consider it disturbing. At least I always feel so...

Christine's repeated expression of how incredible it is that no one reacted to her sickness history as it presented itself from early childhood, is underlined by and mirrored in the syntax of her speech, in the repeated use of subjunctive. In a way, the syntax attests to Christine's refusal to accept the systematic disregarding of domestic violence and sexual abuse in the health care professions which she herself has experienced. Since she is not willing to suppose that all adults are either blind or insensitive, she has to consider the explicit lack of attention to child abuse as caused by rationalization. Her proposal is very simple. "It ought to be possible to say what one thinks." Not addressing what is suspect contributes to maintaining what is hidden. However, were the suspicion to be proven wrong, "it does no harm to be mistaken." So easy, and so difficult, as she admits in the following sentence to being very well aware of. Christine does not claim that the thematizing of assault experiences is easy, neither for the one molested nor for the helper, but it is crucial, because, as she states, *"one should at least not try to treat something which one doesn't understand."*

Did she herself receive treatment for what was not understood? This topic is addressed in the interview in connection with the question of improper approaches in therapeutic relationships. The excerpt is as follows:

C: I've met many different (therapists), but I can't recall anyone who misused my wish for help. I don't believe it to have happened. Rather the opposite has happened, that they've been cool and in a way disinterested. But I suppose that's better, in one way, than that someone had tried to abuse me in such a situation. But I've never experienced something like that. I was rather desperate when I saw the doctor I went to about my allergy. I even gave her some indications that I had things to struggle with, and that it might be appropriate for me to see a psychologist. *But she didn't catch it, she didn't understand that I was talking about something which ought to come out.* I was the one who did it. I, somehow, managed to speak it. *The system still failed.*

A: Deceived you and all the others?

C: Yes. (her voice fades, she trembles and begins to cry, silently) Please switch off the recorder, I can't stand it right now. Wait a little while.[249]

> (long pause, she cries a long time, swallows and attempts to control her breath)
>
> This negligence everywhere. The home, the society, the school, yes, and the so-called helping professions. Not to mention the childcare authorities. The time when I walked the streets in Oslo. I'd suppressed all this, that I and another girl stood alongside the road and picked up men. I wasn't more than thirteen-fourteen years old. I'd forgotten all that. I was so busy with performing and functioning, to keep the pain under control and to adapt to the eczema and the asthma which only got worse during those last years. When I think that no one took hold of me. *No one saw what happened.* There must have been adults who saw what happened, two small girls standing there, children, who picked up men, old men, in order to ... my goodness ... *If I'd heard this story and not lived it, I'd have refused to believe it.*

Once again, Christine addresses the inability, or unwillingness, of a doctor to understand her as attesting to a systematic failure to listen, hear and comprehend. She has a lot of experience with doctors behaving with detachment, which she deems preferable to behaving exploitatively. After having reflected upon the topic in a way people do who feel no real confidence in their own competence to judge the distance which is proper in various human relations, she is able to affirm that she has never felt approached in the sexual way. Her statements concerning her experiences with the systematic negligence of the medical profession evoke her tears. But she insists onto telling more. And when again able to speak, she addresses the systematic betrayals of the society and the societal institutions entrusted with the role of caring of children. Once again, as in the previous excerpts, the wording mirrors a feeling of amazed disbelief that the world of the adults in her society can have been so blind. Again she cannot imagine that no one had seen. And again, the underlying acknowledgment and conclusion must necessarily be that the adults did not see because they did not care. Only thus could her life have developed as it did, and as she herself most impressively put it:

> *If I'd heard this story and not lived it, I'd have refused to believe it.*

Christine's longing for someone to see her and to listen to what she might have to tell was never met in the health care system. Only one female physician stands out in her memory as different, as one by whom Christine felt seen, adequately met, and taken seriously in her need to receive information about what was being done medically and why. She describes a situation in which her interests were regarded and her integrity respected. She has been given the explicit right to control the situation and this engenders immediate trust, provides great relief and makes it possible to submit to the medical procedure which Christine had assiduously avoided until then. She remembers the physician's comments as having helped her to conclude that she ought to have searched for help much earlier, as there was evidence of previous inflammatory or infectious pelvic diseases. This, too, seems to have been interpreted adequately, as indicating Christine's great resistance to attending gynecologists. The doctor, in her way, sees something and this shows in her conduct and her non-verbal communication with her patient. The impression is so strong that it is still in Christine's mind and will remain so, standing out as the only situation where she felt invited to entrust herself. Her background of vain attempts to explain makes her wait for the doctor to be verbally explicit. The inviting remark, however, is not spoken. The supportive hand does not give her the push she needs.

> *She had a very strange facial expression I still remember it, actually. I'll never forget it. She was certainly kind and sensitive. But she didn't help me to get going although I was as close as I'd ever been before.*

Was having been "seen" and then left once again to herself even worse, in Christine's mind, than not having been seen at all? She mentions no immediate emotions after this particular encounter, but proceeds to comment generally upon all those physicians to whom she intentionally gave clues, for example by crying. Christine probably encountered a disparity between her perception of the message potential in her own tears and the doctor's perceptions of the meaning of women's tears in medical encounters. There may have been a culturally constructed incompatibility at issue rendering her tears non-messages in the view of the doctors, or rather a kind of message which might represent the presence of a 'Pandora's box' none of them wanted to open. Nonetheless, Christine, having intended to give clues, feels betrayed. She clearly states the ethical implications of "a professional way" of being either blind and insensitive, or sensitive but defensive toward the needs of abused human beings:

> *It's in a way a betrayal not to register things which are visible,*
> *or not to do more when really having understood something.*

In her final remark, Christine names what she thinks she encounters, for instance in her comments about the waiting list for psychological help and the efforts necessary to be heard: the systematic resistance of the helping professions to meeting her needs. Although aware of the resistance against knowing about, and the systematic negligence of, a particular kind of health problem, namely "suffering from violence," she has made her choice.

> *to be insistent a little longer, not to run away from it. It is serious. I wonder whether I can bear it. I might as well, after all.*

6.2 LINE AND RUTH

In a variety of ways, in shorter or longer narratives in all interviews, such experiences are formulated as a common experience, though linked to quite different situations. There is a range of responses or reactions from health professionals to a patient's information about or disclosure of sexual abuse experience. These extend from non-reactions to the message on the one hand, to its transformation into proof of the patient's insanity on the other. Some such responses shall be mentioned briefly; others have to be elaborated by interpreting how they are presented. Telling about sexual abuse in Line's case was as if not telling anything at all. This is addressed in the following excerpt:

A: Concerning the rape when you were twenty-five, what happened then? Were you injured, were you examined, did anyone do something?

L: I reported him to the police. It was why he'd been imprisoned. I'd been examined. But I hadn't mentioned who he was. I didn't go to the police then. But the examination showed that force had been used. After the rape, I started having trouble with nightmares. And then I got problems with falling asleep. I was unable to sleep. It took me a long time to make up my mind to do more with this case. I fantasized committing suicide. But I had my son, after all. I felt uncertain what would happen to him. But during those years, I saw a few physicians for sleeping pills. I got them very often, for many years. Nowadays I take nothing, no drugs at all.

A: Did you tell any of the doctors why you couldn't fall asleep?

L: No, I didn't tell right away. Only later. My doctor found out finally. He said that he understood my problems. But he didn't say anything else. He didn't give me any advice.

A: You told him about sexual abuse and rape ... what did your doctor do, what did he say?

L: (extraordinarily long hesitation)[250] I can't recall anything, nothing special. What do you mean?

A: Some form of reaction or another ...

L: (long hesitation) No, I can't recall any. He said something about that I should be careful with these pills ... and alcohol.

A: There was no danger in the other ...

L: No. I don't know. It didn't seem so. In any case, he didn't mention anything about it, somehow. He didn't react to it exactly. Well, it was the first time I mentioned it to anyone, and perhaps I didn't make myself so very clear, I didn't tell the details, I mean. But he didn't ask about anything. Not as far as I remember. Only after I'd seen the social worker, then I ... he advised me to report it. We've talked a lot about it. And he advised me to consult a lawyer.

Line had no memory of her doctor's response or advice concerning the rape experience. She repeated it several times, and for her this appeared to be what she could expect and what she had to accept. She did not even share the interviewers distress over the doctor's non-responsiveness, and so had to assure herself what the repetition of the question was intended to express. What she then was told reconfirmed what had previously been said; her response to that might be what had come to her mind other times she had been confronted with what felt like disbelief, or distrust of her honesty. She remembered the doctor's advice concerning the use of pills and alcohol. That was how he responded to her disclosure. She was immediately willing to blame herself for not having elicited another reaction as she certainly had not made herself clear enough, given that she had withheld details. However, she had to admit he had not explored the topic any further, at least as far as she recalled. Feeling sufficiently confident to dare to speak of sexual abuse, after years of struggling with her doubt, and although she had spoken within a frame of trust to a doctor she knew, she was not heard. What she had said remained unheard, as if unsaid.

Ruth described several occasions on which she had dared to express her own conviction that sexual abuse in childhood had caused adulthood problems with which she could not cope. These occasions, all of which concern her psychiatrists or psychologists, shall be reported in the order in which they were mentioned in the interview. The first comment is related to her report of experiences, which she has found offensive, of being improperly approached by men - experiences about which she has told the psychiatrists in the institution to which she was admitted:

> Yes, and what disappointed me so, was - when I told this to the ones who were treating me, the psychiatrists, they defended the men, so it became me that there was something strange about, you see?

> I remember that I told him how I used to react when men whistle at me in the streets. Then he said: 'It's not so surprising, Mrs. NN, that men whistle at you in the streets, because you are such an elegant woman.' It was meant as something defensible, that it should be taken as a compliment. I felt so cheap, I hadn't been out fishing.

> Now I've put in a claim for compensation for damages because of the harm it caused me, and my doctor asked (for the hospital record), and in this record from the hospital there was nothing about what had happened. There was nothing about it. What I had *emphasized* so much during all those years was *not even mentioned* in the record (she puts very much stress on the words in italics). It's really fantastic that they could be deaf to something like that, I mean. Little by little there are a few, perhaps, who begin to understand, I believe, some will understand.

> The psychiatrists understand very little, in my opinion, because when I told my therapist that I had problems having intimate relations with my husband, guess what he said? 'And what if I tried?' he said. (pause) I answered, No.

> As to my part, they've misunderstood completely what has been done to me, and they've placed the source of the problems elsewhere, in my family conditions, in everything else ... this way, I really became sicker than I should have been.

> I always felt that I wasn't heard. And very many times, I stopped talking about things which felt uncomfortable for me, because it in a way meant nothing to them. The way the psychiatrists failed with me was that they belittled something which for me was very - well, perhaps it'll help you to get a different view of it, but to belittle something which was very serious from my view, that wasn't the right way, I believe, at least not right for me.

> I must say that those around me have been very unaware, and I can't understand it shouldn't be possible also for men to understand what such (an experience) means for a little girl, without even having done research on the topic. They ought to understand this out of plain common sense.

The summary of Ruth's statements from her encounters with psychiatric medicine reveals a consistent professional strategy toward messages about sexual assault: *undoing*, which occurs in different ways, and is informed by one perspective: *the male.* Ruth addresses the premise of the normative male gaze on women as informing the responses of the male professionalists she, a female patient, has met. She presents the core of this constellation, and its consequences, in the very first statement. No matter what she communicates about her own feelings about certain kinds and expressions of male attention to her, her judgment is invalidated.[251] It is not the behavior of the men who approach her which is wrong, but her perception of the approaches. The male offenses she addresses are not offensive; it is she who is defensive. The approaches were not insulting; it is she who merely feels insulted. The men's conduct towards her as a woman is judged as correct, thus her reactions must be wrong, which proves that she is deviant. Her perspective is invalid, her feelings inappropriate, her emotions strange. She experiences being defined as "strange," being shamed, having her primary concern completely ignored, having a therapeutic relationship sexualized, being silenced, her illness being made chronic, and having her experience made light of. The male norm is so predominant in the therapists' minds that they listen more as men than as professionals; they are more preoccupied with male standards than the patient's perspective; they judge more according to culturally constituted gender roles than to socially constructed roles for therapists; and their behavior is more in accordance with men's privileges in male dominated societies than with the professional obligation to meet a patient on her own premises. She sees their conformity to a male view as so consistent, however cloaked it may be in professionality, as to cause her to doubt their common sense.

Ruth, herself, is aware of her "career" as a psychiatric patient having been influenced considerably by psychiatry itself and the professionals she has met. From the very moment she recognizes how the destructiveness of the silence she has imposed upon herself out of fear for her father's reaction, and of the shame she's felt over her involvement in an improper relationship, she has tried to tell. She recognizes that her own mental balance, as well as her marriage, are at risk. She herself does not hesitate to tell, although the responses she receives make her feel the effort is futile, or even damaging. Her visible indignation when recalling the responses she found offensive, concerning men's attention and her sexual problems, is quite vivid and not to be mistaken, though the incidents occurred decades ago. These simultaneous violations of female and patient integrity, make clear that the psychiatrists' gaze was, primarily, a male gaze. However, the male and the professional gaze seem quite compatible. Nothing in what Ruth recalls indicates that any of her male therapists is aware of the impact of their "consistent" view of a female patient who herself believes her mental problems to be grounded in sexual assault, in alienating male interference in her primary relationships, and in male disregard for her integrity. Rather, they offer more of the same, so to speak, as Ruth explicitly and repeatedly reports. She thematizes the men's defense of male offensiveness as so obvious that she does not doubt its normative meaning and effect. The norm so thoroughly accords with male premises that her report of sexual abuse is not even mentioned in her psychiatric hospital records. She concludes from the resistance implicit in ignoring, not listening, not responding, and undoing, that her experience has been judged not worthy of being written down. It is rendered not said and unheard. Ruth's exclamation "it is really fantastic that they could ignore it," as in "not hear it," addressing a collective "they" as the many to whom she has tried to tell it while institutionalized, echoes Christine's conclusion that "the system failed." Ruth is aware of having experienced a structurally and systemically constituted and maintained negligence, which is ultimately expressed in a demonstration of the utmost power of the profession: *silence.*

To see this "silence" in the total absence of professional words about her most central concern as a patient is what feels so unbelievable that Ruth would not have been able even to fantasize it. And taking into consideration the impact of an absence of evidence of abuse in the terms considered to carry documentary weight in matters of justice and human rights defense, this absence is more than a mere lack of words. It is the betrayal of a person who has entrusted herself to an institution socially authorized and mandated to give help.

The male gaze as the internalized norm in psychiatrists' judgment of what is and what is not to be considered relevant regarding women's mental problems, and how this may be linked to sexual assault or other conditions, shall be explored. Marcie Kaplan, in her critical "woman's view of DSM-III" from 1983, has examined whether the male gaze has structural and systematic impact, having affected not only male psychiatrists, but also the very classification system of mental diseases, which, by definition, ought to be based on gender neutral, objective, scientific grounds. After having discussed a group of diseases to which data on sex ratios were available in 1983, she concludes:

> ...I turn to the past literature on and conceptions of women's sexuality. Freud's classic theory that vaginal orgasms are different from and more mature than clitoral orgasms caused clinicians, their clients, and the public to believe that women who experienced clitoral orgasms were arrested in their psychological development. It was not until Masters and Johnson published Human Sexual Response in 1966 and claimed that there is only one female orgasm, and it is clitoral, that women who experienced clitoral orgasms were considered cured. In other words, women who before 1966 were immature or even dysfunctional were suddenly, in 1966, mature and functional. The women's sexual behavior did not change, but diagnostic criteria did...Evidence discussed in this article should give one pause when one is considering women's pathology in general, *sexual or otherwise*. Our diagnostic system, like the society it serves, is male centered. (italics mine) (Kaplan 1983).

This 1983 view of DSM-III has been followed up by Joan Busfield with a 1996 view of DSM-IV. In the sixth chapter, entitled Gender and Constructs of Mental Disorders, she argues that "gender plays a pervasive part in official constructions of mental disorders, in the elaboration of normal cases and case identification, because gender is such a key feature of social relationships and a major dimension of social difference, (that it) ...inevitably features in constructions of mental disorders." Since both gender and mental disorders are socially constructed, one has to consider a mutually constitutive effect among these, tautologically conceptualizing the one by the other. This, however, is rendered invisible by a pretense of objectivity and value neutrality in scientific research. Busfield concludes:

> Constructs of mental disorder are, therefore, inherently gendered, and gender and mental disorder is reciprocal. For whilst gender is embedded in constructs of mental disorder, the constructs of disorder which are developed and elaborated by a range of mental health professionals, incorporating ideas about causation and treatment as well as about symptoms, in turn contribute to the way in which gender is itself constructed. (Busfield 1996:118).

And Denise Russell,[252] when reflecting on the most recently defined of the gendered mental disorders, Pre-Menstrual Syndrome (PMS), which in the DSM-III version still was only proposed as a possible disorder, writes:

> The need for a new generalized diagnosis, from the point of view of biological psychiatry as the moral guardian, is becoming particularly urgent with the failure of the biological research on depression. Another factor is that the 'female role' is now under severe challenge in the Western world and conservative forces need to mount strong resistance if that challenge is to be met. The PMS diagnosis must look promising. It has the appearance of being quite specific. It is said to apply to a very large percentage of women and it seems perverse to deny that the cause is biological. The reality is that it is totally non-specific and has not been shown to be biological. (Russell 1995:70).[253]

6.3 RUNA

The presence of a norm implicit within a medical judgment and serving as an instrument of reservation, rejection, suppression or denial, was also to be found in Runa's interview. She had consulted the health care system for many years, primarily for generalized pain in her body which increasingly incapacitated her. From the numerous encounters with therapists in traditional and alternative medicine, she could not recall a single situation wherein the therapist had addressed the topic of adverse life experiences as the possible source of her chronic pain. When recollecting the reaction of a general practitioner to her disclosure of sexual abuse in early adolescence, she mentioned the following:

R: The first doctor I talked about, when I told him that I'd been sexually abused, he said it simply, he was straightforward: 'I'm not very familiar with this. I suppose I can't help you so much. You have to try to search for help somewhere else. But I know one thing for sure, that you aren't only - you aren't only someone who's been affected by sexual abuse, you're so - you're much more, all over,' which meant that I ought to look at myself as much more, because you very often become a loser, you have no more space for anything else, you have no more space for all your positive abilities, you're less good at thinking about the positive in general, you only think of yourself as very odd, very weird, and things like that. I remember he understood this, and simultaneously he said that he couldn't help me so much, that he had little experience and little competence.

A: Did you regard this an honest statement?

R: Yes, that was said very directly, I think.

A: But do you find it reasonable that he told you this ought not mean so much to you? Or don't you think this is what was implicitly in what was said?

R: (hesitates a long time) It may have been reasonable, but it was maybe not so easy to understand, because if you yourself feel that your life is very much shaped by the abuse, it's not so easy to get ... to get the good things in yourself into the foreground, perhaps, and I ought to ... what you're coping with becomes overshadowing. So it was ... yes I think it was straight forward, but still it felt a little hopeless, but I think it was a reasonable way to say it ...

A: ... he didn't pretend to have competence ...

R: No, he didn't.

A: But did you feel he had understood its far-reaching influence on you? (stress on the last word)

R: No, not quite. Because I remember well that he said: 'you have to be aware that there are very (stresses the word) many who have been exposed to something like this. So he - he sort of made it a usual problem. And quite clear, it's not so easy to say how many there are, but this doesn't - it didn't help me so much and still doesn't.

A: I understand. Was he shaken or was it almost as if he had said: 'don't make such a fuss over something so common?

R: Yes, it was a little like that. Yes. He made it quite clear to me that it was not unusual - that may be not correct to say - because he didn't say (stress) so, but it sounded as if it was high time I got finished (stress) with it. That's how it is, it's so much more convenient for everyone else that you get finished with it.

A: But you had just gotten started.

R: Yes, that's so. (long pause)

A: Can others know when you will be ready to say: 'I'm through'?

R: No. Of course not. And you may wonder whether you will ever be through with something like this, because it probably will influence you in one way or another. (pause)

This excerpt has to be commented upon both on the methodological and theoretical levels because, as a whole, it represents: (a) a validation; (b) a silencing of personal voice; (c) a covert opposition; and (d) a professional undoing. The four levels are present and broadly thematized in the first narrative and will be marked in order to make clear the point of departure for the parallel considerations. In the sense of Kvale, the first, the second, and the third phase of interpretation are present, and the first, second and third level of analysis. The first passage is Runa's description of her disclosure and of the physician's reaction, representing the first phase:

R: The first doctor I talked about, when I told him that I'd been sexually abused, he said it simply, he was straightforward: 'I'm not very familiar with this. I suppose I can't help you so much. You have to try to search for help somewhere else. But I know one thing for sure, that you aren't only - you aren't only someone who's been affected by sexual abuse, you're so - you're much more, all over,'

In this text, Runa's statement seems non-controversial for her, she uses an indicative form and recalls what sounded "straightforward." Level three of analysis, the theoretical consideration of professional undoing techniques, will be anchored in this statement, as will level two, concerning an analysis of what Runa says about an experience, and what this may say about Runa. She continues:

> which meant that I ought to look at myself as much more

Here Runa has left the physicians direct speech and interprets what he, in her opinion, intended to say about her, who was abused, and what she, as an abused person, ought to do. The shift from direct to indirect speech, introducing her interpretation of the doctor's statement, is simultaneously, again seen from the third level, an introduction of a norm, a statement about what Runa ought to do. When read in connection with the first sentences the doctor has spoken, as Runa recalls it, the second step of undoing emerges. The physician has, as an immediate response, set boundaries between him and any further information about the topic of sexual abuse in general and Runa's experience in particular. This is done, however, covertly. The rejection of the topic is convoluted into an admission of limited professional knowledge. As a consequence, the physician's possibility to help is likewise limited, as he admits. The strong resistance to any further exploration from Runa's side about whether she wants help and, if she does, what kind, is blocked at the very start of her very first attempt to confide in a representative of the health care system. Most likely, the physician is not familiar with the topic of abuse. Equally likely, however, is that he does not want to become familiar either. Disguised as professional honesty and concern, he not just symbolically, but actually cuts her off. Not only abuse as a topic, but also the abused as a patient, is rejected. Although he admits a profound lack of knowledge, he can still be certain of something which, once again, is formulated as a concern, but which holds a normative imperative. Runa, at any rate, interprets it that way right away. She "ought to look at herself as much more," she is told. The advice may express the physician's presupposition that abused people are completely absorbed by thoughts about the abuse. How can he know about that, however, he who has stated clearly his unfamiliarity with the impact of sexual abuse and therefore refusal, his "legitimate" refusal, to discuss it? Runa has not been given any possibility to tell how she really looks upon herself, thus the origin of her next statements in an impersonal and generalizing "you," are not easy to locate:

> because you very often become a loser, you have no more space for anything else, you have no more space for all your positive abilities, you're less good at

> thinking about the positive in general, you only think of yourself as very odd, very weird, and things like that.

Is Runa speaking in her own voice, from her own experience and knowledge about how "you" think when having been abused? Would she herself call herself a loser, a term from the winner culture of competition rather than from the inside of an abuse experience? Would she say she was completely consumed with thoughts about the abuse such that there were nothing else, while taking care of her three children and a job? Would she not know that she made space for her abilities in her caring for a family home and volunteer work in the parents' groups at her children's schools? Would she persist in negative thinking generally, even as she was actively planning and supporting her children's future development? Would she name herself "odd" or "weird?" It was highly unlikely that Runa voiced her own view, but that she indirectly voiced the physician's judgment about how he thought abused people think and feel, although admittedly unfamiliar with both. This passage, and Runa's short comment,

> I remember he understood this,

reveal a striking discrepancy between his initially admitted lack of proper knowledge, and the consecutively given advice for proper conduct. The emergence of this very discrepancy is what allows the confirmation of the pretense of honesty and concern addressed above. This primary conclusion during Runa's talk, emerging as a matter of confusion in the interviewer's head, seems to be shared by Runa, who, without any indication of a critical stance, concludes her report of her first attempt at disclosure to a professional person like this:

> and simultaneously he said that he couldn't help me so much, that he had little experience and little competence.

The confusion of voices and the confusion of motives emerging from the first part of the text initiate a prolonged third phase activity of interpretation throughout the consecutive text, initiated in the next speech turn, concerning the apparent honesty of the doctor. Runa confirms that his initial remark, which she repeats in the last sentence, has been spoken honestly. Thus, the next step of validation is taken. Does she find it reasonable that the physician tones down the importance of her experience, or does she not share the interviewer's interpretation of the meaning implicit in making sexual abuse a common thing. Runa hesitates before answering, then begins from a "neutral" perspective: that, possibly, he has been reasonable. Then she dares to voice her opposition in a "but," which is immediately arrested and she says it is her fault as it was not easy for her to understand. However, she sees this differently, although she feels obligated to formulate her own experience of a heavy impact in a "you" form of speech. She continues in this "you" modality, although making clear she means herself when confirming what "I ought to" but which she can not. The sentences contain a repeated shift between her own voice and that of the other. The last sentence is still confusing, but her opposition against the authority is more visible. The doctor was straightforward, indeed, but what he asserted felt hopeless to her.

The next validating step concerns the interviewer's insecurity about the interface between the doctor's general statement, as absolute as it was, and Runa's personal experience. Provided she feels that the doctor has made allowances for his limitations of knowledge by refusing to care for her yet still asserting that his authorized advice be accepted, does she let his opinion invalidate her experience? Her answer is clear: his attempt to make sexual abuse a rather common experience, so frequent that it does not deserve much attention, did not "help her so much and still doesn't." She invalidates his advice. She has heard him say, although correctly adding that this was not in the wording but in the meaning, to "get finished with it." She has heard his refusal to listen, his rejection to come close, and his resistance to know. Here, Runa reveals that she perceives his real motives. It is not for her sake that she is advised to overcome it, but for his own. He shares the discomfort about the topic with "all the others" who want abused people to overcome it in order not to be confronted with its reality. Runa has met this reaction before, in much closer relationships and with different consequences. She has learned to keep silent, and she is used to giving others' voices and opinions priority, rendering her almost unable to speak her own interests, demands and needs. As she tries to do so, she immediately becomes aware that her demand to come to terms with the abuse by opening up and understanding it is invalidated in the guise of professional concern and advice. That which she needs to overcome the impact of sexual integrity violation, is not at all a topic. That which *the doctor needs* to avoid its presence in and with her, fills the entire patient-doctor interaction.

The professional resistance against certain topics in a medical consultation, rendering a variety of people's life-world experiences non-topics, is very clear in this recollection. It is equally clear that it still does not represent a stain on one's medical reputation to be found ignorant of a body of relevant knowledge, namely that of sexual assault and domestic violence. Furthermore, the professional conduct of admitting lack of knowledge, which is legitimate, combined with an explicit invalidation of the topic itself, seems contradictory. And finally, to shield normative professional advice behind admitted incompetence, is at best paradoxical. This is addressed in Runa's final remark without any reservation, and with convincing certainty. No other person can know when enough has been told, done and understood to heal the assault which will remain in a person's corporeality forever, just as every other life experience does.

6.4 BJARNE

Does this professional resistance against knowledge of an adequate response to sexual assault, expressed in normative rather than scientific statements, affect only women and become visible only in constellations involving male therapists and female patients? Several passages in the interview with Bjarne, regarding implicit messages about violation, address the male patient to male therapist relationship in a medical setting. Bjarne recalls impressions from encounters with several physicians, as in the following excerpt:

B: I got a job in Sweden for two years. I worked hard, it was fun, but this inner restlessness began to drain me again. I took Valium during these two years.

A: Your restlessness had still not received its name?

B: No. Well, anxiety, that was the name, *I had this anxiety, which I had to learn to live with, as I was told. But I couldn't reach behind it, behind this anxiety.* Then I returned to Norway, my mother died, my father got sick and recovered, I had to help since my brothers and sisters were scattered around the globe. After having returned, the turmoil went on. *I consulted doctors, got prescriptions for medications, walked around almost asleep, but I kept teaching. I was depressed, but I couldn't describe the core of it. I was asked very few questions - and new drugs were prescribed.*

A: It was always on this level, 'purely cosmetic,' so to speak?

B: Yes. Every time, purely cosmetic. I asked to be referred to a psychiatrist, which was done. *There, I experienced the same, less medication, but no real work.* He was interested in seeing me work in a group to have my aggression triggered because he thought I should have it exposed as I was such a ... I don't recall the term he used ... and I thought I probably would have exploded ...

A: But no question about where it might come from?

B: *Never any question. I touched upon it. And the same reaction as before, that that couldn't be so serious.*[254]

A: You brought it up, but it wasn't regarded as possible grounds for your rage.

B: Yes. *Again and again. I was astonished. Nowadays, I understand that they were without any knowledge about the topic.*

Bjarne had repeatedly offered an opening to discuss the experiences behind his restlessness, which was called anxiety, which he could not access by himself but needed someone to whom to tell it and with whom to explore it. No one was willing to follow him there, however, or to guide him into the emotions which fed the reactions he could neither cope with nor modulate. The following statements concerning his encounters with doctors are presented in the sequence in which they are mentioned in the interview:

> **B:** I started to search for contacts (with men) myself. I became more and more confused, this study of theology on the one side and the Christian faith and everything - and then this. I began to wander the streets of Oslo during the nights. I crossed the whole town, night after night, got too little sleep, my whole life disintegrating. But I didn't know ... I saw a doctor, got sleeping pills, I presented myself as being sleepless, I got the prescription, and nothing more came out of it.

> After this encounter with the young man - it was short - he left to go abroad - my life changed - I tried to accept that I was homosexual. I saw a psychiatrist because now things became uneasy ... he was a professor at the University clinic ... he gave me tranquilizers. I saw him every week.

> At a seminar, I watched a young woman fall down the stairs, swearing like a flash of lightening - in this anthroposophical milieu - I found that marvelous. I fell in love with her on the spot. Two months later we got married. (pause) I went to the professor and told him what had happened, and he spoke the following, wingéd words: 'Well, let us state then that you are cured.' I never saw him again.

> In this period I saw several doctors, among them Dr. R. 'You know, I'm homosexual,' I said. He replied: 'You are not homosexual, you are only frightened. I'll test a gas on you.' I fled. I thought I'd received enough pills and what have you. He scared me deeply. He wanted to try a gas, and then he suggested to try - I was there twice - to try - I don't remember the name, I don't think it was LSD - but something similar, but there was a boundary I wasn't willing to cross. I guess I was right. I had preserved a little of my health.

> I didn't think I'd succeeded at anything. Everything crashed, I got an anti-depressive drug and was given very high doses - I guess I've tried the whole spectrum of these medications.

> I worked at school, ate pills, drank a lot of beer. I understood this couldn't work. That I never would get out of it. I went to an American psychologist. But he left after a few appointments. From my doctor I'd gotten anti-depressant drugs. A had a large bottle. Then I phoned the SOS service of the church one night, I had to talk to someone, I had no one to talk to. I said my situation is so-and-so, I guess I can't stand this any longer. I got enraged - there was a violent rage inside me. Just at that time, there'd been a discussion in the newspaper about homosexuals, it was rather revolting. And then I met the same thing from the pastor I

> talked to. I just got furious and said: 'I don't want this any more, I'm finished, I don't want to take part any more.' 'One doesn't do such things,' he replied. 'Yes, I'll do it' I countered. Then I swallowed the whole bottle of pills - and don't remember anything more. (pause)

These statements, and the initial excerpt as their conclusion, covered a little more than thirty years of countless attempts to have a traumatic experience narrated. However Bjarne asked to be listened to, presenting himself as restless, sexually confused, sleepless, depressed, suicidal (twice), enraged or anxious, and, in which ever way he told about an unhappy life he could not master and conflicts with which he could not cope, the answers were drugs and norms. He had been prescribed the whole range of available tranquilizers and anti-depressant drugs. And he had been met with the norms of sexuality. When he revealed feeling attracted by men, he was judged as needing treatment, and when he told about his heterosexual marriage he was regarded cured. When receiving counseling about his conflicts from a pastor, he was condemned. When making public his ultimate refusal to go on, he was reprimanded. Whenever he addressed the topic of his childhood sexual trauma, however, it was ignored. Sexual trauma was non-information in contexts of counseling for life problems and sexual ambiguity, presented by a man who worked frenetically and drank large quantities of alcohol. Manic phases ended in exhaustion and depressions where the world lost all its colors. None of the therapists ever came to know this, which would have demanded that Bjarne told from the time the colors were unmade for him.

Not until Bjarne heard a voice familiar to him on a radio program, dealing with the topic which had concerned him since he was ten, could he speak about the violation and its impact. Not until he had spoken about it, could he reunite with those aspects of his physical being which he had abandoned in his childhood. The social dogma of sexual abuse as either nonexistent or at least not harmful, and the cultural norm of adult heterosexual relationships, were what guided the judgments and actions of the professionals he encountered, and by no means the value and gender neutral knowledge on which the professionals purported to base their advice.

There is still very little knowledge about the health consequences of sexual assault on boys during their childhood. Only a very few comprehensive and selected studies on boys have been conducted as yet. In an increasing number of studies, however, both sexes have been taken into consideration, although not primarily from the gender perspective of abuse embodiment, but rather in order to generate estimates of general occurrence and medical or mental health care service utilization.[255]

This is most likely due to the cultural patterns, gender roles and social presuppositions which have informed the theory of sexual abuse research and the methodology applied. This topic will be dealt with more extensively in chapter 3, linked to victimization and revictimization.

6.5 SYNNØVE AND ANNIKA

Were there other normative elements than "male" and "heterosexual" in the professional advise given to people who had disclosed sexual abuse to health care workers? In the interview with Synnøve, the topic emerges in connection to an admission to a psychologist, and it was thematized as in the following excerpt:

> **S:** I've seen doctors very often throughout the years because of pelvic problems, I've had considerable trouble there, and finally there was a doctor who suggested that my problems were psychological. Then I could loosen the bonds. Actually, I understood that was so, myself. He was the one who sent me to a psychologist, while others haven't been willing to talk about it, it seems to me. It had a lot to do with the pain during intercourse. When I went to the psychologist, what I really felt was that he too was trying all the time to run away from my problem. But I kept seeing him because I thought it just took some time to approach it, although I tried to be quite explicit. He insisted there had to be other origins to my problem. I stopped before the summer vacation, and I'd intended to continue after the holidays, but then I learned that he'd left, and I felt this as a betrayal because he hadn't mentioned it to me. I tried to write to him, I got his address, I wanted to ask him for a written report about our work together, but I haven't heard anything from him. I feel cheated. I felt a kind of sexual pressure seeing him, although it surely wasn't so, but I felt he wanted to get out of me a lot of things I didn't like to talk about, such as sexual fantasies and sexual dreams and the like, I thought that really wasn't any concern of his. I wanted to go into my everyday problems. I really think I received the best help after making contact with the Center.

Synnøve enters psychotherapy after experiencing the relief that, finally, she encounters a doctor whose judgment of her sexual problems and pelvic pain is that they have a psychological base, although he does not try to explore of what kind. She does not mention if he makes any suggestions or asks her questions about her own thoughts, or about her reflections evoked by his proposal. Apparently, Synnøve has not expected such reactions, however. She considers the decision to refer her to a psychologist as adequate, since the origin of her pain is attributed to her feelings.

The psychologist, as it turns out, represents a new problem for her which she obviously has not been prepared for. As previously experienced in her encounters with physicians, who avoided the topic of hidden abuse, the psychologist also "was trying all the time to run away" from what she presents and perceives of as her problem. Perhaps Synnøve is unaware of her psychologist's having certain ideas about the nature of her problem. She is referred to in her doctor's notes as, 'a female patient with chronic pelvic pain mainly accentuated by sexual intercourse and without any clear somatic pathology.' The possibility that her pain and adverse reactions to sexual approaches from her husband might be grounded in childhood sexual abuse, is not mentioned as it probably has not been considered by the referring physician. The psychologist is presented with the fact of the abuse by Synnøve herself, because she considers it relevant. In her view, it is the valid point of departure for her interactions with the psychologist. However, what she encounters is not the open-mindedness she had hoped for, but a familiar avoidance, albeit expressed differently. He wants her to make explicit her sexual fantasies and dreams.

From this particular focus of his attention, one may conclude his theoretical framework to be classical Freudian, and his model of analysis and intervention to be grounded in the Freudian theory. He is trying to place Synnøve's sexual and health problems in the frame of reference wherein a report of sexual abuse in childhood is a manifestation of a girl's overriding, yet suppressed, desire to mate with her father or his substitute. He does not accept Synnøve's reality, namely that she was desired by her father's substitute, and used, against her will, to fulfill his sexual desires. Synnøve's story seems irrelevant to him. Although she supposes that he needs time to approach the topic, attesting to her experience that most therapists are reluctant to address it, she persists in presenting her primary concern: to talk about how this abuse experience interferes with her everyday life. He, however, does not hear her voice. He hears one of the most central voices of his profession, the voice of a theory about female sexuality, developed a century ago and radically challenged and overruled since, but still valid in professional approaches to the analysis and understanding of female patients' sexual problems.

This makes a paradox appear. A woman dares, after decades of silence and for the first time, to thematize her sexual abuse experience. This happens out of her own awareness of the necessity of a disclosure in order to open for healing solutions. She feels confident that a psychologist will hear what she has dared not say previously, neither to other people nor other professionals. She trusts that psychological problems, though expressed in the body, can be entrusted a psychologist. But he seems to flee. He avoids the topic. He focuses on what he insists she must have: sexual fantasies and dreams. She focuses on what she has: a sexual violation experience. His insistence makes her feel violated. She suspects him of sexualizing their relationship, although she immediately underlines that she certainly must be mistaken. A theory denying the real experience of abuse by attributing to all women an inherent desire for taboo violation, affects a relationship where real abuse is the main concern. The initial perversion of reality into fantasy, formulated in a psychoanalytical theory and informing its practice, affects and offends a woman who speaks the reality. Based on theory, she remains unheard, and by theory she feels offended.

Yet another constellation of not being heard is to be found in the beginning of Annika's interview. Her report of previous diseases is recorded, and her presentation continues in the following excerpt:

A: ...I had urinary tract infections when I was in primary school, and I had problems with controlling my urine - and was admitted to the children's clinic for that. I remember the sort of examination where they insert an instrument through the urethra, that wasn't pleasant. (laughs a little, somewhat resignedly) At that time I'd already had intercourse. Later I wondered why they didn't see that. It was the same year my brother had started to abuse me. He was six years older, that means seventeen. He performed intercourse from the very beginning. I was also examined later, when I went to high school. But they didn't notice anything. I still had infections and the like...

a: ...and the abuse just continued ...

A: Uh-huh. I had some bad ovarian infections.

a: What happened then?

A: Well, I went to a doctor, I had such pain at school that I almost fainted, they called a doctor. I was examined, and given medication.

a: But no one asked how you might have gotten these infections?

A: No. I still lived at home. No one suspected anything.

From the age of eleven, Annika suddenly gets repeated infections in her urinary tract, becomes incontinent, and several times has an acute pelvic infection. She is examined and treated but is not asked about any possible origin of such unusual changes and clusters in a previously healthy girl. No one sees that she has been penetrated vaginally although she is examined several times in this area, or, if anyone has seen, he or she does not address that issue. It is not clear whether Annika would have dared to tell at that time. When she does tell, however, several years later and under heavy pressure from her partner, who has sensed something being wrong, the following occurs:

A: (after having made an appointment) I returned to the doctor and told him, sort of stuttering, what had happened - and he said, no, this was very (stress on the word) usual between siblings. Ohh. I really would have liked to kick him. (stress on the last words) It was so usual, and I'd over-dramatized something, that's how I interpreted what he said. It

was so usual, really. And I made it quite clear that my brother had had intercourse with me since I was eleven years. I told him that I refused to believe that this was so common among siblings. I didn't believe that ...
A few years later I was at the children's ward, with my son. There, I met the same doctor. He saw that in the boy's record there was a remark that the mother had mental problems and that she needed help. So he wondered what that meant. Then I replied that I'd been incested. 'My goodness, have you been exposed to that, you as well?' he exclaimed.

a: Was that quite new for him?

A: Yes, it seemed so.

a: What you had told him previously about your brother wasn't dangerous, but when you mentioned the word incest it became dangerous?

A: Yes. There were only a few years between those two occasions when I spoke to him. It hadn't become so different to mention it in the meantime. And I've imagined now and then going back to him and telling him what he really did to me the first time. There was so little left that I might have collapsed, it wasn't thanks to him that it didn't happen.

Once again a cultural presupposition about a norm is introduced into a professional relationship and informs a professional judgment. What older brothers are allowed to do with younger sisters, which has to be accepted and is not worth mentioning, informs a response to a desperate young woman whose marriage is at the brink due to relational and sexual problems. Annika cannot be mistaken about this message, although she initially thinks so and consequently makes quite explicit what her brother has done to her. The doctor undoes her confidence on the spot by invalidating her experience, its impact, and her quest for help. In another setting, when confronted with the same quest for help, but in writing, and formulated by a third person as a statement about her, the mother of a boy he is meant to examine: only then does he wonder. Annika now uses the term incest for what she has earlier described much more adequately. When the term is mentioned, he hears it, but obviously does not connect it with what he has been told previously; possibly, he is unaware that this term and the description she provided concern the same violation. Now, at least he hears. It remains uncertain, however, just what he understands. Annika has fantasized confronting him with his unethical conduct and unprofessional response. However, she has determined that would be fruitless.

6.6 EVA MARIA AND JESSICA

Eva Maria's way of telling about abuse was not to be explicit, but rather to present a covert message in the form of an excess of strain. Motherless from early age, she was sexually abused by her father from the age of seven to thirteen. She got pregnant in late adolescence, but before she had given birth to twins, their father had left her. The children were sent to a foster home, where one of them died. Later, she married a man who, she understood too late, was a violent alcoholic. She regained the right to have the remaining twin with her in her marriage. Fairly soon, she had another child. During the following eight years, she got pregnant seven times, and, each time, had a spontaneous abortion before the eighth gestation week. This rather unusual health problem is thematized as follows:

A: These miscarriages between your second and third delivery, do you see them in connection with what you suffered in your marriage?

E: Yes. That's for sure... (voice trailing off, she hesitates)

A: Violence and battering and forced intercourse - you couldn't keep the child?

E: No. (soft voice) It was a strange time, he was so violent, I had my job, and I had two children at home to care for, it was a lot, I was exhausted. It was too much...

A: I understand. This means: you had a job, two children and a violent, drunken husband. Did anyone know about this?

E: I suppose some people did. I mean, some must have known. But I said nothing.

A: The doctors you met every time you were admitted for a miscarriage...

E: I was admitted every time, both to our local hospital and twice to our regional hospital. But no one reacted. Later I got these pills. I'd mentioned that, after all these miscarriages, I didn't want any more. Then I got pills.

A: But never a question whether you were under stress?

E: No, but I've thought about that ... it was my fault, that I didn't say a thing. I thought I might several times, but I didn't, I couldn't make myself say it ... (pause)

A: What stopped you, do you think?

E: (hesitates, voice soft) I don't know exactly.

A: You didn't want to tell that you were married to a man like that?

E: No. (whispers) No. But you may say, there were certainly people who knew about it, we lived in a large house, one could hear when he was raging, there were neighbors phoning the police, without me saying a thing.

A: The neighbors knew and the police knew, but the doctors didn't.

E: No. They didn't get to know about it.

This is certainly not a narrative about telling about abuse. It is rather quite the opposite. It is a narrative about the collective silencing of abuse which is both heard and witnessed, but not reacted to. Eva Maria knew nothing else than to be objectified and offended, and she let it happen.

She tried to direct the attention of the childcare authorities to the abuse she was exposed to in her childhood. That was futile. Her voice was not heard. The abuse terminated when she moved out, which rendered her vulnerable in a new way. When meeting a friendly young man, she allies herself with him hastily, and they talk about marriage fairly soon. They consult a pastor about their wedding plans. Then Eva Marie gets pregnant and the young man flees from his responsibilities. Though, thanks to evidence given by the pastor, he is declared the father of the children and ordered to pay for their care. But he never helps Eva Maria; she has to cope alone in order to earn a living. The man she marries does not start to abuse her until their first child is born. Although she now understands his character, she feels obligated to stay because of the child. Thus, the violence and humiliation continue. She never speaks out, however, both due to earlier experiences of not being heard, and because of feeling shame about her husband. She hopes someone will intervene so she does not need to tell. This hope, however, proves to be in vain because of a high social tolerance of domestic violence as belonging to the private domain; her neighbors telephone the police but no action is taken. In the meantime she (her body) continues to reject the children he forces upon her.

Her inability to voice her interests becomes a health problem. The doctors she encounters treat her condition, but never question her general situation, her home condition, the amount of her responsibilities, or the presence of threat or violence in her life. Her state must have been rather critical several times, as she is admitted to a tertiary gynecological ward for what was considered to be habitual spontaneous abortions. Eva Maria does not mention this as an indication of her impaired somatic state, which it surely was, and neither does she mention injuries, which she certainly received. She is so effectively silenced that she does not tell until a chronic pain incapacitates her. After many years of frequent sick leaves and many changes of employer due to instability at her work, she receives a disability pension for her sickness. The final diagnosis is fibromyalgic pain. And still she does not tell, until she is challenged in a rather unusual way. This topic was thematized as follows:

A: When did you begin to tell about your experiences, when did they come out?

E: I went so far that I got problems with my legs and my hips. I couldn't walk any more. (strong accent on the last remark)

A: So. You simply stopped yourself.[256]

E: Yes. That was quite clear, yes. I had pain in my hips and was supposed to get help from a physiotherapist, I thought. For the pain in my hips and legs. The physiotherapist wrote this, you may read it, you understand this since you're a doctor.

* A record from a psychomotoric physiotherapist gives a detailed description of a pattern of muscular tension and defense in the upper parts of the body, strongly contrasted by lowered muscular tonus and impaired contact to the lower extremities, the pelvic area and legs, which seem 'deadened.' The therapist has addressed the topic of incest which the patient has confirmed.

A: That was it . . . she just asked you directly, and then it felt all right to tell her?

E: Yes. It was because she was a woman. I hadn't seen women before, only male doctors. I'd consulted physiotherapists before, but not about my legs, only my shoulders. And then they found tensions. They were traditional physiotherapists. Then I got massage and the like. But this was a psychomotoric therapist.

A: Psychomotoric therapists think . . .

E: . . . different, yes.[257] She grasped it by . . . she began to touch me . . . and then she said this was similar to so and so . . . and then I started to get going, we'd found a match . . . she was really good.

Eva Maria goes on in silent and silenced suffering until she actually cannot walk any more. Her ever more incapacitating pain is the focus of a variety of medical interventions, and she is sent to therapy, but only parts of her body are treated at a time. The treatments are in vain, the pain continues, and so does the abuse. As Eva Maria is unable to voice her suffering and abuse, she carries it in her body in an ever increasing burden of lived experience. Her corporeality is filled, quite literally, with inflicted pain, and she becomes almost deformed by it. The most basic functions fail, and her legs cannot carry her. Finally, someone regards her corporeally, touches her whole body, and begins to "read" the pattern of defensive and vigilant presence in the upper part of her, and the withdrawn and powerless absence in the lower part. The therapist, capable of sensing the muscular tonus and its differing messages, skilled at perceiving a Gestalt, and capable of telling what she perceives and thinks, helps Eva Maria out of the silenced room. This accords perfectly to Christine's claim that, "it must be possible to say what one thinks." By taking Eva Maria's corporeality seriously, the therapist can tell the connections by which Eva Maria lives.

In her description of how such an examination, based on the principles of psychomotoric therapy, is performed, I quote from a study by Eline Thornquist about encounters among patients and physiotherapists. She writes:

> Information elicited through verbal dialogue and clinical examination were woven together in different ways and viewed in relationship to one another. The therapist did not operate out of any kind of hierarchy in relation to information sources; verbal and bodily information were considered equally

> relevant and valid for understanding the patient's complaints. Thus, direct connections were established and there was continuity between the different elements of the examination, *and* between the patient's verbal and bodily messages and expressions. There was no talk of eliciting information by words, and *then* through the body in traditional terms, wherein the 'subjective' is linked to history-taking and the 'objective' is tied to the body and the clinical examination. The patient's body was not treated as more 'real' than her words. (Thornquist 1995)

The tension and pain, the powerlessness and lack of strength, were the lived impact of silenced violation. As they were spoken, the violation was addressed by another and to tell explicitly became legitimate, Eva Maria was able to do so. She was one of several who had "started to get going" (also Eva Maria used the ski term 'å komme på gli') due to the insight and directness combined with empathy and acceptance of a psychomotoric therapist. By interpreting the lived body and suggesting the appropriate context for the corporeal presentation, the therapist had opened the way for telling, releasing and understanding.

Finally, the interview with Jessica will be used to explore yet another kind of unheard message. The consequences of her abuse embodiment have affected her in a way that prepared an assaultive contact with the health care system in adulthood. She describes her abuse history as follows:

J: I experienced abuse from the age of seven or eight until I was sixteen. The abuser was my stepfather. Apart from that, I was raped by an adult male neighbor at the age of fourteen. There was also another man making approaches. That was almost the same year. For several years, I belonged to a group of young people, this means I approached young men and invited them to be sexual with me, and then I 'let them down' or ridiculed them. There is a lot of shame linked to memories from that time. I kept this up until I was eighteen ... I moved out of the house when I was sixteen, from then on I had to take care of myself. (voice lowering) But then I started with drugs and alcohol, in my teens, I started in the eighth grade, but it got worse when I got out of high school. I used marijuana and pills and things like that.

A: You mixed whatever you got?

J: Yes. And when I was drugged I had these affairs with these guys I mentioned, it never happened when I was sober. Never when I was sober. I had anorexia when I was eighteen, during my pregnancy. In that pregnancy I was so intoxicated. High blood pressure and protein in my urine.

Jessica's child is born premature and demands a great deal of care, which she cannot provide alone. She is dependent on the help of her parents, which forces her to meet her abuser regularly, to entrust her child to her mother's and stepfather's care, and to express gratitude for their help. She is confident that her stepfather will not touch the child improperly since he is a boy. Still, she remains dependent on someone she abhors. Jessica has a heavy crisis with a new anorectic period at twenty years old. A girlfriend and her mother intervene and get her functioning again. While attending a school and caring for her child alone, she meets a man, gets pregnant, has two miscarriages, and is left by her lover. Then she meets her future husband and her life seems to stabilize. She tells him everything, and he accepts her past, though unaware of the implicit impact. This is addressed in the following excerpt:

A: How was it to live together, I mean sexually?

J: It has become worse and worse during the years. After the last birth, it terminated. I can't manage any more. I can only tolerate intercourse when I'm almost drunk. But this makes me suffer. I should be able to tolerate him when I'm sober. I have to drink quite a lot to get over that threshold, so I can. (soft, worried voice)

A: Does this mean, among other things, that you're frightened you'll develop an alcohol problem?

J: Yes. I had one in my teens, you know. So I know what that means. (talks again in a very soft voice, mumbles, almost as if to herself) What is so difficult when I'm sober is my shame, I feel shame. I feel shame when I'm aroused.

A: Arousal and shame have gotten linked together for you earlier?[258]

J: Yes. (long pause)

A: Does this mean that your stepfather has made you ... that some of the abusive situations were such that you felt arousal?

J: He did it in such a way that it felt good. This is what I can't cope with anymore. ... When I'm sober and start to feel aroused, and when I really am drawn in, then (claps her hands sharply, making a 'cutting' sound) it turns and I withdraw. Little by little, I haven't dared to go into it. (long, long pause) I don't cope. It doesn't work. And the man who lives with me is gradually starting to loose patience ... (long pause) I've suggested we consult a psychologist. But he doesn't want to. I don't know what to do. (pause)

Jessica knows that her marriage is at risk because of her withdrawal from sexuality with a man she loves and who arouses her when she's intimate with him. She knows what happens and why. Her only option for not withdrawing from him is alcohol, which frightens her even more as she was on the brink of dependency in late adolescence. Her husband knows her history, but has no possibility of comprehending the link of shame and arousal he, unwittingly, becomes a part of. He cannot see his role in it, and can consequently not see the validity in Jessica's wish for them to attend therapy together. He is not the problem, as Jessica has to admit, but his influence on her in intimate situations is the trigger which she can only avoid by avoiding him. Her withdrawal started during her second pregnancy. She became aware of the problem as increasingly and potentially destructive for her relationship. Receiving pregnancy care from the physician who knew her from the time she was seventeen and who had always been correct and attentive, she decides to entrust to him her worries. Her description of the situation was as follows:

J: I began to tell him that I had increasing problems with my sexuality. He suggested I might quit using the pill for birth control after the delivery, because many women lost their libido when taking the pill. I myself knew it wasn't there. I said I believed it wasn't that, but that it might be a consequence of my stepfather's having abused me as a kid, I said. 'You have to try to cut it out, it isn't of any use to go back to the case,' he said. I went out, and since then I didn't mention it any more, you know? Now I'd shown him my cards - and I should just cut it out. That was that. I was disappointed. I was really, really disappointed when I left his office.

A: A complete rejection.

J: Yes. Complete.

A: Didn't he understand what you said?

J: I don't know - or he didn't grasp it.

A: He heard what you said, after all. He weighed what you said against the pill, and gave the pill the highest probability.

J: Yes. He understood what I said ...

A: But not what it meant ...?

J: I've never mentioned it since. If you're rejected in such a way *by someone I liked so much and had been going to for so many years ... he ought to know who I was.*

Jessica had trusted the long-standing familiarity of her doctor to create a secure frame within which to divulge her main problem. She wanted acceptance and help to disentangle a problematic link. Who should help if not the doctor who knew most of her life story? He did not want to know more than he knew, however, and thus he betrayed, rejected and silenced her in an act of undoing.

6.7 SUMMARY

I have explored interview excerpts which thematize a variety of unheard messages as a particular aspect of sexual abuse experience. The explorations demonstrate various ways in which health care professionals respond to information about childhood sexual abuse, with normatively rather than scientifically grounded argumentation and advice. The examples make evident that both the topic of the health impact of abuse, and the abused patient her/himself, is at risk of meeting rejection when disclosing sexual abuse experiences. Theoretical frameworks developed by Kaplan, Busfield, and Russell offer options for estimating the impact of cultural presuppositions, and of the social constructions of gender and disease on responses to patients who suffer from hidden violation.

7. REACTIVATED EXPERIENCES

7.1 BJARNE

Each interviewee talks about experiences concerning reactivation of memories from the time of abuse or the abusive acts. In Bjarne's biography, a reactivation of an experience is linked to a violent assault by a young man living in the same building and who had harassed him repeatedly before the attack due to the man's contempt for homosexuals. A comprehensive presentation, derived from Bjarne's interview and reinterview, is as follows:

> This work was brutally interrupted when I was assaulted by a young man living close by. He'd been verbally abusive toward me before, but I wasn't aware of his violent nature. One day, he stopped me in the entrance and attacked me with an iron bar, beating my arms which I'd raised to protect my head. *After that, my well known depression returned, and the pain in the arms incapacitated me for half a year.* I didn't even dare to be in my flat. Gradually, this has been overcome. But recently I ran into him, accidentally, in the market place in our town. He screamed obscene words at me. *And I escaped by shaping distance, as I'm used to doing in threatening situations,* as if putting him under a glass bowl. *This resulted in a heavy panic attack* a week later, the first for many years. I recognize the connection and I'm aware of the work which has to be done.

Bjarne is assaulted on the staircase of the house where he lives. Prior to this, the offender has abused him verbally, but Bjarne is unprepared for a violent physical attack aimed at doing a maximum of damage. He is unaware of representing a personified provocation for a young man who has tried to kill his abusive father, and who harbors a hatred toward homosexual men. The assault is serious. Bjarne's forearms bear large contusions, but no open wounds or verified fractures. He cannot use his arms for weeks. While recuperating at his brother's, he becomes depressed, anxious and suffers a heavy, lasting pain, although his arms gradually seemed to heal. He is completely incapacitated, as if paralyzed. This regressive state of disability lasts half a year. The pain, however, is still present a year after the assault.

While recalling this process, Bjarne is moved by strong emotions, as is mentioned in the interview transcript. He consents to the interviewer's interpretation of the event as reactivating his childhood assault memory, "almost violently," as he adds. The assault has repercussions on his overall conduct. He is on his guard and attentive, watching the surroundings whenever he is outside, prepared at all times for a new attack. His nightmares about the rape and the shriek have recurred, as he reports prior to the recollection of the assault. This coincidence makes clear beyond his doubt, where the threat and the unexpected pain have placed him mentally. His feelings, as he states, had totally overpowered him. After Bjarne's narrative of the post-assault year, the interview continues as follows:

> **A:** May I ask you: People like you who, early in life, have experienced the world as a dangerous place ... may it be that experiences later in life in a way - not similar in circumstances or details - that these break into ... a room of experiences, where the prior knowledge of the world as dangerous ... becomes reactivated ... then there might emerge a very strong reaction to what happens here and now. This reaction may seem quite unreasonable in relation to what's happening now, but you're in a way in a double room, the past and the present. Can you see this perspective?
>
> **B:** Yes, I can. I can understand this. I can describe it. Maybe I would have chosen different words for it, but what you say fits very well with what I perceive, because in the beginning I didn't even have words, only feelings. First now, I can begin ... having gained a little distance ... now I can begin ... *I recognized it, after all, didn't I, this good, old, familiar figure which has reemerged all through my life, it relates to everything that's happened all the way ahead ...*

Indeed, Bjarne can share a perspective on the recent trauma to reactivate the prior. He can confirm by his own perception that *when an old trauma experience is renewed, the new trauma is aggravated.* The simultaneous presence, invisible to everyone other than the one having previously embodied the trauma, can render the present trauma impact incomprehensibly and disproportionately huge. Only from a perspective of a corporeality composed of former traumatic temporality, spatiality and relationality, may the present experience be evaluated as to its impact. Such a co-presence of past and present offense, however, is not visible in the medical documentation of present damage, leaving a problematic discrepancy between objective "findings" and subjective "feelings." The medical, instrumental gaze cannot see what Bjarne ironically calls, "the good, old, familiar figure," in his corporeality, which had accompanied him throughout his life, and which had reemerged under particular conditions. He is aware of this connection to such an extent that he steers the talk in the interview back to the basement room in China. First here, after having talked about this reactivation, does his most complete description of the childhood rape experience emerge.

"The familiar figure" deserves exploration. The Norwegian word *'figur'* which Bjarne uses, may mean quite different things in different contexts. From a dictionary, one learns that it may designate a body figure as outer shape, as a form. It may mean a character, a role in a piece of literature or drama, or a social role as, for example, a position. It can designate a shape of any form, an elaborated line, a pattern, a prescribed path as in ice-skating, a contour, a silhouette, a proportionality, a drawing, a geometrical scheme. It may signify a group of notes in a piece of music, a change of sound in a sequence of chords, a defined passage in a dance, a movement, a contrast, a composition, or a grouping as in the flight of birds or words in a certain order, a turn of the body as in gymnastics, a human type as the clown or the one who is depressed or the one who is stubborn, a constellation in a change of light, a rhythm in writing, moving, swimming, flying...

Can a word which denotes so much and such different things, which means almost anything or nearly everything, be useful for any relevant purpose in the context of trauma embodiment and reactivation? Yes. Indeed. Or rather: precisely. The word takes its meaning from the context more than many other words do. What can be a figure depends on the perceiver. A figure and its significance is shaped in-between. It is perceived by recognizing, and is created by living, being or acting. The figure as such is not constituted as a figure in itself; its existence as a figure depends on the perceiver who recognizes it.

What is *'figur'* in Bjarne's context? Concretely, it is a male body of literally or symbolically overwhelming size or power, the configuration of a danger. Bjarne looks out for it wherever he goes after the assault. In a way, it is present in his eyes as a latent and immediately activated "seeing," and with a high degree of specificity. This echoes Synnøve, her eyes searching for a name in the obituaries and for a male figure in the downtown streets. It echoes Beate's seeing her father's way of handling knives mirrored in her son's hands. It echoes Veronika's perception of "the movement" in the shower curtain. Annika hears the "sound" of her brother's sexual arousal in her husband's breathing. Pia is cast back to the past by a movement of her partner's mouth which recalls her stepfather's "tongue" during his sexual arousal. Any "water flowing over [any] mouth" brings Judith back into her past. Berit cannot cope with the "too big" head of her child being born. Line's abuse reexperiences, too, are configured in a highly specific constellation.

Can there be as many figures or different patterns as there are abuse experiences? The patterns or figures inscribed in the senses during abuse, can they even be reactivated by entirely different situations, that is, ones which are emotionally positive, ostensibly non-traumatic, personally sought or desired?

7.2 KARLA

Karla has been abused by at least eight persons, among these her father, an uncle, and friends of this uncle. The systematic abuse began when she was between six and seven years. There was, however, a situation before this, when Karla was four years old, which she only partly recognizes at the time of the interview. Its recall has been reactivated by the sudden emergence, during pregnancy, of an allergy to eggs. She has never been allergic before, but suddenly she reacts with a florid eczema whenever she eats eggs, and the smell of eggs evokes attacks of nausea. During one such attack, she catches sight of herself as a little girl in a strange situation. She describes it as follows:

> I don't know whether this means anything to you, but there's something which has made me understand that there's some connection to a hen house I ran into because someone was running behind me. Yes, there is. And the head doctor at the psychiatric ward (where I was admitted several times) has tried to make me turn round in my head to see who comes from behind. I manage to remember everything else ... as... that I run into it, there are several behind me, I stop, *but then it turns into a kind of sauce,* I can't see what happened. I don't manage to - *I really don't dare to turn round to see who's behind me.* I can see the striped shadows the sun makes falling into the room, and the shadows of those who come in behind me, stopping behind my back ... I get so far. I know the others come in after me. But I don't manage to stand calmly and turn around, *but I know that I did it.* I see myself my daughter's size as she is now, I must have been four. I don't dare go deeper into it. But one day I'll be strong enough to dare to turn round. Of course, this is something I can work with. Apart from this, I have to be aware that there are reasons why I'm not able to yet. I want to know who it is, after all, but I don't manage it. It's quite clear to me that there's a psychic barrier. (excerpt from pg. 2)

This is a narrative echoing Bjarne's and Elisabeth's, and as in the case of Elisabeth, one that emerges during a state of consciousness which, judged from outer appearances, is named psychosis. There is a part of a "film" in her memory which is in need of a context, but there is no one who might be willing to identify the place where it belongs. The picture of the hen house emerges first. As is addressed in many of the interviews, the state of pregnancy may be the trigger for a retrieval of childhood abuse memories. It is during pregnancy that there come: a nausea elicited by the smell of eggs, a skin reaction from eating food containing eggs, and the memory of herself at four years old, running into a hen house where she feels caught. She does not succeed in turning around to face her offender(s). She recalls details, the striped shadows made by the sun and the people, but Karla is unable to turn around, although "she did it," back then, as she says. There is a blending of consciousness levels in the description of abuse situations, wherein different fragments of a situational perceptions are placed on different levels of awareness, some emerging quite clearly and others being rendered "sauce-like," as Karla says, or "foggy," as Berit and others express it. Although it feels distressing not to know what is so important, namely literally to face in order to know it anew, Karla accepts that she is not ready for this particular recognition.

Karla's sickness and abuse history is extreme, as is her tendency to be accident prone. She was in contact with all kinds of health care workers and social workers until late adolescence; she has been anorectic and amenorrhoeic, an abuser of drugs from the age of nine, and in need of special educational assistance at school. All kinds of health problems escalate during her pregnancy and she is admitted to the hospital twice since she can feel no fetal movement. In late pregnancy, due the suspicion of ulcus ventriculi because of abdominal pain, she is scheduled for a gastroscopy. She tells the following about this examination:

> I was admitted to have a lot of tests, among others, a gastroscopy. And I thought, all right, this is okay for me. After all, I've had a tube in my stomach before, when I was intoxicated. But I'd never had this strange test. I was put in this examining room. The head doctor of this gastro-ward who was going to perform the examination came in. *He just starts, but I hadn't settled yet. I was pretty tense. I didn't know what was going to happen.* I imagined a lot of preparations would have to be done first. But he went right ahead. I thought "OH-oh,"[259] okay, to open my mouth is what I can do, but the very moment he came that far that I was supposed to swallow to get this tube down my throat, *when he was so far down, I panicked.* I drew the tube out in a hurry, and then I cursed and yelled at him. I didn't know then that he was the leader of the whole stomach-department there, but it wouldn't have mattered. I was completely pissed off. (almost shouting) 'You,' I shouted, 'now you have to listen. *If I'm going to get this into my stomach I have to be allowed to push it down my throat myself.'* He got very angry, said something like this, that that was just impossible, he was just forced to do this examination, we had to, because I might have a serious disease. 'Well,' I said, 'you can get it done, but I have to push myself. When I let it go, you can continue further down. *But the part where it goes through my throat I have to push it myself.'* He had to give in, in order to get me examined. *And just the fact that I was now in control myself, made it so I could swallow it without any problems.* I just relaxed and concentrated on my breath. That was why I managed it. I just breathed and thought of nothing and fixed on the nurse with my eyes. (excerpt from pg.1)

Karla has not been given any information and is not mentally prepared when the doctor begins the examination procedure. However, she expects a phase of technical preparations by the staff, and a kind of presentation of the steps of the procedure and her role in it. She thinks she will receive some sort of instructions as to how cooperate so she can be "ready" for it. She is not given the possibility to learn what shall be done, however. She is expected to adapt and submit without her prior and explicit consent. Karla reflects briefly, during the very first phase of the procedure when she recognizes that she must be mistaken. When she is just simply told to open her mouth, she obeys in silence, all the while perceiving an "alarm" getting started. The announcement of threat is given from herself to herself, and then explodes into an eruptive and desperate physical and verbal reaction the very moment she feels the pressure of the tube far down in her throat and is asked to swallow it. Karla's fully blown panic is, within the space of a second, transformed into refusal, rage and protest. She is so desperate that she curses and scolds. (So vivid is her recall of this situation during the interview that she almost shouts out.) She seems in a way to be impressed by, or actually, very content with, her reaction, expressed in a change of level in her narrative, which introduces a complex meta-comment about roles and emotions. The doctor whom she refuses to obey and offends verbally is the head of the department, which means, that she is actually refusing to obey the person, the man, of highest rank and authority in this clinic. She admits to having been unaware of this, but stresses that she would not have reacted differently had she known. With this statement Karla may be emphasizing how intense her rage was, and, at the same time, expressing that she had no time to reflect upon who it was she opposed. She just did it. She formulates her conditions for cooperation, and in a way which makes her demand an ultimatum. She maintains her position although the physician presses her, scolding her for her impossible conduct, her unacceptable refusal, and her irresponsibility considering the possibility of her having a serious disease. She dictates how the examination has to be performed if he insists on doing it. She guides the tube, she pushes it through her throat, she lets it go when she feels ready to - and then, voluntarily, leaves the rest to him. She succeeds, both in making the physician obey her instructions and in getting the tube into herself by regarding her breath, disciplining her thoughts and fixing her eyes on a woman besides her.

According to Karla's description, wording and intensity, the invasion of her mouth in the context of a medical examination, although deemed necessary despite her advanced pregnancy, is too similar to something else. Lack of mental preparation, and an abrupt approach to her body without even regarding her as personally present, creates immediate insecurity, expressed in high attention:

> *He just starts, but I hadn't settled yet. I was pretty tense. I didn't know what was going to happen.*

A warning emerged right before a sensory perception became a reactivation:

> *when he was so far down, I panicked.*

The panic emerging attests to a degree of similarity which is unacceptable and which provokes Karla's defenses. She can only accept the procedure provided there is a crucial difference from the first attempt:

> *If I'm going to get this into my stomach I have to be allowed to push it down my throat myself.*

The moment her demand is accepted, the necessary procedure can be performed since it now is perceived of as a voluntary and cooperative insertion of a medical instrument, and not as an invasion of the mouth:

> *And just the fact that I was now in control myself, made it so I could swallow it without any problems*

Karla, by demanding to be respected and having that wish prevail, breaks the similarity which in her perception has made the procedure something other than it appears to the medical staff. She guides the tube in the rhythm of her regular breath and she swallows what she knows is an instrument for a medical purpose.

> *That was why I managed it.*

This narrative suggests a forced oral abuse experience, in accordance with the explorations in some of the previous chapters, and attests to the abuse embodiment of an alienated mouth. Is this reading justified by any remark about oral abuse in Karla's interview? The topic emerges on page six as a brief comment in the context of consultations due to secondary amenorrhoea:

> It began early in my teens. I went to our family doctor, the doctor my mother consulted. He reacted to my very rough voice. He also noticed that I was kind of boyish in my body, kind of boyish in my use of my body. And he reacted to my beard. It doesn't show so much any more, I've removed it. I pluck it out, like this, instead of shaving my face, that's so irritating. So I pluck it out. 'Isn't that painful?' I've been asked. But I don't feel it. *I have no sensation around my mouth. But that's perhaps due to all this sucking that the feeling has disappeared, and that I've switched off the sensitivity around my mouth, I suppose.*

The reading of Karla's active transformation of a threatening, because too similar, situation by means of instructing and controlling, is supported by her descriptions of other situations in a health care context, addressed when talking about an ultra-sound examination at her gynecologist's:

> I'd never had an ultra-sound exam performed before. There was a lot of equipment around which I didn't know. 'Hell,' I thought, 'what does this mean?' 'Where should I lie down?' I asked. 'You lie there. Then we shall do this ultra-sound. You'll get a kind of stuff on your stomach, I call it stuff, it's a jelly I have to use to be able to move this sound-sender around. It's

> not dangerous, only sticky and cold.' 'Well, yes, okay then,' I said. 'Are you ready?' he asked. He's learned this, that I have to breathe, that I have to feel calm, and when I say: 'now it's all right,' then he can do whatever he wants. Then he can do whatever he wants, really. I'm away from it in a certain sense. I'm there and feel what he does, but I'm not quite there. How to explain this in another way, I don't know.

Karla has taught her gynecologist what he has to do to enable her to tolerate what he does. He knows she needs to prepare and to know. He respects her demand for information, and he provides it as a part of the procedure. Whatever she can understand the purpose of, she can tolerate, given the time she needs to concentrate upon her breathing. The function of breathing in a calm and regular rhythm is a precondition for relaxation, which again is a prerequisite for Karla's entrusting her body to others' hands. It may also have been how she coped with oral sexual violation; now however, she uses the strategy for mastery. Only when she is "ready" does she know the situation to be appropriate, which means under her control and on her premises - and, as such, *made not dangerous.* Her actively detaching the current situation from whatever might evoke the perception of similarity to being threatened and overpowered, results in her calm cooperation with her sensibility preserved. She can allow herself to perceive in a state of attentive control of her breath and her bodily tension. This is literally "being there" and "not being there" simultaneously. She need not dissociate. Therefore, even if she distances herself to a degree, she still feels, hears, understands and remembers what is done, - as the text following the quotation above makes obvious.[260]

7.3 ANNIKA

At sixteen years of age, Annika experiences a double reactivation of abuse experience, during a medically indicated physiotherapy session after the third surgical intervention on her left elbow, which she addressed as follows:

> He was the chief therapist, he was instructed to treat me. *I really experienced an abusive situation.* It surely wasn't like that - but for me it felt like. I was sixteen, lying on a flat bench. The therapist was a tall, adult man, he must have weighed at least ninety kilograms, and he placed himself on my arm, he broke it into pieces, at least it felt like that, *and I was lying there on the bench and screaming it hurt, but no, I had to bear it.* I had all these pins and screws inside. Then I had to have an x-ray taken, and they saw a dislocation they wondered about the cause of. This was the last time I went to physiotherapy, it was much too tough. (pause) Since then, I've had a whiplash injury, I was involved in a car accident, and then I needed treatment for that. *I really don't know what happened,* the therapist was rather nice, I mean he was friendly and talked a lot about a lot, but I got exhausted by it. *He surely meant to*

distract me, but the more he talked the more tense I got. For me, it would have been better if he had just done his job and shut up. (pause) Everything happens to me. Operations and whiplashes and accidents and the like.

* She tells about several situations where *a man has held her and she's felt deathly afraid and simultaneously completely weak, unable to move.* What she neither can cope with nor tolerate in these situations, e.g. during physiotherapy, is to be held, immobilized. *She acknowledges this link to the abuse.*

Annika has an accident at four years old, causing a fracture in her left elbow. The fracture dislocates. Corrective surgery is performed when she is twelve, resulting in chronic pain. This pain becomes the indication for the third operation, when she is sixteen, an attempt to fix the joint in a more correct position. Afterwards, she is told to begin to mobilize the joint, but is injured anew. In Annika's perception, which probably corresponds with what happens, the relocated joint is dislocated again by the therapist's bending it. He not only ignores the screams of the recently operated girl, but tells her that she has to endure the treatment.

I really experienced an abusive situation.

I was lying ... on the bench and screaming it hurt, but no, I had to bear it.

Although she gives space for an "objective view" to contradict her own, as in previous passages, she insists that she perceived the treatment as abusive. As a consequence, she later avoids physiotherapy at almost all cost; this may in fact have caused her more pain and functional reduction in her left arm than necessary. However, when injured once again - indicating a proneness to injury which she herself thematizes - she has to accept physiotherapeutic treatment. This treatment is certainly quite counterproductive, or even anti-therapeutic.

He surely meant to distract me, but the more he talked the more tense I got.

It almost seems as if Annika feels alarmed by the talking of the man manipulating her neck, which, of course, means to be held. She dislikes being intentionally distracted as she must direct her awareness to what is being done, not to what is being talked about. A physiotherapeutic setting carries the threat of being held, and the danger of being hurt.

This being held by a man, however, and the corporeal reactions it evokes, her weakness and inability to move, does also occur in other situations quite different from these. Annika knows there is a connection to her abuse experience. Just why is Annika unable to move if held by a man? The background of this connection is thematized on the last page of the interview text. Annika reports how much attention she pays to other people's breath, mainly at night and in the same room. Thus, her husband's breath disturbs her sleep and, during intercourse, reactivates abuse memories. After this, the speech turns occur as follows ('A' means Annika, 'a' means the interviewer):

a: Your brother did quite a variety of sexual things to you?

A: Yes. It sometimes happened that I was awakened by his licking me. Worst of all was that then I felt arousal. I got furious with myself and felt guilty, and ever since, it's been like that: whenever I feel aroused, I find myself disgusting, I can't perceive it as good, I feel shameful when the arousal comes.

a: Arousal and shame are one.

A: Yes. He did what he pleased, I didn't dare oppose him, *because I was deathly afraid that my sister would wake up, I slept in the same room as my little sister.* I thought I had to protect her. I did whatever necessary to prevent him going to her. I was the eldest, after all. I had to prevent it. It took me many, many years before I could talk to her about it. Because I felt not totally certain that he hadn't abused her as well. But now I feel confident that he hasn't. (pause) Quite incomprehensible that she never woke up, so often as he came, and with everything that he did. But it feels good to know for certain that not she too ... that it was only me.

Once again, the emotional synonyms of arousal and shame or guilt resulting from the linkage of sexual arousal during sexual abuse are addressed. In Annika's particular abuse embodiment, however, there exists yet another link: to be held by her brother meant to be forced into silent passivity due to her sleeping little sister. In her corporeality, this link expresses itself as, "feeling too weak to move away," when held by a man, regardless the setting. Annika, self-defined as responsible for keeping her brother away from her little sister, endures her violation submissively. Feeling responsible is still one of her most consistent traits, as are keeping quiet, doubting her own judgment, and subordinating her needs to those of others.

7.4 HANNA

Hanna's interview provides a very complex reactivation history which must be elaborated step by step through the complete text. In the first passage of the interview text, transcribed in summary, Hanna mentions pain in her legs during childhood, and certain memories concerning intestinal worms. Hanna recalls a situation where she, as a little girl, is lying on her stomach across her mother's lap, her pants taken off and her mother examining her anus for worms. Hanna is highly uncomfortable because she feels her uncle's eyes on her from somewhere, as he often is to be found peeping into the windows of neighboring houses. She is uncertain whether he is there or not.

A: Do you visualize the scene? Do you see someone near you?

H: I've asked whether my uncle might have come into the room on such an occasion. My mother denies this, but I have a feeling . . . you see, the doors in a farmhouse are never locked, at least we children couldn't imagine a door being locked. But whether he came in or not, I definitely have the feeling of not being safe there which is concrete. Whether it happened or not, what I do know for sure was that I wasn't safe in such situations, someone could come in and see what felt so humiliating for me. But what could be done? It was just this way things like that were handled during my childhood.

A: Well. I immediately think that you had already experienced that you weren't safe, or that someone went around and peeped, or that someone's eyes were at improper places. I wonder if children don't have a well developed sense of eyes directed at improper places.

H: There's no need to guess concerning this uncle, because he's known as a strange person who peeps. It's not only me who's been bothered by him being allowed to trespass freely, you might say. One of the problems with living in the countryside is that so much is accepted; people showing deviant behavior are more easily tolerated than they would be in town; there it would have been different; that I'm convinced of.

Hanna's uncle, an incompetent according to public opinion, is known for his indecent habits. She connects a strong discomfort to his gaze. She answers affirmatively when the interviewer questions whether she personally experienced his gaze as disrespectful, as undressing. Almost five pages later, after having reported the video setting and her psychologist's reaction, Hanna says:

H: I have to add something here. Yesterday, I stood in front of my parents' house with my mother. I stood there with my mother in the dark and examined my sister's car, she'd had an accident. Suddenly I felt - I feel two points inside here (turns around and points at her lower back on both sides) - that my uncle is close. It's just incredible. There are two points I can localize precisely. There I felt it. He was near by. He shouldn't have been because he's been told not to show up near the house. He hasn't done so the last month. I haven't seen him. But yesterday I felt: there he is. And there he came, with his laundry. He doesn't do his laundry himself, my mother does it for him. He comes and brings it once a week. So he came legally, so to speak. But he'd moved silently close to us, and I felt it in the dark. There must have been fifty to sixty meters distance when I sensed he was near. Perhaps this is what you've addressed with your previous question, I sense him, I've known these eyes before, there's something which clings.

A: Yes. It seems as if you've integrated his gaze bodily. As if you've stored memories that he's watched you from hidden places, that his eyes have been behind you.

Was the "imprint" of eyes in a certain area of Hanna's back a way toward an understanding of a corporeal reaction of discomfort which is felt before its source is registered? The points Hanna has marked as the field of reaction when her uncle is around, are addressed once more on page thirteen when she talks about her legs, which have been painful since pre-school age. Hanna says:

> I have a problem with my legs. At night, my legs are usually very unsteady. When I wake up at night, I find myself in fetus position. My therapist says that it looks as if the muscles in my legs are in tension all the time, there's a strong tension deep, deep inside, it results in me walking with my knees bent together, and it's very painful for me to really straighten my legs completely. That's correct, I understand that, I feel constant tension in my legs which is deep inside, and this is something which I've been asking about and searching to get treatment for, God knows how many places, and for years, because I've recognized that there's a connection between my legs always being painful, and cold. My legs and my lower behind are always cold; everything is cold. To be able to achieve a different condition, also bodily, is now one of the most important aims in my life. It's uncomfortable to feel like this, never really to calm down, never really to relax; just imagine falling asleep and feeling refreshed when you get up. It's really fascinating, what I've always known: there's something inside, and that it's a nightmare which plagues me. And it's quite concretely in my muscles.

When challenged to reflect upon whether this might express some kind of defense, in the sense of her legs not really belonging to her, she says:

> That my legs are not mine, you mean? Yes, yes, completely correct. When you say that, I realize that I've used such words, when I've gone to see a healer and God knows who else, I've told them then that there's a long distance to my legs, there are miles between me and my legs. It feels like that, from my back and all the way down. I can show you. (turns around and indicates a line) [261]

In the next passage she addresses the forces she feels hold her down, and which she has begun to recognize as her uncle's hands holding her down on his bed, although she still has no memory as to what is done to her on this bed in these situations. She feels an enormous "drain" on her physical and mental strength and energy. She feels counteracted by these forces she cannot properly name as yet. Hanna has understood that the pain in her legs, caused by the tension of self-defense and vigilance, drains her of the strength she needs for other purposes and which she misses when engaging in her personal dream: becoming a professional painter. This is addressed as follows:

> Twice I've made myself an atelier, with a big investment and a lot of work, but I didn't use either of them ...I work like a hero in the initial phase, I get ready - then nothing happens. This causes me great pain, I've invested so much, in so many senses. I've a tremendous desire to express myself; so much is not accomplished. And the years just pass by. I get so easily disturbed, I can't calm down, it doesn't turn out well - I have a real grief there.

* We talk about the grief and the strength she must invest to cope with it. We talk about her ambivalence concerning being "seen" because she links

so many shameful and humiliating situations to knowing "herself seen by others." To be seen and to feel shameful are closely tied together. She tells that, right before her first exhibition was meant to open, she gathered all the paintings and placed them under her bed, while she sat on the bed disapproving of them. She grasped that that was no "sane," healthy way to behave. She could by no means do any thing about it, however. We talk about the meaning which the words "sane" and "insane" carry when people's reactions are judged, and that it might help to ask for the intention behind or the meaning of an act in order to glimpse its purpose, no matter how irrational the act might seem.

* She addresses some problematic tendencies of hers related to paintings. During the latest years, she has succeeded in convincing herself that she really does not like paintings, that to paint is foolish, and that she really does not want to. She has transformed her longing for painting and creating, and her horror of exposing herself shamefully to the gaze of others, into: "I really dislike paintings." But she has many dreams about painting and paintings, and these dreams are filled with sadness.
* Finally she formulates her utmost aim: to let herself be seen, unashamed.

The path of a particular discomfort, evoked by another's gaze, leads from her uncle's eyes upon her, which equates with shame, to others' eyes on Hanna's paintings, which reactivates this shame. To expose her art means to expose herself indecently, improperly, shamefully, and dangerously. The necessary strength for creating and exposing is not available due to a consuming defense against a force which is identified only in part. The riddle continues to be more complicated than Hanna can cope with. There are pictures seeking a proper place, there is a distracting uncertainty, and there is an obligation to be grateful which makes rage and distance illegitimate. But the embodied reactivating, abusive, shaming, and limiting gaze is identified.

7.5 FANNY

Quite a different reactivation of abuse experiences emerges in the long interview with Fanny. She was abused by an uncle from the age of five or six years, and by an adult male neighbor during the same period. The continuous abuse by her uncle terminated when she was twelve years old due to external circumstances. He raped her, however, when she was nineteen while she was involved in a sexual relationship with a young man. When she recognized that she was pregnant, she could not be sure whether the pregnancy was a result of the rape or of her love affair, but she aborted spontaneously. Some years later, she had an imaginary pregnancy which created a great deal of confusion. Then, she was involved in a traffic accident and had a whiplash injury. Other health problems she mentioned were an allergy of increasing intensity from early childhood, a chronic urinary tract infection, and a long period with a hemiparesis in the right side of her body. She had been admitted once to a psychiatric ward. Fanny reports:

F: I was admitted to a psychiatric ward after my third delivery. It was because after this delivery I reexperienced something of what had happened before. I had pushed it away as long as I could, I haven't been willing to admit this to myself. I didn't believe it had affected me so heavily. I knew about it, in one way, but I tried to convince myself that it wasn't so serious to me. This is how I think. *But I regard it as very important to talk about, as right now to you. Because I believe it may help others. And this is so very important because of what happened to my daughter.* She's eleven years old right now.

A: Do you want to talk about her now?

F: Yes, I might as well. She's a very good girl, we've been talking together, the same way you and I have been, you might say. I don't know how to say it, but she doesn't get adequate help.

A: What has happened to her?

F: *She's been abused by a neighbor boy. Then she was four and a half. She's told me pieces of this earlier, but little by little more has come out. I've been blamed in the town because I've made it public. The abuser has done whatever possible to break us down.* The situation is this: a divorced woman with her two children and her brother, at that time nineteen years old, moved in next door to us. It was this nineteen year old boy. I should have thought so, but I didn't. They played a lot together, these three kids. (Fanny's daughter and the two neighbor children) It's occurred to me - when I think about it now. They were in love, she said. She was so very proud of being in love with a nineteen-year old. I have a son who was as old as this young man was then. And my daughter had a very good relationship to her older brother. It didn't hit me, in a way, that a nineteen-year old could do something like this...

A: You mean that another nineteen-years old could abuse your daughter?

F: Yes. They had a strong relation. They were in love. I laughed a bit at this, I found it sweet.

A: It was this she talked about? Did you mean this with the pieces?

F: Yes, I don't believe it was like this from the start. He surely invested energy establishing real trust between them. And then she was meant to spend a night together with these kids. They often slept over at each other's house. Once, my neighbor called me and said my daughter wasn't well, that she had a fever and wept, and that she felt pain in her genital area. She (the neighbor woman) wasn't allowed to take a look. Then it appeared, in pieces. She told a little more, and a little more. But inside of me a bell rang at once; immediately I understood what had happened. When I asked my neighbor about it she got angry at me. Later, they understood that it was serious because the girl got an asthmatic attack, she couldn't breathe, she was hysterical, so she had to be admitted to the hospital. I couldn't tell about it the same evening, but some days later we went to the allergy section and there I talked to a nurse. And then I talked to the allergy doctor. He only replied: 'But what are you going to do about this?' She had to be examined whether it was true or not. Then he sent for the hospital social worker. We had a discussion. The social worker wondered what should be done. It ended in nothing being done right away. Then we were referred to the child and adolescent psychiatric ward. I felt they understood what was at risk. They did. We talked together. *But we had such huge problems, my husband and I, that we felt like wrecks. It turned out to become enormous.* But I didn't report him, because he was a neighbor, and I thought perhaps it wasn't so serious. Gradually, we understood more and more. Then I confronted him with what I knew. *And then he reported me to the police. I was reported to the police, accused of having insulted his honor.* His sister and he stood together in that I'd attacked him harshly, that I was an intolerable neighbor, lots of dirty linen washing, that I didn't take care of my children. *The case went to court - and he went free.*

A: There was actually a case against you?

F: Yes. I went to a lawyer. *It's been horrible. These years, everything that followed, it's drained me. This, this has drained me more than my own case...*

Fanny was the third of the women selected among those volunteering for an interview who was known also to have an abused child. However, several others of the interviewees did reveal sexual abuse in multiple generations, as Fredric and Tanja, among others. The interviewer was, however, not aware that it was the daughter's abuse, and the very stressful circumstances of the disclosure, which was Fanny's main motivation for talking to me about sexual abuse. She wanted to talk on behalf of her daughter, and only indirectly on behalf of herself. She made this a topic when leading from the initial report of her own sickness history to the inadequate help her daughter received:

But I regard it as very important to talk about, as right now to you. Because I believe it may help others. And this is so very important because of that which happened to my daughter.

Thus, when asked whether she wants to start with telling about her daughter, she immediately does so. After a few sentences, however, it becomes obvious, that the daughter's and Fanny's abuse experiences were densely interwoven, the narrative about this web starting with:

She's been abused by a neighbor boy. Then she was four and a half. She's told me pieces of this earlier, but little by little more has come out.

As a result of a close neighborliness with a somewhat unconventional family, Fanny's daughter is exposed to sexual abuse which she initially reports to her mother "in small pieces." Little by little, these pieces make a pattern, which suddenly finds its name. When her daughter falls sick one night while staying with the neighbor's children, Fanny knows what has happened. One may wonder how a woman, herself a victim of sexual abuse, does not prevent her daughter from being molested. Fanny blames herself for not having imagined the possibility, but she can only admit that she failed to do so. Or rather, the abusive adults she had encountered rendered a boy the age of her own son a male not representing any danger. She simply did not think of this young man as a man, as a potential abuser. Her frame of reference did not include a general perspective of the gendered social world, where females are endangered by the presence of males due to structural conditions. She only saw the parallels in the relationship between her daughter and her daughter's beloved big brother, and her daughter and her daughter's beloved big friend (at the age of the brother and more a child than a man in the girl's mother's eyes). Therefore, she was not suspicious, but instead, inattentive, to "the small pieces" the little girl tried to mediate about what this young man did to her when she was in his home.

The disclosure causes problems, partly affecting Fanny, but gradually involving the entire family, expressed in adverse reactions from the local society and from the abuser.

> *I've been blamed in the town because I've made it public. The abuser has done whatever possible to break us down.*
>
> *But we had such huge problems, my husband and I, that we felt like wrecks. It turned out to become enormous.*

Fanny confides in the health care staff during her daughter's hospitalization. She does not, however, receive much support. The physician turns her question back on her, thus communicating that he is not "in charge" of such a case. He mentions an examination of the girl as being necessary for further intervention, but he neither initiates it nor consults an expert about the measures for securing such documentation. Consequently, nothing is done in the actual situation directly related to the girl's sickness after having spent a night in the neighbor's house. The social worker is empathic, but defensive, as is the child and adolescent psychiatry staff. The topic of a legal proceeding is not addressed by them. This professional defensiveness indirectly heightens Fanny's insecurity, or her reluctance to report the young man to the police. She hopes the harm to be minimal, and the question of guilt to be private, a matter of unofficial clarification between her and her husband on the one side, and the offender on the other. When confronting the young man once more, however, he takes a step beyond her imagination:

> *And then he reported me to the police. I was reported to the police, accused of having insulted his honor.*
>
> *The case went to court - and he went free.*

From being the accused, an offender having molested a child, he turns the situation around so that he is the accuser, of a mother having insulted her neighbor's honor. And as no evidence of the sexual molestation of a child is provided; he gains his satisfaction, and the mother is sentenced for making false allegations. Fanny's perception of this judgment blends her own and her daughter's abuse:

> *It's been horrible. These years, everything that followed, it's drained me. This, this has drained me more than my own case...*

Although indicating that conflicting emotions burden her current life, she gives no more details, but reports how the molestation of her daughter shows in her occasional "spacing out." She has become very tough and does not want others to know. The officials have withdrawn after the law suit, rendering Fanny the one who "blew up a case which never happened." She becomes more and more desperate on behalf of the girl. The case and the sentence break Fanny down, and she asks for psychiatric help.

> I'm specialized in premature infant care. I had to cut out the job. *I met my own pain when this came up. I became so very aware of what had happened to me. I began to think about abuse in relation to every one of the women giving birth.* I thought: 'has she been abused, has she experienced something like this?' That doesn't work in a job. I became disabled. Then I started in a half-time job on a medical ward. I've been there until now. *But now I'm at the end.* I don't function anymore. I don't cope anymore. I'm a hundred percent disabled. I'm hesitant and indecisive, feel like eighty, don't manage a thing at home; I sort of prepare some food, nothing more. The rest is just turmoil. *I feel all the time that I'm persecuted.* (her voice grows indistinct, the next sentence cannot be understood) But I've received incredibly good support at the hospital. (her former place of work) I've gotten a considerable amount of help from a psychiatrist. I go to therapy every two weeks and have become familiar with the psychiatrist, after he encouraged me to talk about myself. That helped. I couldn't be in a group. It only got worse. *There I reexperienced every assault the others explored, and I got really mad. It was my daughter's and my abuse on top of each other.* But regardless how I feel, the worst is hers. I think of her every day, how things will turn out, how she feels inside ... oh ... no ... (long pause).

Fanny's abuse experience being reactivated by her daughter's starts a gradual breakdown. Thrown back into her own past, Fanny wonders whether she frequently encounters hidden abuse in her everyday life and in her work. She herself has been able to reconstruct that her first pregnancy had resulted from a rape. Thus, every delivery she assists, and every women she meets in the ward, revives the question of sexual abuse. Fanny perceives of abuse being covertly present whenever women address the medical system with gynecological or reproductive questions. This awareness paralyzes her in her work. She has to quit. But she can neither escape this awareness, nor the fact of her own repeated violation and her daughter's abuse. It is everywhere, yet unseen and disguised. She feels persecuted, which must not be interpreted as an indication of insanity, but rather as the logical consequence of Fanny's reality and the current social conditions she knows so painfully well. Thus, she cannot relate to a group of abused persons: all abuse becomes hers.

Fanny, as a professional, presents several other considerations which contribute to her general and professional disability. She has learned about the sexual abuse of several members of her family through witnessing third persons. Several women in her generation were molested as girls. Therefore she is more and more concerned about children:

F: I believe there's a lot hidden in conditions we don't understand as yet. I suppose among the MBD children there are abused kids, that something else is hidden in behavioral disturbances. I believe there are many who receive treatment and who are admitted and sent home again without anyone grasping what may have happened. But I think that becoming aware of it may lead to a different approach; then one may be able to give things a name, so to speak. I myself have not been aware that my anxiety linked to this. It's not overcome as yet, but I try.

A: Anxiety?

F: Yes, I get these breathing things, breathing problems.

A: Say a bit about that.

F: Well, whenever I think, when I come close to these things, like last summer, I get sick. It's not the right name, this word, but I can't express it properly. I get sick. I get distressed. I get such a distress reaction. This affects my breath.

A: Other things as well?

F: Yes. I get nauseated. (voice very soft, tuneless) [262] I feel the pain of all the trauma experience I come in touch with. Nightmares and dreams and the like. It's something I have to live with. It's cruel to have these very often, it's exhaustive. Then I feel like eighty. It takes a lot of my strength, it takes an enormous amount of strength. ...

Fanny makes explicit how the reactivation of abuse experience, in whatever form, drains her. She is speaking about coping with ever-present and everlasting invasion. The distress of embodied abuse, kept alive and alert by the awareness of violence being imminent in society, consumes all Fanny's strength. As she tries to confront it where she meets it - in her family history, (and as such, in talks and meetings with relatives), in her daughter, in her patients, in the disabled children she knows of, in therapy settings - it exhausts her. She knows more than she is able to contain. She sees more than she is able to adapt to. *Her abuse reactivation has become constant* since she recognized her life in her daughter's abuse.

A: When your daughter's abuse came up, was it first then you understood what had happened to you?

F: I'd thought of it before, after all, memories had returned. But I hadn't comprehended how much it was. I only remembered a little, I thought of it as negative memories, repulsive to think of. I believe I hadn't understood how much I'd collected *before I entered the group and recognized things the others were telling.* I began to think that others might know something, people I knew, that they had known this for a long time and what they might think about me. *I felt dirty, ugly, unworthy and the like. I think others think of me like this.* It's violently aggravated. And that's a burden to bear. It's easier to meet people whom I know can't know about me from earlier. (long pause) I've always felt like this. It affects the quality of my life in all respects. When I remember my school years, this feeling of being incompetent, being stupid - it's been with me, always.

A: Did this affect situations where you were supposed to succeed in something, or to prove something - that you withdrew?[263]

F: Yes. I fled. *I've always been convinced there must be something wrong with me. This feeling that something was wrong, regardless. And that it was with me, in what I did.* It has accompanied me all my life. It's still like this.

As in Bjarne's interview, recognition of other people's recollections of the abuse they suffered reawakens Fanny's memory. She gradually knows anew what has only been diffuse and distasteful. In contrast to Bjarne, however, her recognition does not bring a reconstruction and thereby a comprehension of the narration of life. Fanny's context is different. Her well-being is affected anew by proxy, through her daughter, the legal system and social prejudices, being in touch with abuse at her work and in her family. Consequently, for her, to know anew does not occasion relief but rather a re-living. A prevailing presence of sexual violence overwhelms her. Wherever she turns, she catches sight of it. When turning to herself, she only can see with others' eyes, judging herself as she quite recently had felt herself judged by a court of law: as an evil and unreliable woman who accuses good neighbors of criminal acts. Just as Elisabeth expresses being and doing evil, Fanny "knows" that she and what she does are wrong. Fanny's narrative is that of a *revictimization in a triple sense.* Her reaction confirms the above mentioned hypothesis of there not necessarily being a directly proportional relationship between the actual trauma and its embodied impact. Fanny breaks down; she cannot adapt any more.[264] How does this show, concretely, corporeally?

A: May we return to the whiplash injury (mentioned initially). When was that?

F: In 19xx.

A: Did you have the hemiparesis in this connection, was that what you wrote about in the sickness list?

F: No, it came during the preparation of the court case. It was when our neighbor's brother, that young man, was always persecuting me. He harassed me, he slandered me, he made trouble for me in all respects. *Then I got this paralysis.*[265] *I believe it was caused by the violent strain on my back.*

A: It was a paralysis which came during this distressing situation?

F: Yes. *And of course this should have been addressed in the trial. It showed how unreasonable everything was.*

A: Does it remind you of something? Have you ever before experienced something similar, that you in a way switch off parts of your body?

F: *Yes, I've had this before.* Not exactly that I was paralyzed, but that I switched off, that I disappeared, in a way. That's how it is, when things get too heavy, I switch off in a way. Then I don't register what's going on around me...

This passage demands comment on two levels. The first is that of hypothesis-generation and theory-building towards an understanding of assault embodiment; the other is that of method or technique. First as to hypotheses: a sexually multi-traumatized woman (both in childhood and in adulthood), is, related to one complex setting, multi-traumatized again as her daughter is abused sexually. Her integrity is violated by being harassed and made the object of accusations. Reactivation, reexperiencing and actual experiencing coincide. The confrontation is overwhelming and partly paralyzing to her, literally. She is unable to cope, and tries to shield herself by using a previously acquired adaptation method, dissociation. Her response to a current experience of powerlessness confirms the previously formulated theory that adaptation may turn maladaptive. But Fanny, herself aware of the origin of her reaction and the current trigger of it, introduces a new element. She imagines her corporeal response, though not addressed in the trial against her, as holding the potential function as proof of abuse. She thinks of it as demonstrating convincingly the rightness of her accusation and the wrongness of that of the young man. To see her, paralyzed, ought to convince the court that the assaultive acts had indeed occurred. How can she think this? What makes her believe a paralysis in a mother could be proof of the sexual assault of the daughter? How can she, apparently easily and without any further explanation, formulate this the way she does? A highly probable answer is: she perceives of her own corporeal logic as so concrete, unambiguous and self-evident that everyone must simply comprehend it as a reality and a proof of the truth. The court, however, rules against her. The visible proof, the evidence of assault, is not seen at all. The man who molested her daughter and violated her own integrity, defeats her twice. Her gradual breakdown makes her enter therapy. There she becomes increasingly aware of her previous experiences. Recently judged wrong by law, the wrongness in her becomes more and more all that she is. Her career terminates; her life is in chaos.

With regard to method or technique, this passage makes evident two serious limitations of the interviewer's skills: The interviewee, the eighteenth in the sequence of interviewees, addresses a topic never mentioned before, namely the conviction that the particular sickness in one person, which in addition in a medical sense is defined as non-somatic, could attest to sexual abuse experience of another person, such that it could provide evidence for or argument in a legal proceeding. This topic is, unfortunately, not elaborated. The dialogue moves on to the next pattern addressed, about dissociation as a technique of mental self-defense, an interpretation which the interviewee validates without any hesitation. However, further exploration of her bodily response, medically proven and diagnosed, is not done because of the immediate introduction of a new topic by the interviewee, one concerning a childhood teacher. The interviewer's attention is distracted by what she perceives as a clue to yet another abusive relationship. This, however, is not confirmed in what follows. Consequently, a possibility to explore a so called psychogenic paralysis (previously referred to as "hysterical paralysis"), is missed.

7.6 BEATE AND GRETHE

What Beate reports may be seen as reactivation by understanding retrospectively. Her abuse history and recollection have been described earlier. Apart from becoming aware of the abuse of her mother, to which she was a witness, of her son, and her own abuse by a relative, she has become aware of her abuse by a professional. Her memories of this experience and the acknowledgment of its true nature, first occur to her during visits to general practitioners and gynecologists as an adult, which is addressed as follows:

A: All these infections when you were thirteen to fifteen, did no one wonder, did no one ask why?

B: No. I went to the doctor for pain when urinating. I was given pills, all these Sulfa-things, it was almost continuously, all the time. I went to the school physician. He actually liked fondling quite a lot himself, so that wasn't so easy. (laughs a little, voice unclear)

A: Once more, what did you say?

B: The school physician liked fondling a lot, (very clear voice, stress on every word). He wanted to see the breasts of all the girls, he wanted to see whether they had grown since the last time you'd been there. And then he had to touch them.

A: The school physician screened breasts?

B: Uh-huh. Every time. (laughs)

A: He's not hired to do that. You girls knew that too.

B: No, not then. This is quite a few years ago. We thought he was within his rights.

A: Indeed. You said he also fondled you elsewhere?

B: Uh-huh. He did ... well, it was just to pull down your pants and look, and then touch ... well, around ... but he never touched in the opening. Only around. He was my school physician, yes, I actually had this doctor until I left the ninth grade.

A: The doctor did several things which don't belong to the procedures he's obligated to perform.

B: Yes, that's what I thought, later. Because when I nowadays, as an adult, go to a doctor for a check-up, very few physicians examine my breasts during such an occasion, although it might be appropriate.

A: Well, and none would ask you to take off your pants and look and touch...

B: If it isn't what you came for. These gynecological examinations are the worst thing I know. I don't know how often I've changed doctors because I can't stand this examination.

Beate and the girls in her school are exposed to legitimized violation, a tactical preemption in the sense of Riches. The fact that the school physician "explains why" he must see and touch the girls' breasts at every visit regardless the reason for the encounter, attests to his awareness of illegitimacy. Framing this act as medically purposeful makes the girls cooperate. Therefore, he need not fear their opposition or their reporting it to others. Why report what is usual and correct? His tactics work, since he can even legitimize an inspection of the genital area, "touching around it," though never invading the vagina, as Beate explicitly states. In her recollection of the situations, she syntactically reveals her discomfort, which she also expresses when initially stating that visits to the doctor were not so easy. She knew, of course, what the doctor would do whenever she consulted him, especially when attending for urinary tract infections and genital pain. The discomfort in the actual situation is also expressed in a link between her hesitating description, and her lack of options to avoid these covert violations before the age of sixteen.[266]

The interview with Grethe provides insight into both a reactivating of abuse at her place of work, and a "legitimate" abuse in a very particular medical setting. The interview text starts as follows:

G: I was eleven when the abuse began, and I was fourteen and a half when it ended. The abuser was an adult man, a close relative.

A: Later in life, did you ever experience that someone came closer than you wanted, or that you've been touched improperly?

G: I have a male employer, who's been very troublesome; I've repeatedly gone on sick-leave because of that.

A: Does this mean your boss is molesting you?

G: He comes too close. When he approaches me to talk about something, he comes too close. Then I start perspiring, I get very tense like this, (elevating her shoulders) I go around constantly with pain in my shoulders and my neck when I'm at work.

A: This means that you react to him? Are there no other men?

G: No, there's only one man where I work.

A: Have you experienced something similar with other men, or is there something strange about your boss, in his conduct?

G: Yes, it's his conduct. It's in a way his conduct. Rigid. Rigid.

A: Do you feel threatened, is this what I'm to understand?

G: No, not threatened. But he comes too close. I've talked to him about it, but he's careful in the usual way, he says, he's watching himself, at least he says he is, I've tried to talk to him in a calm manner, and he's aware that I'm very much bothered, I've had sick-leave many times each year. So now, I see a psychologist for therapy. And I talk regularly with the social worker in our county. Because I also have problems with one of the others where I work, there's one who harasses me. She finds a way to get at me while I'm doing the cleaning. Then I get pain. (pause) I'm very vulnerable this way. I feel it's just as if she picks on me all the time. Goes after me and corrects things and says that she can see I've been there, corrects the curtains and the like.

A: She's after you and tries to find things she can blame you for?

G: Yes, she really is.

* She tells that she feels very insecure all the time, and that she feels a constant tension, also when out in the city. Quite often it happens that she meets her father, her abuser, when out shopping. That is a problem, because she never knows where and when it may occur the next time. (She tells even more about the problem of meeting him at all the places she may possible go to, the post-office, the bank etc., but due to her very soft voice, this cannot be transcribed verbatim.)

The initial speech turn pattern of the interview text shows that Grethe is very reserved; she does not easily tell about herself, and almost every one of her statements demands a validation in order to grasp what the problem is at her place of work, and in the relation to her male employer and a female colleague. The dialogue is concentrated around three topics which create so much trouble that she responds with pain, causing her to take sick-leave many times each year: proximity, rigidity, and criticism. This may even be made more specific: male proximity and rigidity, and (female) criticism. She cannot tolerate her employer within a certain distance without feeling distress, although she denies feeling threatened. She knows he will not attack her physically. She has even dared to talk to him about the distance required to make her feel comfortable. He has accepted this, but there seems to exist a discrepancy of measures for what *is* too close and what *feels* too close. Proximity, rigidity and criticism are the basic ingredients of her tension, awareness and anxiety at work, causing pain and resulting in sick-leaves. Elsewhere, however, the terror of chance encounters with her abuser is another basic substance of her tension, awareness and anxiety. Wherever she turns, she feels at risk of encountering him. The sum total of this situation means not only the reactivation of abuse situations now and then, it means constant vigilance and a generalized reexperiencing, at all times.

Grethe is five years old when her parents divorce. She remains with her mother and, when her mother remarries, gets a stepfather and siblings. The stepfather terrorizes her with battering, scolding, and shouting. Her mother does nothing to prevent this, and Grethe is the only one of the children to be abused. She is eleven when her mother dies. Therefore she is sent to her father, who immediately begins to abuse her sexually. This abuse terminates when Grethe runs away and hides in her grandmother's house. There she tells. The grandmother takes action, reporting Grethe's father to the police. The police demand a medical examination:

G: I was terribly afraid when this was decided. He (the physician) was the type who is very rigid and very heavy-handed. I was held down while he examined me. He's a pensioner by now. But all the women who've been to him get completely hysterical when talking about him.

A: Why do you say 'they get hysterical,' and not 'he's not suited for the profession he had?'

G: Yes. (short laughter) That's more correct. But *I* didn't want to say that. He frightened me so much. I couldn't relax. But he just said that if I didn't calm down he would give me an injection. Because I was so afraid. He said so when I was sitting in that chair. He said if I didn't calm down he would give me an injection.

A: As a threat? You just had to be relaxed although he was harsh and heavy-handed?

G: Yes. It was really hard. The same happened when I was to give birth to my first child. It was the same doctor.

A: And you had no choice, he was the only you could go to?

G: Yes. There was no one else here. (long pause)

A: You said you were held down the first time, when the police demanded ...

G: Yes. It's also written in the protocol. I was held down. They tied me (to the chair) because I was too upset. And this examination reminded me most of all of the first time, when everything began, when the abuse began. It was precisely the same in later examinations, I couldn't relax, I felt like I was having cramps.

A: That's not strange at all.

G: Yes. So I wanted to have it over. I got sterilized after the third delivery. (long pause)

This was Grethe's description of her first gynecological examination, which was not only compulsory but executed by physical force and under threat of forced sedation. In order to secure the medical proof of vaginal rape she was violated again, in a medical setting, in the name of help and for the purpose of justice. This medical examination which, according to her perception, was directly comparable to the first rape by her father, became the frame for all later gynecological examinations. She met the abusive physician again during her first delivery, an absurd constellation of dependency and vulnerability, threat and disrespect. Consequently, she deliberately eliminated her fertility in order to avoid any further examinations of this kind.

7.7 BARBARA

Finally, Barbara shall be mentioned. She was abused by an uncle from the age of ten to thirteen. The uncle, a bachelor at that time, lived in Barbara's home. Suddenly he left, and for unknown reasons as Barbara did not dare to ask. She merely felt relieved, although she later considered the possibility that someone had intervened, though certainly not her parents. They insisted she participate in his wedding some years later. Futilely, she tried to refuse, but she was not allowed to break the traditional rules in her strict Christian family. *She fell sick.* She, nevertheless, had to join her family on the journey, but she was far too sick to participate in the ceremony. These events led to her decision to attend a boarding school far away from home the following year. As she had chosen a very strict school, this was accepted. Barbara reported as follows:

* She tells about this school in the North and a rather free summer holiday before entering the next grade.

B: I drank and went to parties and behaved as never before. There I met the one who I started going with for a while and who, after all, raped me.

A: Why did you say "after all," why did you somehow twist it?

B: No, I don't know, really. I mentioned this relationship in the start (of the interview).

A: But this was what it was?

B: Uh-huh. He was - I was sixteen and he was twenty-six, so he was much older.

A: An attractive man?

B: Yes, until he raped me, then he wasn't attractive any more. But this, just this I haven't told my husband, he doesn't know about this. I haven't dared, it's too touchy.

A: Is what makes it touchy that it might possibly seem as if you had wanted it, as if you had invited him?

B: I'm quite sure that my husband would not end our marriage because of this, but I have problems with it myself, I wonder whether... whether I myself did ... (pause)

And after this trip I entered the class. Since I was brought up in a very religious home, I went to a Christian school ... (long pause, deep breathing)

A: That must have been very difficult.

B: Yes, it was very difficult. I was meant to be there for two years, because then I could pass my final exam. The first year was simply horrible. I had no control over myself. It wasn't me who was there, it wasn't me who was there. I really didn't want to remember that year.

* She tells that everything she had experienced became a source of shame and guilt according to the norms she was surrounded by, and which were supposed to apply to everyone. There was no room to talk to anyone about what she had experienced. The conviction that she was a sinner became ever stronger. All discussions which occurred at school and among the pupils confirmed that she was a person having committed grave sins.

B: During the next year I got into more and more trouble. I ate and ate. By close to Christmas, I'd become so fat ... I was really huge, really ... and so I decided to get thinner, so I accustomed myself to throwing up, to vomiting, so it went on, but just before Christmas I couldn't cope any more, I told the Dean I had to go home because I was completely out of balance. I was given permission and went home. I couldn't tell everything, but they must have understood something. I said I had problems sleeping. My mother got afraid when she understood I really was in trouble with my nerves. She took me to a doctor to get me something to sleep. It was an elderly doctor my mother trusted very much, she had known him a long time. But he was very strange. He just looked through me. I don't mean that something was wrong with his eyes, but he looked at me without seeing a thing. I remember I thought: 'Why can't you ask mother to leave, I'm old enough to speak for myself.' And I also thought: 'Can't you ask me why I don't sleep?' Because, had he asked, I'd have told him, I'd gotten that far by then. ... But still I didn't tell, so it was my fault, it was my fault, but had he asked, I'd have told him. Yes.

Barbara's childhood is informed by a religious codex which renders sexuality a non-topic. She is abused in her own home and estranged from her parents in whom she cannot confide. She has no options at all for telling. When her uncle leaves, the abuse terminates. The feeling of guilt, shame and sin, however, remain incorporated. In an earlier passage of the interview, she mentions a suspicion that her uncle himself had been abused. She also wonders why her grandfather, sleeping in the next room and never closing his bedroom door, had not heard the uncle visiting Barbara's room. She even considers there might have been a conspiracy between her uncle and her grandfather. Such thoughts and reflections, however, make her feel even more sinful and guilty. The family norms regarding obligations and traditions are much stronger than the parents' interest in why Barbara tries to refuse to participate in her uncle's wedding. She is neither seen in the despair over having been abused, nor heard in her opposition to obedience. Her silenced and ignored rebellion turn into acute illness, in the sense of Scheper-Hughes. Though not freeing her from the journey, it does free her from witnessing lies being spoken in the House of God during the wedding ceremony. Barbara is so desperate that she considers warning her future aunt prior to the ceremony, but dares not, although she feels great pity for a woman who blindly enters into marriage with this abusive man. Barbara's inner image of him is presented as follows:

B: I remember little from this time. I mean after the abuse. I don't remember how it happened. (her voice lowers, hesitates)

A: You mean that you've wondered whether you've been seduced or threatened, that you might have a memory of that?

B: No, I don't think he used force... In a way, I don't understand ... I've always felt that I was stupid, I understand that I feel stupid, mainly from childhood, perhaps I can't remember ... everything around ... (hesitates, searches) ... I can't comprehend how I could do it, take part in it ... I mean that I ... that it could happen... I was a little afraid ... there's something strange ... I mean when I think of him, he never has a head ...

A: He only has a body?

B: Yes. Uh-huh. He just has in a way ... the head is not there ... and his eyes I don't recall, I don't see them. I only know he's old. But that's not correct either, he must have been thirty, but I think of him as an old body, old, ugly pig of a guy, I've seen him like this always, and he only has a body and hands ... and a penis, of course, so I can't comprehend how he could get me ... how ...

Her abuser is in Barbara's memory an "old, ugly old body with hands and a penis, but without a head." She cannot comprehend how he made her cooperate, as she can remember neither threat nor force. The next man whom Barbara lets touch her was also much older than she. Although twenty-five years have passed, she is still doubtful concerning her own responsibility for both, the abuse and the rape, since she did not resist her abusers and thus feels she may actually have invited them. This feeling of probably bearing responsibility because she did not resist and perhaps fears she may even have "cooperated" seductively with both men, was constantly addressed and indirectly talked about in the school she attended. Wherever she turned, she heard the voice of moral authority condemning her as a great sinner. Her involvement in premarital sexuality attested to her personal sinfulness in general, while her lack of explicit resistance to sexual acts, and even with several men, testified to her female sinfulness in particular. The voice reactivated her experiences unceasingly, and it aggravated her shame and guilt. There was no room for telling in order to spell out or explore the questions of guilt and responsibility. The answers were provided in advance: she, an unmarried girl, having practiced sexual acts with two men, regardless the circumstances for these acts, deserved condemnation according to the religious codex. Thus, her sinful burden became ever more unbearable until she could not adapt any more and had to ask for help. However, as a result of social constructions, the measures to release the impact of silenced sexual violation and of implicit moral condemnation, had to be sought in a medical institution. These measures, by necessity inadequate, maintained Barbara's conflicts and problems into adulthood.

7.8 SUMMARY

I have explored interview excerpts which thematize a variety of reactivated experiences as a particular aspect of sexual abuse experience. The explorations make clear that silenced trauma, embodied in a variety of ways and influencing individual corporeality according to the interpretation of the abuse-context, can be reexperienced in many other kinds and types of contexts. The triggers are configurations, perceptions or constellations of high specificity and particularity. These are characterized by bearing resemblance to abuse memories to such a degree that they become nearly or entirely indistinguishable and inseparable for the abused person. Thus, any situation, any innocuous context, may be transformed in an abused person's perception and interpretation into one which seems dangerous or offending. That this includes contexts defined as offering help, consultation or support, may contribute to an exhaustion of the person's adaptability to perceived or actual invasion or offense.

Chapter 3

EXPLORING THE MEDICAL MAKING OF PATIENTS

INTRODUCTION

The seven phenomena emerging from the interview narratives about violation, sickness, and treatment, were shaped by the dynamics between abuse and health problems, and by, intertwined with the former, the dynamics between the abused person and the professionals representing the health care system. These dynamics held specific characteristics which shall be elicited in the following.

The first characteristic was the limited validity of *linear time* as a basic, structural element. Although the primary assault, in most cases, had to be regarded as an origin or a driving force, sickness could both precede and facilitate violation, as in Oda's and Annika's cases. Sickness of unknown origin, occurring simultaneously with undisclosed abuse, could terminate the ongoing abuse, as in Nora's case. Violation, sickness, and treatment very often occurred rather simultaneously than successively, as in Oda's, Fanny's, Eva Maria's, Nora's and Mary's cases, just to name a few.

Just as the concept of linear time was overruled by simultaneity, so was the concept of *causality* challenged. Medical interventions "caused" by a pain, became the "cause" of successive pain of different kind which "caused" repeated treatment, which, in turn, "caused" chronicity, as in Nora's, Annabella's, Line's and Annika's cases. This chronicity, however, could imply a maintenance of a certain kind of pain presentation, evoking stereotypical medical responses which failed to be adequate solutions to the problem. It could as well represent the artifact of previous medical treatments which, meant to release a health problem, in fact caused yet another.

Additionally, the concept of *individuality* was overruled by identification with, and responsibility for other persons, or a reactivation of own violation experience by means of violations hitting other people. The assaulted person's health problems could be decisively affected by conditions in another person's life. This could be read from the tics of Annika's daughter, in Hedvig's abuse by her father as Nora's "stand-in", in Beate's by-proxy reassault by way of her molested son, and in Fanny's and her daughter's simultaneous though different violations by the same man.

Finally, violation, sickness, and treatment did not present themselves as discrete *categories*. Medical settings could be experienced as non-therapeutic in nature. Consequently, medical interventions or counseling could, in effect, represent acts which were alienating, blaming, offending, rejecting, insulting, sexualizing, assaulting, and promoting chronicity. Acts defined as help were experienced as a new, and as such, a primary violation. However, they also could feel so similar to a prior violation that they had the impact of an offense, and thereby evoked a violent reexperience. This was the case in a variety of modalities in, for example, Veronika's, Eva Maria's, Mary's, Bjarne's, Line's, Ruth's, Christine's, Runa's, Synnøve's, and Grethe's histories.

Grethe's experience of a medically performed violation by a male gynecologist was, in her perception, both a repetition of rape experiences, the memory of which she had tried to escape, and a new assault. An act of help, a medical examination, was unmade, in the sense of Scarry. A professional became a molester while securing medical proof of sexual abuse on the demand of the police and for use in a legal proceeding.[267] The medical abuser could not even be avoided later; Grethe encountered him involuntarily during her first delivery. She had no other option than to entrust herself to the care of the same man who had violated her "legitimately" before. These repetitions of encountering abuse, even in a medical setting, finally led to Grethe's voluntarily terminating her fertility in order to avoid the intolerable gynecological examinations.

Such voluntary termination of fertility in the wake of, or in anticipation of, omnipresent or latent sexual violence was also thematized by Oda as her only possible option for protecting herself, as she could not protect her body. Sterilization as a means to prevent a threat to health during pregnancy, albeit based on an erroneous medical supposition, was thematized by Nora. Annika chose not to have more children as she felt so very incompetent regarding her mothering of her eldest son and daughter. Eva Maria, having miscarried in seven of ten pregnancies, decided never again to risk pregnancy and requested sterilization. Elisabeth, having suffered from pelvic pain for years and being suspected of having a malign process in her abdomen, was partially ovarectomied and advised to have her fallopian tubes cut in order to influence her chronic pain. Gisela also agreed to this intervention in the hope of being released from her incapacitating, chronic pelvic pain. Camilla was the eighth woman reporting a termination of her fertility, by tubal ligation. In her case, the intervention was a means of curing excessive, chronic menstrual hemorrhaging. Their abuse-sickness-histories will be explored in the following.

The presentations make evident that each of these eight women had a highly individual context and background for the same, final decision. Simultaneously, every one of them addressed the overall influence of the sexual abuse of girls and women on questions related to female fertility. Implicit in these eight narratives was the contribution of the health care system to the consequences of sexual abuse as regards female fertility and reproduction. This contribution reflected the blindness of gynecological theory and practice towards sexual violence. Finally, each woman, while presenting in her unique way, contributed to making visible the general elements involved in every process of a sexual assault experience which turned into sickness.

Sexual violation, having occurred for years in a child's life, could suddenly "by chance" or apparently by coincidence have been terminated by a medical intervention for a pain of unknown origin. This particular "solution," a presentation of pain and a demand for medical intervention, could be chosen again and again years later to avoid ongoing abuse in an adult relationship. Consequently, a treatment became the "cause" of a sickness, and the maintaining force of a chronicity. A treatment would represent a silencing force, since labeling a health problem in diagnostic terms effectively prohibited all other approaches to understanding. Thus, medical treatment promoted chronicity, and contributed to the general silencing, and, of course, to the maintenance of ongoing abuse.

Treatment as the "cause" of sickness could blend with treatment as an assault, in the sense of iatrogenic harm through interventions based on dubious indications, and in the sense of assaultive acts by the therapist. Treatment as violation necessarily occurred in therapeutic settings where information about sexual violation given by the patient was ignored or undone, and whenever therapeutic relationships were sexualized, symbolically or concretely.

A sickness would "cause" new violation when the medical intervention chosen was perceived as resembling a previous violation. Thus, the intervention pattern itself became the driving force of a sickness history. On the other hand could repeated ideational reexperiencing and recognition of the assault heal the assault's impact and at times be the only effective treatment. The same approach could, however, due to other circumstances, accelerate a personal breakdown. A refusal to comply with a therapeutic treatment plan, or opposition to a therapist's advice, sometimes showed be the first, or a decisive, step in the reintegration of salient memories.

Consequently, in an analysis of the impact of hidden and socially silenced sexual abuse, the relationship between the elements violation, sickness, and treatment would change since their meanings showed to be interchangeable. They even would mean something different in consecutive situations in the same person's narrative, and they could carry very different designations in the different people's lives. Consequently, their impact was, by logic, different, which happened to contradict what is perceived of as logic if the terms are taken at face value as if they were unambiguous.

In other words, in the context of hidden sexual abuse, the designation of violation, sickness and treatment was situationally constituted and determined.

1. CAMILLA

In the analysis of one single interview shall be demonstrate how, on a background of hidden abuse, what seemed to be a somatic disease resulted in a treatment which was an abuse. Treatment and abuse, repeatedly interlinked in medical settings, were consequently driving forces in the promotion of chronicity. Since the treatment was experienced as abuse-like, it both maintained the sickness and rendered it incurable. Both sickness, treatment and abuse were then subsequently terminated by an unintended reactivation of multiple abuse memories in a non-medical setting.

Camilla's interview was the penultimate, and very rich as to information about somatic health problems. Her abuse history shall be presented in relation to her sickness history, as documented in hospital records from nine hospital admissions and six out-patient clinic examinations, during thirteen years, due to gynecological problems. This abuse-sickness-history will be explored in the same way as Camilla approached it herself a few years prior to the actual interview. This is commented upon in the end of the interview, where Camilla attempts to summarize her emotions after having talked for three hours about abuse and health problems.

> **C:** I've avoided gynecological examinations during these last years. I used to go, earlier. I was so frightened, afraid of everything that could be wrong. I had a lot of pain in my stomach. But now I feel - I guess I don't tolerate this (gyn.ex.) anymore nowadays. (long pause)

What I feel now, what I feel right now, is that although it was horribly painful, it feels good, although it was painful. Because it's the first time people believed me. I've tried to tell all this. But I can imagine that the reason why I haven't received feedback was my way of telling. Finally, a sudden breakthrough happened, it happened in connection with problems where I work, I participated in a psychodrama group, it really concerned difficulties I had in my work. It concerned men, I cooperated with three men. But there, the rapes suddenly emerged. Suddenly, I was sitting there telling about them to the others. One of them got angry at me and said it was disturbing that I smiled while I told about it. Like I was making light of it. So I take into account that earlier, when I tried to tell, I've presented it the same way ...

A: ...and made it 'not dangerous' while trying to tell...

C: Yes. I guess so. So no one thought it was serious.

Suddenly, while playing out a *conflict between herself and three male members of the staff* at her place of work, Camilla tells about rapes. In her narrative about this situation, she herself uses the word 'suddenly' twice, expressing her own astonishment about an almost involuntary, or at least unplanned and unprepared narrative. What she tells seems in contrast to how she tells it, which is addressed by a member of the psychodrama group who dares to express confusion or irritation about this discrepancy. For the first time, Camilla is made aware that she herself may contribute to not having been heard during former attempts to tell. "Laughing away" the pain of abuse experience is her technique, which, as she admits, may have been counterproductive in situations where she wanted others to understand. Does she recall such situations?

C: I had a period where I suffered from sudden memory-loss. I could be driving along, and suddenly I didn't know who I was and where I was meant to go. I always managed to pull off and stop and calm down. Then it returned. It felt like a movie stopping suddenly. There were long passages which disappeared, and I couldn't remember how I'd gotten to the place where I found myself. I got very scared. I believed that I was about to lose it. It lasted a complete week. I drove the wrong way every day, I came too late every day to the course I was supposed to lead because I drove an entirely different way than I should have, and when I found myself there, I understood that it was the wrong place to be. But people didn't believe me, they thought I was joking and just saying this to escape from something.

A: Did you receive some comments that they found you strange?

C: Well, I was very self-ironic, I told what had happened and ridiculed myself and made jokes about it.

A: You made it 'not dangerous'? You took the message out of it by saying it was only a joke?

C: Yes, I did. *I took the severity away, I was so scared.*

A: You distracted the others so they wouldn't find the traces of what was so serious that it overwhelmed you totally?

C: Uh-huh. Yes. I distracted them. But I saw physicians several times, because I was so scared. I told them about the memory losses and the fainting and all that, and the headaches and the pain in my neck and all that, and the constant pain I had in my stomach, but they had been with me since childhood.

A: You described the situations and told that you were afraid of going mad?

C: Yes. Yes, I said it, I told about it. What I got back was that I wasn't believed. I guess they thought I made up these situations, that I went to doctors ... well, that it felt satisfying to go see doctors ... I don't know. I thought very many times that they didn't believe me.

A: What did they do?

C: Well, they told me I should calm down, that I was nervous, my mother was nervous as well, I don't know how much she has contributed. I don't know what she knew, but she has overprotected me. I have an older brother and an eight year younger sister. They always were told that I had to be taken care of, that I had to be protected.

A: And now you start wondering whether she had reasons to say so, whether she knew something.

C: Yes, I really do wonder. Because this protecting me, it would have been natural to think it rather should be my younger sister. But she's dead now, I can't ask. I can't ask my father, because then he'll die. He has heart-trouble, he takes drugs, but I guess he uses his sickness to protect himself...

A: And you respect this?

C: Yes. Certainly. If something happened, I would have trouble with my conscience. (long pause)

Perceptions of unreality, irrational behavior, memory disruptions, fugue states while driving a car, all these serious indications of severe distress, when presented to health care professionals, are partially ignored and made light of by the patient herself as she talks ironically about these frightening experiences. The physicians Camilla consults let themselves be distracted by her mode of telling. They do not consider that the very manner of presentation may be an accentuation of a message about life-threatening experiences. Although the patient expressed her fear of going mad explicitly, this is not heard as it is said with a smile. The joking is not seen as a part of a defensive and protective strategy against knowing and acknowledging what might be at issue. Camilla could have been killed in an accident while driving without being mentally present, yet no one asks her *why she feels forced to "space out" in a car in her present life situation,* as she had been driving cars for years. Her presentation is interpreted as indicating "nervousness," without any specification of what that might mean; and Camilla's nervousness is linked to her mother's, as if the existence of a logical connection to familial conditions could render this term part of an explanatory system, similar to the way Hanna's migraine was "explained." The more or less random choice of explanation engenders the rather random advice she is given. According to a concept of "nervousness", expressing exclusively excitement without a proper cause, she is told she ought to "calm down."

Camilla is in a state of excitement, yes, but of a very special and potentially destructive kind, that of reexperiencing previous trauma. The emergence of *a conflict between her and her three male colleagues* at her place of work burdens her. The conflict is finally addressed in an interactive approach by a psychodrama including the complete staff. Here, and to her own obvious surprise, Camilla "suddenly" tells about rapes. *The first rape she mentions occurred in a car,* performed by her driving teacher when she was nearly eighteen. Thus cars were unmade. They represent dangerous places which, during a state of conflict of interest linked to dependency on and cooperation with relevant men, become places of reactivation. Thus, invasive trauma memory engenders Camilla's "spacing-out" while driving, endangering the lives of herself and others, and scaring her profoundly when she becomes aware of what happens. However, the breakthrough into awareness of trauma starts with the most remote or least proximate of repeated sexual molestations, which, as with all of her violations, has never been mentioned hence not been overcome. *The next rape she mentions, was performed by three men she knew well.* In the interview, her report of this rape is linked to questions concerning her main gynecological problem, as follows:

C: I believe, but I'm not so sure about this - I've had bleeding, bleeding all the time, bleeding on sheets, and in beds and on clothes, and it happened that I'd been sitting on a bus and then standing up and everything was red, leaving a bloodbath behind me - for a long time I believed that first started after the rape by these boys, but I'm not so sure anymore.

A: But after all, it seems natural you linked it to that since the other one was even more dangerous for you to think about?

C: Yes, I connected everything to that. But when recalling - this history there, he was a boy some years older than I was, and I was very much in love with him. We dated. But we hadn't had sexual intercourse. I held back, I was afraid. And then his two comrades decided that we had to have intercourse since we were together. On a certain occasion, they locked the door to the room where we were, all of us. I remember that I couldn't understand why no one came to help me since I screamed as loud as I could, at least I believed I did. But later, when remembering, *I recognized that I didn't scream. It was only a scream inside me. Now I know it didn't come out.* And later I've thought that - although I know it's nonsense - I can't help but think it was my fault ... (her voice has softened gradually, has become hesitant, searching)

A: ... you ought to have screamed?

C: Yes. But I can see it now. *I had my reasons for not doing so, I know it wouldn't have helped.*

Camilla is sixteen years old at that time, the young man she is in love with and his comrades are nearly twenty. While being raped by a man she trusted, she perceives of herself as screaming, feeling despair as no one comes to help her. The young man she is in love with takes advantage of his friend's initiative and rapes her in their presence in a locked room. Why is the scream only inside, why can she not voice her refusal to be profoundly betrayed, humiliated and overpowered? Why does she know it won't help to scream? Is there a parallel to Bjarne's suppressed shriek as "everything suddenly was dangerous" the very moment his most beloved man becomes assaultive? Camilla gives no clues to the danger associated with either screaming or the active suppression of a scream. Rather, she expresses her fatalism, a conviction of futility as regards the effect of voicing that an improper subjugation is taking placing. From where does this conviction of futility come? The answer is given in a passage in which Camilla tells about her second abuser whom she met when she was eleven years old, a young man working on the neighboring farm, who, that same autumn, is dismissed due to his improper conduct toward several girls or young women in their small town. While discussing this man and some, as yet, rather unclear memories, the primary abuser is introduced. This is presented in the following excerpt:

C: I don't remember exactly how everything turned out, but he came to our house, he came several times. I know I was afraid him, and once, he came and I fled and hid in the house, and he came in, behind me. But I don't remember more. Not exactly. During the last few months, I've tried to find a way to remember more. And here is something concerning *my father, because he let the man follow me into the house,* they were present when I ran into the house to hide. There was even more with my father. *Because I remember I told him something about this friend of his, I asked for help. But Father wouldn't listen to me, rather the opposite,* his friend was the only one who was allowed to enter my room to say good night whenever he was at our house at bedtime, and I tried to hide myself from him, *but father found me and then I had to say good night and was scolded* because his friend was so fond of me that it was so nasty of me to hide.

A: It wasn't enough that you didn't want to?

C: No, and *when I hid, he came and got me.*

A: So he almost presented you to this other man?

C: Yes. Yes, that's what he did. But I'm uncertain about whether I'd told Father what he did to me. But I'm confident that I said I didn't want to be

> with him, that I disliked it. *The next, about the man from the neighboring farm, I didn't tell about.* It was something which had disappeared for me in a way. But when talking about him, I remember that later, I learned he was dismissed that winter because he couldn't leave the girls in peace. And I know he peeped through the holes in the wall of the lavatory several times when I was in there. This came back to me right now, and I can clearly see how much he frightened me.

One of the best friends of her parents, an adult, male neighbor, abuses Camilla in her own room and in her bed since pre-school-age. She hides whenever he comes in order to avoid a good night ritual which she hates. She tries to voice her feelings about this man to her father, though most probably unable to describe what feels so wrong, as she is too young to know the proper words by which to make adults understand. Her message of disliking and flight is neither heard nor "seen". Rather the opposite occurs. She is offered to her molester, and scolded for being nasty toward a man who loves her. *This repeated experience of her father, to whom she has confided her resistance, offering her to her molester,* results in a very effective silencing of a child's protest against abuse. When her next offender approaches her, she even does not try to tell; she knows it is in vain. Her father's repeated conduct confirms her supposition. He lets a man, from whom his daughter flees before his eyes, follow the girl into the house and into her room. There is no protection to expect or to hope for. The father does not consider that her reactions should guide his own; how she feels seems unimportant or irrelevant to him. Therefore, when raped by acquaintances a few years later, she only screams inside. She has learned that men have the right to offer women to other men - and no one will intervene. By the time she is sixteen, she has been victimized by three men, violated countless times, and offered and betrayed by the most significant men in her life until then, her father and her first love. Silenced by her father's ignoring her attempt to communicate fear and revulsion, and even more by his active delivering of her to her abuser, as she experiences it, Camilla is in a constant state of sexual and personal violation, which gradually leads her into a sickness history from which she does not escape for the next twenty years. Camilla reports her first admission to the gynecological ward as follows:

> **C:** Shortly after this rape, I was admitted to the hospital for the first time. We were celebrating a cousin's confirmation, we'd been having dinner, I rose from my chair and saw my mother's face besides me turn pale, I'd been unaware of it, but my entire dress was stained with blood. Since it happened on a Sunday, I didn't go to the out-patient clinic before the day after. I was taken into the examining room and checked. I didn't know what would happen, and I was scared to death. I was lying there, watching the instruments, when I suddenly felt a pain beyond my imagination, and

then there was a lot of blood and I was completely confused - the doctor had done a D. and C. on me - (long pause)

A: ...no ...(long pause, deep breathing can be heard)

C: And later the same was repeated, almost once a year. (long pause)

A: (soft and rather terrified voice) What have they done to you? Was it the same doctor every time?

C: No, after all, it was done in the out-patient clinic, but he was the head of the department. He decided...

A: ...And the others agreed to it ...once a year without a positive indication. Some of them must have understood that it didn't help.

C: Yes. More or less. I had an impression that the doctors couldn't cope with it in a way. Because very often when I met physicians, they would contradict each other. And I suppose I could see that they wondered, and that not all of them found it easy to know what ought to be done. I could register this from their reactions. But I was in and out again. Mostly, I was told that it was due to my nerves. Gradually, I wondered, myself, s it like this, am I nervous, is it as the doctors say it is¿ I couldn't make sense of it ...

The performance of a curettage without anesthesia of a sixteen-year old, completely unprepared girl is confirmed by the records.

A few weeks after having been raped by friends, Camilla is molested again, but this time in a medical setting. Like Grethe, harm is inflicted on Camilla and pain under physical force during a gynecological examination. In her first summary while filling out the sickness list, Camilla had mentioned this situation briefly, as follows:

C: I had the first D. and C. [268] when I was sixteen. It was done by a doctor in the out-patient clinic, he wouldn't tell me what he was going to do until he was ready.

A: My goodness. An abrasio at the age of sixteen...

C: Yes, I was lying there and was held down, I was held ...He was a doctor who's well-known. I see it now, I'd repressed it, it was abuse, that's precisely what it was. But no one has asked, no question mark ever. There were many who told me that I was nervous. And I may have been so.

A: What were abrasios intended to help?

C: To stop me from hemorrhaging.

A: But your bleeding was ostensibly caused by nerves? Where is the logic?

C: Yes, well you may ask.

The explanatory system of nervousness as a cause for uterine hemorrhaging in a young woman, demanding an abrasio of the uterine lining, seems to hold inconsistencies. But still, the same intervention is performed on this "indication" six times at the same hospital, and four times in another hospital from which no records could be obtained. Between these,

Camilla becomes pregnant, although she has been told that she would not conceive due to the frequent prior interventions. The hospital records confirm two very late and one early miscarriages over a span of four years, and an abrasio for excessive bleeding and pelvic pain in between. In the collected records from the hospital where most of Camilla's medical interventions have taken place, mention is made of uncertainty regarding the origin of this pain and bleeding. "Nervousness" is addressed, without any comment upon what that means or what the physicians in charge may have been thinking of. Furthermore, no measures are taken which could be more adequate, such as a consultation with a psychiatrist or psychologist. The text of the records also supports the conclusion that none of the physicians has tried to elicit, together with the patient, information as to what may be fueling her continuing "nervousness." The "nervousness" has been mentioned to the patient, but still, the conclusion is: *Abrasio muc. uteri. Dilatatio canalis cervicis.* Half a year later, the same intervention is made, based on the same medical reflection. Again, the uterine lining is removed and a cervix-dilator inserted for twelve hours. Histological and hormonal test results are described as normal. The year after, Camilla has her first miscarriage, and an abrasio is performed in the fourteenth week of amenorrhea. The next year, she has her second miscarriage after having been hospitalized for almost three months due to bleeding during pregnancy. An insufficiency of the cervix cannot be verified, the fetus having died several weeks in advance. The same process occurs the year after, when Camilla again has to be treated for a miscarriage in the eighteenth week of gestation. Doing the abrasio, the physician believes that a myoma has been registered, a finding which is given the status of being a probable explanation for what is, by now, labeled as "habitual abortion." Half a year later, however, a gynecological examination cannot verify this previous finding, and a myoma as being the cause of habitual abortion must now be excluded.

After the next abrasio due to hemorrhaging, Camilla chooses to have an IUD inserted as she fears yet another miscarriage in case of conception. When controlled after the insertion, a note is made in her records concerning a possible uterus bicornis as the cause for her habitual abortions. The gynecologist considers a hysterography which, however, is not performed as this hypothesis holds a very low probability. The year after, Camilla once again attends the out-patient clinic due to frequent bleeding, but no surgical intervention is performed. She has decided to terminate her fertility and has adopted a child. Two years later, she is admitted for a tubal ligation, which is combined with yet another abrasio. In the records, mention is made that Camilla wants her uterus removed in order to avoid the almost continuous bleeding. The attending physician cannot verify any myoma or other pathology in the reproductive organs. As in all previous notes from the hospital admissions during the years, Camilla is described as being in good somatic health, and nothing which might be regarded as a cause of pelvic pain and bleeding can be verified. The term "nervousness" does not reappear in the notes of the latest admissions. The measures taken are described briefly, and no other considerations are mentioned. What has Camilla herself been told, and how much is she informed about the actual indications for the various interventions, or the findings? Most of her encounters with gynecology are characterized by uncertainty. This feeling is most concrete whenever she meets the physician who did the first abrasio. She describes such a meeting, occurring in connection with her third miscarriage, as follows:

C: The last time, when I entered the hospital with fully active uterine contractions, he was the head of the department. He was the one to see me, and two other doctors were present and the chief nurse. I knew her very well. He was to examine me, with me sitting on that chair, he checked me and then suddenly he said no, here there was no reason to keep examining anymore, threw the instruments down and walked out. Just walked out. *He left it to the others to comfort me, because I was totally in despair.* The chief nurse took most care of me, and she told me a little ...

A: ... what they thought about this man?

C: Yes. Then they gave me a single room, and *he came later and apologized. I suppose they had taught him a lesson. But I didn't believe him. It didn't look like (an apology).*

A: Did you regard it as an obligatory act of penance?

C: Yes, it seemed like that. Later, years later, a new doctor came in his place. He talked about that there might have been several organic reasons why I miscarried every time, *they talked about myomas and a split uterus and other things. But they didn't find out. But finally, five years ago, I got sterilized. It was also meant to help with the bleeding.*

A: Did it?

C: Yes, *it improved for two years, I had less hemorrhaging.* But what happened were two or three menstruations very shortly after each other, and then nothing for seven or eight months. Totally unpredictable. But that made me insecure again, I wondered whether I again had become pregnant. *The doctors said I'd become pre-menopausal.*

A: But you weren't?

C: *Of course not.*

A: The irregularities have something to do with your emotional life?

C: *Yes. After all, it was three years ago I began to work, and feel, and find my way back to the threads from my childhood. So I took for granted that it had something to do with that.*

A: And since, there haven't been any medical interventions?

C: No

A: Concerning the split uterus or the myomas, how did you find out about this, did they perform a laparoscopy?

C: No. They checked it when sterilizing me. They looked for it. But I don't really know what's done and what the answer was. And that's what makes me really desperate. I may keep going and fight for all other things, *but concerning my own health, I really get, I really get ...*

A: Have you given up?

C: Yes. I haven't managed to fight, I've just accepted whatever they've told me without putting in any question marks. Now, I understand the whole process, now. *I see that I've experienced everything as abuse, what they've done.*

A: Please, say it once more.

C: Yes. I said that *I've experienced the manner in which they've done things on the same line with the other abusive acts, I feel this now, and that's why I haven't managed to fight.* (very long pause, weeps)

Thirty-five years old, Camilla has had the last of fifteen gynecological interventions which, in most cases, have been seen as indicated due to a meno/metrorrhagia, and which have been linked to three miscarriages, and which finally ended with a sterilisatio tubarum. Cryosurgery is performed in between. She has had excessive bleeding from early adolescence after her menarche at the age of ten. The first intervention is deemed medically indicated due to bleeding shortly after a rape about which no one knows. This rape is the third abusive relationship Camilla experiences, having been rendered unprotected against abusive men by her father who ignores her protests and literally offers her to her childhood molester, and who lets an adult stranger approach his daughter in his house. Camilla has learned early about the futility of protesting against males deciding over her body and her reproductive organs. This experience is confirmed, intensely, during her first encounter with the male-dominated and male-conceptualized world of gynecology, which Emily Martin has explored in depth. A medical violation is Camilla's debut as a gynecological patient. She has to accept the same doctor over and over again, either indirectly, as he guides the policy of the ward she is admitted to most often, or directly, as described above. Once again, he leaves her in despair, which is witnessed by others who intervene on her behalf. She never dares ask, however; she never dares protest; she never insists on knowing and deciding for herself. The repeated conclusion of her record notes, that no pathology can be verified, has not really reached her. She believes, again and again, that the proposed intervention is correct and medically indicated. She is told it will help and will stop her bleeding.

In retrospect, Camilla can see the core of her experiences connected to the medical interventions. As the first experience is an abuse in a medical setting, what constitutes an abusive act and what constitutes medical treatment is rendered indistinguishable for her. Consequently, each later intervention feels so similar to the previous sexually and medically abusive acts, that she re-experiences sexual abuse. In her corporeality, sexual abuse has been repeatedly linked to being offered, betrayed, left unprotected, and overpowered by men she trusted or into whose care she was entrusted. Thus, fundamentals of security and confidence have been unmade in the continually repeated constellation of men in power approaching her body. She never dares voice her protest, her doubts or her questions as to the correctness of the medical judgment, or the logic behind the medical conclusions. Therefore an invasive procedure is repeated ten times, although never providing the kind of effect it is supposed or presumed to, a stereotypic ritual more than a rational intervention. The "explanations" given to the patient deviate from the medical measures taken at all steps of a nearly twenty-year long sickness career. From the records one can read how this chronic health problem is gradually constructed by medicine, starting with a treatment which equals rape to such an extent, that it unmakes a gynecological setting, converting it into an abusive situation. The unmaking is worsened by the perpetration by a doctor who personalizes disregard, assaultive conduct and an abuse of power, influence and trust. Camilla, after repeated previous violations, is made sick by professional intervention which she experiences as the expression of an over-all patriarchal power and a male sexual offensiveness, which is even integrated into medical acts and clinical reasoning.

Medical interventions, guided by an inconsistent medical theory, create and maintain sickness. Treatment, acting as violation, causes the sickness it purports to cure.

Finally, Camilla herself breaks the perpetual trap of treatment which equals violation which causes sickness which requires treatment which... She finds herself "suddenly sitting there, telling about the rapes" in a non-medical setting meant to solve her conflicts related to three male colleagues. This particular conflict constellation has actually reactivated abuse memories as evidenced by the emergence of repercussions from the last and most remote of the sexual abuses, a rape in a car during a driving lesson. These repercussions are strong enough to endanger her life in the period prior to this meeting. Fifteen years after a 'forgotten rape,' she dissociates whenever driving a car in a certain direction. She has tried to voice her despair and panic to several doctors, although not really daring to show how deathly afraid she is of her own absence in presence. She cannot explain her particular corporeality, she knows it sounds crazy, she ridicules herself for being stupid enough to miss the right road every day. She tries to grasp what happens, aware of a threatening state of being which may presage death or madness. Not until she is challenged in a psychodrama setting, does she suddenly comprehend the present threat to be a threat of the past in the present conflict. The configuration of these three men relating to her in a particular way feels too similar a "figure." However, in encounters with the health care professionals, she once again lives the repeated lesson about the futility of voicing her rights and speaking out her concerns when feeling her boundaries endangered. As her father has "taught" her from early childhood, and as she frequently has had confirmed, no one will listen.

Has she never felt heard in her countless encounters with the health care system, apart from by the comforting chief nurse?

C: I've met two female gynecologists during my hospitalizations. I know for sure that the one could only stand to be there for half a year. She was very kind when I met her. *Those were the only doctors I've ever felt tried to look a bit underneath my skin. But they had no possibility to reach further, they stayed for such a short time.*

A: How did you feel this, how did they express this?

C: The one, B.S. was her name, *she came to my room and sat down and talked with me,* and she tried ... I can't explain it, but *she came to see me several times and took her time talking to me.* I felt that she saw there might be something she ought to come closer to. She was very vital, but she told me that *she couldn't work there, she couldn't stay there.*

A: When did this occur?

C: Around my last miscarriage, in (year) I suppose... *She's the only doctor I've met before to whom I might have told it.*

A: ...because you felt that she had sensed something?

C: Yes. *But I didn't tell her either.* I'm not even quite sure whether I was aware of it all. I knew about the rapes, in a way, but not what had happened earlier, not then.

A: But you didn't mention them.

C: *No. Nothing. It was impossible.* But I remember her. *She took her time, she took her time and came into my room and sat there.* That's why I remember her name. There was another one, but her name I don't recall. But I remember B. well, she was somehow open about *the conditions in the clinic, it was known,* I'd worked at the hospital so I knew *it was known* that the gynecological department wasn't especially cozy ... (long pause).

A: You consider her as having been concerned in two ways, both the disdain for women practiced in the ward, and because she wondered whether you'd had experiences which might have been the origin of your problems?

C: Yes, I'm sure. Not because she said so, *but because of her way of asking, that's how I was sure she understood more.*

Only once does Camilla really feel seen; only one doctor "came, took her time, sat there, talked," therefore her name, her vitality and her empathy are remembered. A female gynecologist senses there may be information of relevance never elicited from a patient with an incurable health problem. She may have been one of the two doctors who witnesses how Camilla is met when coming in for her third miscarriage. She may also have wondered about the consistent lack of a verified cause for this health problem. She may have felt uneasy concerning the documented failure to cure a sickness which, nevertheless, always is seen as indicating the same intervention. She may have felt uncomfortable to see this obviously healthy woman unable to carry her unborn child more than seventeen weeks into gestation, a fact she herself confirms in the records, while aware that the fetus is not misshapen, and that no other probable defects can be verified. It seems, from Camilla's perspective, that this doctor is the only one who listens to her own professional discomfort when confronted with an incomprehensible chronicity combined with a medical repetitiveness. What makes Camilla confident, is a combination of empathic conduct and careful approach, showing that the doctor is willing to use her time to listen and talk to Camilla on an equal level.

she came to my room and sat down and talked with me
she came to see me several times and took her time talking to me
she couldn't work there, she couldn't stay there
she's the only doctor I've met before whom I might have told it
but I didn't tell her either. No. Nothing. It was impossible.
she took her time, she took her time and came into my room and sat there-
the conditions in the clinic, it was known, it was known
because of her way of asking, that's how I was sure she understood more.

The physician seems to know that time and an interchange of thoughts are needed to approach Camilla's sickness. A mutuality is shown in the physician's comment on the atmosphere in the hospital which forces her to leave. Camilla knows even better than she the hostility and disdain for women integrated into the clinical practice. She understands the doctor's refusal to take part in systematic, professional disregard and lack of respect. This means, however, that Camilla has not the time she needs to be able to confide in this doctor. She cannot just trust this system which has provided her ever stronger reasons not to. The only doctor to whom she might ever speak about sexual abuse and male violence is about to leave because of the abusive and disrespectful tenor in the place, before Camilla is ready to tell. She is left again, though empathetically, to the systematic disregard from which she cannot escape as easily as can her doctor. Camilla does not accuse the physician, unlike Christine, who feels especially betrayed by the only doctor who "had shown her that she saw something," and who addresses the moral obligation to intervene which those who sense the presence of any violent assault or humiliation have. Camilla does not accuse, but accepts fatalistically being left after having been explicitly invited to confide herself. She does not even speak about feeling disappointed. She has been left and sacrificed before by people whom she has trusted, and even loved. Even when feeling betrayed, she will never accuse. This is stated in the very first interview excerpt. She has never dared to ask her mother what she knew. Nor will she ever accuse her father of having offered her to a sexually abusive man, as it might cause his death from a heart disease which, as Camilla suggests, is his shield against hearing her speak his betrayal. However, she heals her sickness herself by speaking the other men's abuse.

2. BJARNE AND ELISABETH

If the process of becoming sick in the wake of hidden and silenced sexual violation is to become comprehensible, both as to it's course and nature, then it must be explored contextually, as exemplified by the interview with Camilla. It is the making of sickness out of violation within a frame of social, cultural and professional conditions, all of which contribute in specific ways to its very particular dynamics. The sickness emerges gradually, and its turns and peaks do not occur at random. When one reads the sickness history through the 'filter' of the abuse history, its intrinsic logic becomes evident, as previously demonstrated in the abuse-sickness-histories of Veronika, Bjarne, Fredrik, Oda, Nora, Annabella and Fanny.

Theoretical reflection is required in order to gain access to the process of becoming sick, and to the making of a patient in a medical context. For this purpose, documentation from sickness histories as they are provided in medical records are deconstructed from the perspective of adaptation to hidden abuse experience. This deconstruction renders visible the process of becoming sick as being a breakdown of adaptability to chaos, in the sense of Pierre Janet. Driving forces behind this breakdown, however, are: medical interventions, which may contribute to the maintaining of chaos; and the silence around socially silenced abuse. Pierre Janet's theory is grounded in observations of the phenomenon of dissociation, reflections about the nature of memory processing, and hypotheses concerning consequences of the failure to narrate trauma experience.

As mentioned previously, dissociation has recently been confirmed to be a central aspect of mental states, or "translations," of trauma experience, which are conceptualized as representing syndromes, posttraumatic stress disorders, PTSD, according to psychiatric classification. The DSM IV definition is built on six criteria related to the initial condition of trauma, whereof the first are formulated as follows:

> A. The person has been exposed to a traumatic event in which both of the following were present:
>
> (1) the person experienced, witnessed, or was confronted with an event or events that involved actual or threatened death or serious injury, or a threat to the physical integrity of self or others
>
> (2) the person's response involved intense fear, helplessness, or horror. Note: in children, this may be expressed instead by disorganized or agitated behavior. (DSM IV 1994:427)

The formulation allows the conclusion that these criteria for the definition of trauma are primarily or predominantly focused on threats to the physical aspects of human life and existence. Some experience modalities of the traumatized person are taken into account, however, as regards children, these are expressed from the viewpoint of an observer, based apparently on the supposition that horror, for example, is a feeling in adults and a behavior in children. All other criteria are linked to psychological or mental aspects, although some of them may be reflected in physical activity, such as hyper-vigilance or outbursts of anger. In other words: trauma is generated by physical threat but reflected in mental states. The state of PTSD, however, is very often associated with a variety of other states which psychiatry defines as being discrete, and which are then conceptualized as independent from trauma. Psychiatry regards their occurring in conjunction with symptoms indicating PTSD as a co-morbidity. The frequency of this co-morbidity has led to the supposition that people suffering from PTSD may also be at risk for these states. This is expressed as follows:

> There may be increased risk of Panic Disorder, Agoraphobia, Obsessive-Compulsive Disorder, Social Phobia, Specific Phobia, Major Depressive Disorder, Somatization Disorder, and Substance-Related Disorder. It is not known to what extent these disorders precede or follow the onset of Post-traumatic Stress Disorder. (DSM IV 1994:425)

When recalling the list of disorders mentioned above, the observable expressions they denote have all been mentioned by the interviewees quoted in this study thus far. Through the narratives about these "states" as they are provided in the interviews, however, the "states" may be seen as related to and evolving from violations and assaults, as well as how these were felt, experienced, interpreted and adapted to. In narrative form, told by the experiencing subject, these "states" represent ways of living under ongoing threat or danger. They are modes of coping with uncertainty and unpredictability. They mirror techniques for surviving and shielding. They represent strategies for the protecting of personal integrity. They are reliable, undeniable and eloquent testimonies to the prior occurrence of serious, molesting and traumatizing assault. Likewise, they bear witness to something which ought to have been told but has not been, and which demands to be understood but has not been - regardless of the reasons. Furthermore, these states are impressive demonstrations of an internalized destructivity which has been "forced into" the individuals' lives, which was not invented by them, and of which they themselves are not the source.

And finally, the "states" reveal the incredibly creative capacity with which human beings are endowed for shielding the self from the forces of imposed humiliation, and from the extinguishing power of having one's life, body or self objectified, which is identical with the privation of human dignity.[269]

The literature concerning PTSD, however, demonstrates a far greater interest within psychiatric research in objective proof of pathology than in subjective accounts as to the different modalities of trauma experiences, the human adaptive potential, or the individual creativity in the wake of assault. Neurophysiologist van der Kolk and psychiatrist Fisler write:

> If trauma is defined as the experience of an inescapable stressful event that overwhelms one's existing coping mechanisms, it is questionable whether findings of memory distortion in normal subjects exposed to videotaped stresses in the laboratory can serve as meaningful guides to understanding trauma memories. Clearly, there is very little similarity between viewing a simulated car accident on a TV screen, and being the responsible driver in a car crash in which one's own children are killed...
>
> ...Surprisingly, since the early part of this century, there have been *very few published studies that systematically explore the nature of traumatic memories based on detailed patient reports.* (italics mine) (van der Kolk & Fisler 1995)

Despite the ever increasing acknowledgment that trauma cannot be simulated, and that 'as if' experiences do not equal 'originals,' the study of trauma impact has continuously been characterized by an avoidance of subjective sources. These are considered to be flawed because verbal testimony is a priori subjective. Among the objective sources regarding traumatized subjects, narrative accounts are considered less reliable than observations of behavior. These, however, are regarded as still more imprecise than measures of physical substances. The proof of trauma experience has been based on, for example, the verification of altered physiological or biochemical states in traumatized persons. Thus, various studies have concentrated upon the measurement of certain indicators for stress. This is referred to in an article reviewing the thoughts and theories of Pierre Janet. He proposed, rather than a biological approach, an evaluation of the gradual breakdown of adaptability to trauma experience as leading to chronic helplessness, expressing itself through psychological and somatic symptoms. He described the process as being guided by an alternation between intrusion and avoidance symptoms, as was later confirmed by other researchers in the field of traumatology.

> Contemporary research shows that in situations reminiscent of combat stress, war veterans with PTSD show chronically increased autonomic arousal and abnormalities of the catecholamine and endogenous opioid systems. Many studies of traumatized populations have confirmed Janet's observation that *they suffer from psychophysiological symptoms of the respiratory, digestive, cardiovascular, and endocrine systems.* (italics mine) (van der Kolk & van der Hart 1989).

The next step of confirmation of Pierre Janet's observations was the acknowledgment of age and mental, emotional and cognitive maturity at the time of trauma experience as being of determining importance. Janet had proposed trauma memory to be stored on various levels, as sensory perceptions, visual images, visceral sensations, or as narratives, quite depending on the perceptivity and the cognitive skills or abilities of the assaulted person.

> Cognitive psychologists have demonstrated that memories determine the interpretation of the present, even when they are not conscious, i.e. encoded on a linguistic level. They have identified three modes of information processing: enactive, iconic, and symbolic/linguistic, which closely parallel the stages of sensorimotor, pre-operational, and operational thinking described by Piaget. These different modes of information processing reflect stages of the CNS development. As they mature, children shift from sensorimotor (motoric action) to perceptual (iconic) representations to symbolic and linguistic modes of organization of mental experience. When people are traumatized, i.e., exposed to a frightening event that does not fit into existing conceptual frameworks, they experience 'speechless terror.' Janet believed that when people are terrified, the usual cognitive schemata are inadequate to create a mental construct which places the experience in the perspective of prior knowledge schemes, causing it to be left unintegrated and to persist as a psychological automatism. (van der Kolk & van der Hart 1989)

When recalling the reports of the early memories which Bjarne and Elisabeth had confirmed by searching for the place where they belonged, the phenomenon of visual memories from a pre-language age ought to be reflected upon. Both persons remembered in movie-like glimpse what they had not stored in words but which they, years later, could easily describe with words. By using a symbolic representation for what, in its original, was stored as sensory perceptions, the pictures were made narratable. Thus, the "told pictures" could be identified by others, whereby they were attributed a meaning they had lacked in the original imprint. In the identification of words for pictures, these memories were acknowledged as true, valid, reliable and no longer just isolated images, but rather as a part of events. The connections between the pictures and the events were brought forth in a narrative about the original circumstances and the authenticity of the images. Only by their being told and then recognized as representing witnessed events could the iconic be transformed into the symbolic. Unless someone else had known them, they would have continued to be "unknown" by the remembering person.

Elisabeth knew by pictures being acknowledged by her mother that she had not been mad and hallucinating when seeing them, but had merely been in a speechless time of her embodied life, an authentic and past phase of being. Therefore she knew her diagnosis rested on flawed premises. Bjarne was able to utilize the confirmation of his recall ability to increase his eagerness to actively recall and reintegrate that which had been disrupted for more than forty years. He regained confidence in his memory, and he could trust anew that which, in his mind, represented lived life and not just some intrusive thoughts coming out of nowhere, created by nothing. Bjarne and Elisabeth had been in psychiatric treatment for years, the greater part of their adult lives. The psychiatric response to their attempts to tell had been to minimize or ignore the information; the result had been an ever increasing sickness. Their sickness had been labeled differently, but neither one had been understood. The gravity of its expression and consistency had been taken into account, but only treated symptomatically, not as a valid indicator of grave, unacknowledged and untold trauma.

Bjarne's sickness career included a variety of states which were all interpreted as being mental symptoms, qualifying, one after the other, for a range of psychiatric diagnoses. From early adulthood, these diagnoses were linked by health professionals to what he expressed concerning: his sexual identity ambiguity; an alternation of depressive and manic states; a profound disturbance of sleep including nightmares; an abuse of alcohol; an abuse of drugs (gradually imposed on him by professionals); and repeated suicide attempts. In addition, he experienced critical disruptions to his private and professional life. Finally, he was disabled by chronic pain and a reactive depression after a violent assault.

As Bjarne's abuse history has been explored in depth in the previous chapters, the course and nature of his sickness career can be decoded as being a result of a social, cultural and professional suppression of the narrative about his childhood trauma and the dramatic context by which it was shaped. Consequently, the sickness career can be understood as a making, and not as a becoming, not as a somehow "natural" development of a psychiatric multi-morbidity in a disposed individual. It emerges as engendered by a social and professional suppression and denial, of a consequent silencing whenever he wanted or dared to tell. Silencing what needed to be told in order to be integrated brought illness upon the person who was forced to hold back what he wanted to share. He was obliged to contain the pain of the past, which forced him to numb himself with alcohol, drugs, and at times a near mania which led to breakdowns due to exhaustion, and to numb his entire body by avoiding any awareness of its problematic sensuality.

Elisabeth's sickness history included, apart from psychiatric treatment, a variety of somatic conditions leading to medical treatment, some of which have been mentioned already in connection with her dental problems, the pain in her jaws, and her inability to open her mouth properly. When reporting her somatic health problems in the beginning of the interview, she had noted pain as a major problem during her childhood, localized in her neck, shoulders, head, and pelvic area from before she was ten years old. This pain remained during adolescence, the pelvic pain increasing and then peaking at every menstruation, evoking repeated loss of consciousness. From early adulthood, the pain gradually affected her entire body, but was localized particularly in her pelvic area. In addition, her steadily increasing problems related to the mouth and a broad array of allergies generated frequent contacts with the health care system. After having given birth to two children, she was admitted to a hospital suspected of suffering from cancer of the reproductive organs, which was considered a possible cause of the chronic and ever more incapacitating pelvic pain. Peroperatively, no malign process was verified. Nonetheless, the surgical intervention resulted in an appendectomy, a partial removal of the right ovary due to what was explained to her as adhesions, and a bilateral tubal ligation, terminating her fertility. A record of this hospital admission could not be obtained, therefore the indications for such a sizable intervention cannot be compared with Elisabeth's memory of the information she was given after the operation.

3. LINE, CHRISTINE, SUSANNE AND MARY

Elisabeth's somatic sickness history is not exceptional, as became obvious when exploring Camilla's, Eva Maria's and Oda's interviews concerning their gynecological and reproductive health problems. Neither is it exceptional as to chronic pain being the major complaint in this group. Nor is Elisabeth's sickness history unique in its blending of long-term psychiatric and somatic health problems. Therefore, four interviews are chosen to allow an exploration into the process of becoming sick, as it is told by the interviewees, *and* as it is described in hospital records. The potentially destructive force of hidden sexual violation on somatic health will be approached by analyzing the development of Line's lower back pain, Christine's pelvic pain, Susanne's arthritis, and of Mary's acute referral for suspected meningitis. With the exception of Susanne, all these interviewees have been referred to in several of the previous chapters.

Line recalls, apart from a broken arm at the age of two when having fallen from a bed, frequent urinary tract infections and an increasingly problematic urine incontinence. This was thematized as follows:

> **L** Of diseases in childhood I can recall the doctor stating that I had an infection of the urinary system. I don't recall precisely how old I was, perhaps nine. I don't remember. Possibly I was nine. It may have happened after the first abuse. This urinary tract problem stayed with me further on. I was very troubled by it. I peed in my pants very easily, so I had to use huge diapers, you see? I was very, very troubled by this. I couldn't get rid of it until I was eighteen years or so.

Line had been forced into performing fellatio and had been fondled genitally by her father once at the age of eight. The next time, when she was twelve, he penetrated her vaginally but without performing intercourse. Both abusive acts occurred in the mother's absence due to a journey and a hospitalization. Thus, Line has reliable dates to relate her own age to. Until she was eight years old, she had neither had infections nor any problems controlling urination. From the time of her first infection she became incontinent. When asked about whether special measures had been taken to treat what was a great hygienic and social problem, she recalled having been examined by a doctor while at the primary school level, but never later. She had neither received any treatment nor been examined after she entered high school. At school, however, she had been allowed to change separately before swimming or gymnastics, and she believed none of her classmates ever knew she used diapers since she had developed a variety of strategies for hiding this shameful fact, such as never sleeping overnight at the homes of her friends.

Her incontinence and frequent infections disappeared after she was eighteen years old. At that time, she left her family and entered into a relationship with a man. The development parallels Veronika's, who also suffered from a profound, chronic impairment to health and social functioning which improved abruptly and without any intervention after she had left her home and her abusive father. Although, unlike Veronika, Line had not suffered from continuous abuse, she may have lived in the realistic and constant expectation of new offenses. She even recalled her suspicion of her father abusing her younger sister and one of her brothers, mentioned in later parts of the interview. The threat of abuse and the threat of disclosure and punishment was paramount. One may even be tempted to consider that, by taking strict control over her mouth in order not to let out what must not be told, she had lost control over another bodily function she previously had mastered.

As stated, Line's regaining of control over her urinating coincided with her moving away from home. However, she entered into another destructive relationship. This is thematized as follows in three parts of the interview, first in the dialogue about her lifetime abuse experience, next in a summary of her statements about the father of her child, and finally in a short remark about the termination of this relationship:

> **A:** Have you later ever experienced situations when someone came too close?
>
> **L:** Yes, but not that they ... well, I had a partner who battered me, but I never have experienced incest from anybody else.
>
> **A:** Your partner was violent. But was the violence ever connected to sexuality, he didn't force you?
>
> **L:** No. It didn't happen. No one else either.
>
> * We talk about how she met her partner and about the period when she got pregnant. She tells that she more or less fled into this relationship, and that she stayed there although she very soon understood that this man had a serious alcohol problem - and that he turned violent when drunk. After some time, he began to act violently even when not drunk. But she thought she had no choice but to stay as she had no other place to go to. She could not earn her own living.
>
> * We talk about the time of her cohabitation with the father of her child and about the violence. It ended by her fleeing from him to a women's shelter where she stayed for some time. Then she went into therapy with a psychiatrist in combination with attending the psychiatric day-treatment center. In this connection, her son was placed in a foster home.

Line had left a situation of being constantly endangered by a sexual abuser, and gone into a relationship with an alcoholic who gradually revealed his violent character, although it took some time before he allowed himself to act violently without being under the influence of alcohol, and thus excused from responsibility. Line then had to face the fact that he assaulted her by will. She needed time to comprehend that she had not escaped but, rather, was trapped anew. However, her options for escaping were just as bad as in her early youth. She had no other place to go as her family home did not represent a shelter. Although employed, she was an unskilled worker, having very limited possibilities to earn a living on her own. At twenty-one years old, she develops continually more severe lower back pain which, after a period of observation, leads to her being admitted to a third-line orthopedic hospital far away from her home-town, due to a condition termed "spondylolisthesis." The entry note from the records is as follows:

> *Present state*: Woman in normal condition. Seems somehow slow, psychically. Answers questions slowly. One feels she has problems finding words for her

answers. Walks a little carefully, but not insecure. Spastic muscles cannot be found. When standing, a hollow area according to L5-S is found as with spondylolisthesis. Mobility in her back is good, although the back is kept fairly straight. Anisomeli of her right leg is found, clinically equaling almost 2 cm shorter than the left.

Sensibility: No certain defect of sensibility found.

Mobility: No pareses to be found.

Reflexes: When examining the reflexes one feels uncertain as to findings. Both patellar- and achille-reflexes show almost O extension. The plantar-reflexes seem to be inverted.

Gen. exam: Blood pressure 115/80, pulse rate 76 regular, auscultation cor and pulm no remarks.

One permits oneself to refer the patient for neurological examination. Today's neurological status is difficult to interpret as to reflexes.

The next day, previously taken x-rays are viewed, leading to a note as follows:

There is agreement to perform a neurological examination and a lumbal radiculography. *A clear indication for an operation is present* since there is a considerable gliding to be found between S1 and L5 (3.degree)

The neurological examination is concluded as follows:

When performing a usual neurological status one cannot verify any defined pathological findings, especially no positive signs indicating any spinal roots concerning the extremities to have been affected.

After also having performed a myelography, the conclusion of the orthopedists is as follows:

The patient's myelography is viewed and the neurological exam reported. The neurologist cannot find any affection of the CNS with roots and medulla spinalis. Myelography shows bilaterally a short root at L5. No narrow spinal passage. One considers an outer fixation between L5 - S without laminectomy being appropriate.

This conclusion, based on several examinations which cannot verify an acute threat to the central functions of this young woman, or to vital parts of her obviously healthy body, seems strange. There is one finding only, representing a relative instability of the spine at one point. It was already known, was not caused by any acute, recent injury, and has to be considered as most probably congenital. At any rate, this is explained to the patient. Line says:

The doctor said I possibly was born with it, but he didn't know for sure. It could just as well be ... if I hadn't been operated on, I could have ended up in a wheel-chair, he told me, due to ... well, he didn't really explain why. But it was serious. Otherwise I wouldn't have been operated on, after all.

Line's pain is explained by an anatomical anomaly of the spine which she may have had since birth. If not, it has to be an acquired defect, although no one knows when or due to what. If it is congenital, the "finding" does not explain a sudden onset of pain at the age of twenty-one. The anomaly does not affect mobility, it does not threaten any vital functions and does not impair sensibility and motility. Why is it still a "clear" indication for a surgical intervention, which is not performed without a certain risk? What seems "clear" is that there is a pain and there is a finding. This combination seems, almost by logic, to equal an "indication." Thus, the operation, a spondylodesis, is performed. Before this, the patient is advised to use a corset, meant to provide stability, but which actually causes a new problem. This is mentioned as follows:

> When using the corset she has a considerable inclination to the right which is caused by the deviation of the pelvis due to anisomely. With 2 cm under her right foot one gets her somehow straight. She feels this a bit unusual.

The first medical decision - a finding equals a cause - engenders the next - the cause has to be corrected, by means of surgery and a corset which actually presses Line's body out of balance. To compensate for this, a special heel is ordered placed under one foot which influences her internal perception of balance thus making her feel "strange," as well as making her aware of the fact that one of her legs is shorter than the other, which she has never noticed before. Shortly afterwards, she is operated on, the corset is put on from the second postoperative day, and after two weeks she is referred to a rehabilitation clinic for several weeks of treatment. Here a circle closes, because Line had been advised previously, by her general practitioner, to attend the same clinic. This is told as follows:

> **L:** I'd begun working as a kitchen assistant. *There were maybe some heavy things to lift. The work was varied,* I did the dishes or prepared sandwiches or different things. I'd worked there for two years. I felt more and more pain in my back. Because of this, I discovered a tumor in my lower back. Here's how I discovered it. I went to my doctor here. He sent me to a specialist. They took x-rays at the rehab-clinic. There they found my spine was curved. There was a curve. But they'd found more, because they talked about an operation. At that time I had problems working.
>
> **A:** But it wasn't due to bad pain that you went to a doctor, it was because you found a tumor almost accidentally?
>
> **L:** *I can't remember feeling pain when I was at the rehab-clinic.* But I remember they talked about an operation, and they referred me to the University hospital. It took its time, at least several months.

Thereby, the making of a patient is completed. A very young woman complaining of lower back pain is examined in the area of her pain. An anatomical finding, though most probably a congenital anomaly, is verified and attributed a causal function, despite the lack of any other kind of pathology. As the "cause" is localized, a surgical intervention is made in an effort to remove the pain. No one realizes that the patient herself does not seem terribly affected by this pain, or even, that she is primarily concerned with a tumor in her back, which she herself had palpated when touching her own back. She cannot remember pain at the time of examination at the clinic. (This echoes Annabella's and Nora's histories.) However, a finding serious enough to lead potentially to a life in a wheelchair, is accepted as a reason for operating. Line is convinced she would not have been operated on if this had not been a serious finding. She admits, however, to having received an inconsistent or incomplete explanation. However, she does not comment upon the other measures, the corset and the higher heel in her right shoe, forcing her into an alien posture. She seems to have accepted this interfering limitation as one of the necessary measures.

When reporting from the time before the operation, Line has adopted the view of the doctors, mirrored in her wording about her work. She talks "with medicine," mentioning the "heavy" things she lifted and the "varied" work she performed. These words can be found in her records repeatedly, where every admission after the first operation includes a repetition of the conditions at her place of work. During the next admission, half a year after the first, mention is made that the patient "has relatively few problems with her corset," which indicates the topic had been addressed. A new problem, however, has appeared: the x-ray examination of the intervention area shows the development of a pseudarthrosis. No further comments are made to this finding, however. The next contact is an out-patient clinic check in order to change the corset to a lighter model. The record note is as follows:

> She has spent a period at the rehab-clinic in order to eliminate the corset and for training. She has continued to use it during periods of strain. She herself feels that *the old pain* has disappeared, *but she gets pain in her back when standing a while.* A light corset is now required which she may use when straining her back. Apart from this, new x-rays shall be taken one year post-op. She receives sick-leave certification for the rest of this year. If she feels her back is well enough, she may begin to work from the start of the new year.
>
> Apart from this she has a paresis of her n. facialis, probably caused by virus. She has attended the neurological ward and shall be checked there.

An "old" pain has disappeared, a "new" one has appeared, and yet another health problem has emerged which, however, belongs to another specialty and therefore is touched upon only briefly. Improvement is mentioned in the next record note. No association, of what kind ever, is suspected. Half a year later, the record text states that the patient has been somewhat incapacitated by pain for several months already and that she has quit her job. Once again, she is examined clinically and with x-rays. The conclusion was:

> *The patient's clinical findings may indicate a pseudarthrosis in the operation area.* It seems reasonable, however, that we use the corset for a time more systematically and she receives a requisition for training of stomach and lower back muscles with instructions for self-training. If this does not help and the patient still has the same complaints when checked after four months, we should conduct motility x-ray examination and planigraphy in the fused area.

Now, the recurrence of the "old" pain is attributed to the "new" finding, the pseudarthrosis, which is an artifact of the operation of the "old" finding. Concerning the consequences of being unemployed, no remark is made, nor is there mention of movement in day-to-day life including housework. Four months later, the patient has not improved, but the planned examinations cannot be performed because she has recently become pregnant. The usual clinical tests do not show any indications of neurological affections or muscular atrophies. As regards her pregnancy, a new corset is recommended which can be enlarged gradually and may support her back. A year later, after she has given birth, a note concerning an out-patient clinic contact states that the patient is pain-free.

During the entire record to this point, covering a three year time span with two in-patient and six out-patient clinic encounters, not one single remark concerning Line's private life can be found. Detailed comments regarding her work are contrasted with a total absence of information as to all other areas of life. Whenever the word strain is mentioned, it is related to physical work and her job, even during sick-leave and unemployment. The pain as it was presented initially, is explained by a congenital anomaly. When it returns, it is explained by an artifact. The strain of pregnancy in a woman incapacitated by a pain which may be the result of an operation due to pain, seems not to be of orthopedic concern. The patient is advised what she has been advised before: a prosthetic "back," and training. No one is aware that she has been battered frequently and increasingly during the last four years, and that she suffers from heavy limitations to her freedom of movement. No one wonders about the lack of a positive outcome of a rather demanding intervention, three hospital admissions adding to several months of hospitalization, a year's sick-leave, physiotherapy, training, and the continuous use of corsets. No one knows that she, when appearing for a check-up after giving birth, has just left a violent husband and lives in a shelter with her little baby. "She is pain free," is the only comment. The earlier mentioned pseudarthrosis is checked for but it cannot be verified by usual x-ray. The previously planned examinations to verify it properly, postponed due to pregnancy, are not performed since she has no pain any more. Thus, the tentative cause or origin of an unexpectedly recurring pain is not proven.

Two and a half years later, Line is once again admitted for pain. The record note is as follows:

> The patient has pain in her back and down both legs. She also feels that her legs become numb when she walks for a while, but it is really very difficult during the examination to get a proper impression of how much she is bothered by this and what kind of complaints she really has... It is possible that she has a kind of spinal stenosis symptomatology, and we ask her to return for control within half a year.

After a long period without pain, Line again is "in pain." Her bodily message of her legs not carrying her, and the lower part of her body becoming numbed and aching, is not heard. There is no question as to events of her life straining her considerably. Therefore she cannot tell that, after she fled to the shelter, started therapy and settled in on her own, her father has raped her in her own home. One year later, she still is in pain, and once again several examinations are performed. But, as no new findings which may serve as explanations for a chronic and therapy-resistant pain can be verified, the physicians decide:

> There is no indication to examine the patient with regard to a new intervention from our side. Therefore we do not make a new appointment for check-up.

A six year follow-up after surgery, a considerable medical investment, and a patient in a totally unchanged state of health, is the status. The hospital records do not contain any indication of reflection on this rather paltry professional outcome. A sickness history is terminated from the side of the therapists. Do they have an idea that they helped to engender it? Do they reflect upon what the substance of this intractable pain was? Does anyone wonder how this woman "in pain" lived?

Christine's becoming sick in connection with a persistent abdominal pain shall be explored in her sickness history and her medical records. From the age of seven, she had been sexually abused by the employer of her father, who threatened to fire the father from his job if she would not cooperate. Christine could recall pain in her stomach as her main health problem in childhood, together with repeated cystitis and frequent infections in her mouth leading to a tonsillectomy. Her stomach pain has been mentioned in connection with her first gynecological examination and her only experience of a supportive doctor. Here she learned that she might have sequelae of untreated infections having affected her reproductive organs. The next gynecological examination was different, and the excerpt from the interview is as follows:

> **C:** It was different the next time I had to search for help. It was during Easter week, I went to the out-patient clinic because of pain, there was a gynecologist who obviously didn't like to be bothered. I remembered the first examination anyway, which I, after all, had survived. But he

> wasn't very attentive and was so unpleasant that *I disappeared into a fog while he examined me.* Then he performed an ultrasound and said he couldn't find anything. But he said it seemed there had been a cyst which had ruptured and that something was on its way out, or something like that he mentioned. But the pain just increased. I had to go to the clinic again. This time I had a female doctor. She didn't examine me, only talked. Probably she could see I wasn't sick but had pain. She just kept me there for observation. Everybody in the ward was all right, of course they had their routines, but they weren't unpleasant. Finally they decided to perform an operation. To be there with all the others felt quite all right. All of them had the same incision and all laughed and coughed in the same way. But I recall that when I woke up after the anesthesia, a male doctor was standing there. The first thing I thought when I saw him was: a butcher. He was standing there in his white coat and I saw the blood stains, although they most probably weren't there. But they were in my head, because he was a butcher. And he said *they had to take my tubes because they were so damaged.* While he said so, I in a way saw the tubes fly in the air. 'Now I can't have children any more,' was the first thing I thought of. And then, he said, *they had taken some other tiny bits.* Then he left. I stayed there for four days, but I didn't receive any more information. *I never got to hear from the doctors what these 'tiny things' were.* I didn't really understand this about the tubes either, that they had disappeared in a way. Had they taken both? I still don't know whether they have done it or not.

From the records concerning the first examination at the out-patient clinic one learns that the physician is aware of Christine having been appendectomized at the age of ten, having had an unspecified injury to her back, never having been pregnant, not using any contraceptives as she lives in a lesbian relationship, having felt an increasing pelvic pain, mainly localized *in the right side of her pelvis,* during the last week. The conclusion after examination by vaginal exploration and pelvic ultrasound is:

> Everything seems quite peaceful now, she will wait for any developments at home. In case of worsening, she will make contact.

The next note, dated four days later, is written after her admission to the gynecological ward. The doctor states that the patient has been sleepless for several nights due to pain, *predominantly in the right side of her pelvis,* that she has no fever, that her general status is without any signs indicating a serious condition, and that the gynecological examination cannot verify pathology. However, since the pain starts anew after the clinically negative examination, an laparoscopy seems indicated. The description of this intervention the same day is as follows:

> 4042 Laparoscopy
>
> 7111 Resectio tubarum
>
> 7010 Resectio ovarii sinister.

> (description of initial surgical steps) ... One finds the uterus normally configured and without striking findings. No adhesions in the lower pelvis. Right side: Right ovarium without pathology. Tube worm-like, bowed and thickened and with nod-like parts. Several diverticles on the tube. Fimbriae partly normal. Left side: Left ovarium slightly thicker with one little cyst, and one finds a crypt like a ruptured follicular cyst. Bleeding initiated in the area when inspecting. Left tube also curved and changed, probably due to previous salpingitis, *but not as much as on the right side.*

When inspecting other organs in the area, a previous appendectomy is confirmed, no pathological findings appear, especially no adhesions to the omentum. The consulted senior surgeon decides a laparotomy to be indicated. The procedure is done in the same session. As it seems, the findings are interpreted as representing the cause of the abdominal pain the patient has complained of. A discrepancy between a consistently mentioned pain on the right side, and the more modest findings on this side as compared to the left, is not commented upon. The interpretations of ultrasound findings with regard to a ruptured cyst as a possible cause of the pain, are not confirmed since the cyst is localized on the left ovarium. According to the description of the operation, a part of the left ovarium is removed. The left tube remains and is not resected. The right tube is removed completely. This contrasts the surgical report, where the plural genitive form "tubarum" indicates the resection of both tubes and not the removal of one. Possibly, this inconsistency reflects merely a problem with Latin grammar, also mirrored in yet another fault, but is consistent with what the patient is later told had been done.

In this part of the records, no remark can be found that the patient, who has never given birth, ought to be consulted before this restriction of her natural fertility. Furthermore, no reflection is noted concerning the patient's complete unpreparedness for this considerable expansion of what was argued for and initiated as a diagnostical procedure. Some days later, the patient is released. The final record remark is:

> Postoperative course uncomplicated. The patient is released with general advice. She needs no sick-leave certificate. Will receive a message in case histological findings indicate the necessity. No routine control scheduled.
>
> Histological report: Tube with lymphectasies and moderate chronic salpingitis. Confer description.

These short remarks stand in contrast to Christine's emotions about the information she is given when still under the influence of anesthesia. A message about *removal of both tubes* reaches her, but not about the partial resection of one ovarium. She recalls thoughts about sudden infertility, but does not speak about this. The topic is not mentioned in the records. A sick-leave certificate is, however, found worthy of being mentioned. A thirty-year old woman has her fertility decisively restricted before ever having been pregnant, without having been informed that a diagnostic procedure might possibly be expanded peroperatively if the findings should indicate such a step. The findings, a change in the shape of the fallopian tubes, and a little cyst on one ovarium, are immediately evaluated as carrying indicative weight. They seem to represent such an urgent threat to health or life that they are adequate reason for their immediate removal, although no other clinical signs are verified. The result of the histological examination, confirming a chronic inflammation in a moderate stage, renders false the emergency-like expansion of a diagnostical procedure into an imposed and non-consensual intervention which holds far-reaching consequences for a young woman's life and future.

How does Christine experience this development? She addresses this in the following excerpt:

> Those 'tiny things' were some adhesions, as they said. A nurse who'd been present when I was there told me this. She came to me some days later and said so. So I partly knew something. But not from the doctor who'd operated me and who came in right after I woke up, because I was so afraid and fairly 'foggy.' The only thing I heard of what he said was that I couldn't have children. There a doctor was standing and telling me I could have no children. And then he said something about that they of course could help me with this in another way. And then he left.
>
> I felt very broken, I was down for a long while afterwards. *But the pain didn't leave. I just continued to feel pain.* Not that it was present all the time. It came and went. And it didn't have the same intensity all the time. But I thought all the time about what the doctor had said, that the tubes were so damaged. I figured more was wrong, *that I was in a way rotten inside.* But I continued to live with it. *Gradually I also recognized that the pain emerged when I felt stressed.*

The incompleteness and inconsistency of information given to Christine is confirmed in the next record note linked to her admission to the same ward with the same problem, now termed 'Dolores abdominis.' The doctor examining her notes a previous removal of the right tube and a partial resection of the left ovarium. A gynecological and general examination cannot confirm pathology. The final note after this process is:

> The patient is admitted for exploratory laparoscopy and she agrees to an eventual laparotomy *if this may help for her pain.* Possibly there are adhesions to the anterior part.

The next step, a laparoscopy under anesthesia, is performed as if it were "ordered" beforehand by the referring physician despite the negative findings of the specialist. The intrinsic logic leading from the description after the examination to the decision to perform an endoscopy, is not really convincing. The gynecologist knows that the patient has been operated on before in order to remove a pain but that the previous intervention has clearly not lead to the result it aimed for. Still, endoscopy with a possible peroperative expansion to laparotomy is deemed indicated. This time the patient's consent is given prior to the procedure, but the reason for her consenting, having been told that her pain could be terminated by more surgery, makes this consent rest on dubious premises. Of course the patient wants to have the pain terminated, but she is not the one to decide whether surgical measures are appropriate measures. Her consent must be seen as guided by the specialist's suggestion that this might help. In addition, it is grounded in her tacit and shameful conviction that what is "rotten inside" ought to be removed. The description of the second intervention is as follows:

> (description of the initial gynecological exploration and the first steps of the operation) . . . Uterus normal size, mobile. Adnexa right side: status as to salpingectomia, ovarium normal. Adnexa left side: normal tube with good fimbria apparatus. Ovarium looks normal. No adhesions. No sign of endometriosis. There is no pathology elsewhere in the abdomen. Senior surgeon consulted. Conclusion: Normal findings. Status post salpingectomia dextrum et resectio partiarum ovari sinistri. No explanation as to the patient's abdominal pain can be found.
>
> She needs no sick-leave certificate since she is a student. Is explained the findings and is advised relaxation-exercise or TNS. Contacts own physician.

No pathology, but the expected status after removal of a fallopian tube is confirmed. The left tube, previously described as deformed, and by the doctor informing Christine described as "damaged," is found to be normal. The same senior surgeon who the first time decided to expand the surgical intervention is consulted and confirms the present normality. The same situation with the same patient with the same problem is responded to quite differently, supervised by the same professor, at one year's interval. The final advice to Christine is to "practice relaxation," suggesting tension in the patient. However, where this suggestion is derived from, what it may concern, and what the term "tension" is intended to designate, is not made explicit. It does not lead to any question or remark as to its kind or nature; the "advice" provided is so profoundly non-specific that it can hardly be called "professional." How shall relaxation be practiced when the origin, nature and kind of tension is unknown, and that question is left entirely unaddressed and unexplored? What do the doctors mean? What are they trying to indicate? Can Christine use this advice? Does she know the origin of the tension which is a possible source of repeated pain? She has mentioned her gradual acknowledgment of the pain being triggered by stress. This is addressed as follows:

A: What does stress you most, have you found that out?

C: Yes, in a way. *Everything which reminds me, everything concerning the incest, then I get pain in my pelvis.* I've managed to localize that. It is related. *I haven't been to a gynecologist since. There was no point to going for the pain.* Now I consult a doctor nearby. He knows everything. I've told him that I use the incest center. He is very fair and cooperative, and when he's forgotten to close the door to the room where the gynecological chair is, I tell him to close it. And so he does.

Christine has found a doctor she can talk to, who follows her reasoning and who respects her demands which are the prerequisite for her feeling safe. It is most crucial to her to shield herself from the sight of a gynecological chair.

Christine has not received help from the health care professions for her chronic diseases, mainly pain, allergies and asthma. She has not given up hope for receiving therapy. She has gradually regained confidence in her own ability to know her needs, and she has acknowledged that medical intervention and advice is inadequate.

Susanne has been abused from the ages of twelve to seventeen by her eldest brother-in-law, in her home. Her extended family lived together in one large household. The father of her best girlfriend abused her several times during the same time span, and so did an adult male neighbor, a friend of her father. Later, she was harassed by several men at her place of work, which is dominated by men. She married a man eight years older than she, when still fairly young. This was her way to get out of the house where she was abused. While pregnant for the first time, at twenty, she had already acknowledged that the man she had chosen bore many similarities to her main abuser. She had to cope with her increasing resistance to being intimate with him, whose hands and body-size reminded her constantly of her abuser and of the abusive acts. She gave birth to four children. Each time, she suffered from severe pelvic pain during pregnancy, termed "bekkenløsning" in Norwegian, suggesting a loosening of the pelvic ring during pregnancy resulting in an instability which causes pain. Pain became her main health problem. This is thematized as follows:

S: I got arthritis when pregnant with my second child. The last time I was admitted for this disease was two years ago. Then I got the diagnosis. It hadn't been so serious before so I didn't know what it was. The physiotherapist mentioned it first. She thought it was arthritis. I had pain in my joints, swollen ankles, mostly the right one. For several years I believed it to be sciatica, in a way it was a continuation of my pelvic pain during pregnancy. But then it localized more and more in one hip - or in the lower pelvic area. So I thought it to be gynecological and went to a female gynecologist. It wasn't especially pleasant. She sent me to the hospital. But there they didn't find anything. Nothing was wrong

> with my pelvis. They checked my back, took blood samples - and the SR was 100. They attributed this to the contraceptive pills I used since no other explanation could be found. It was already rheumatism. But no one mentioned it. Mostly because it felt as if it was in my pelvis - or, anyway, the region of my hip. From there the pain spread.

* She tells about the uncertainty connected to the first hospital admission due to pain. In this connection, mention is made of her brother suffering from psoriasis-arthritis, and that there is RA in her family. Apart from this she has received the diagnosis hypothyroidism and uses a medication to compensate the lacking hormone.

Pain from the age of twenty, appearing at the onset of her first pregnancy and mostly localized in the pelvic region or the hips is diagnosed as "bekkenløsning." Between the pregnancies the pain, still localized in the back and down the hip, is associated with sciatica. Its severity makes it necessary to seek physiotherapists. One of them mentions arthritis as a possible diagnosis for the origin of the pain as it, now and then, is felt in the lower limbs and joints as well. Susanne feels it primarily in the pelvic region, however. As no pathology in the pelvis or the reproductive organs can be found, a gynecological origin is excluded. The only finding, a high erythrocyte sedimentation rate, is attributed her use of contraceptive pills. A constant pain in the pelvic region is aggravated during every pregnancy, mainly the last. At that time, Susanne's marriage breaks up. The first admission to a rheumatological ward does not lead to a diagnosis or a solution. Heredity is considered. The next admission is two years later; she has recently been divorced and is fully employed. The record notes from this admission confirm years of a history of pain, frequent sick-leaves due to pain, a varying localization of this pain in ankles, elbows, hip and shoulder. Previous x-ray examinations of all relevant joints and the pelvis show normal conditions, as does the spectrum of serological tests taken. Pain, stiffness and sleep disturbances are the three major complaints. The patient is described as walking without any problems. Her general status shows no signs indicating acute sickness. The serological and radiological examinations are negative. The only finding is described as follows:

> One can see a moderate tendovaginitis in the right ankle related to the tibialis posterior group. There is also a similar tendency in the left ankle. Conclusion: *the patient has a seronegative rheumatiod arthritis.* During the hospitalization mainly reaction in right ankle and right hand. Starts with medication X and continues at home. In case of insufficient effect or side-effects she may try another NSAID. In the long run, one ought to consider specific medication (gold).

Susanne is advised to seek medical training in a group of RA patients. She dislikes being in groups, however, and dislikes being among people who talk about sickness. Consequently, she has refused to participate. She has been told that she has to accept her disease. Susanne does not feel sick, however, and prefers to play sports with her daughters. Since her divorce, she has felt increasingly healthy and strong. She cannot comprehend the point of adapting to a diagnosis which seems irrelevant, as she almost never has pain any more. During recent years, she has learned to shield herself from pictures showing violence and she still reacts to harsh male voices with trembling and fear, but now she knows why.

Mary's process of becoming sick, the breakdown of her adaptability to continuous abuse, is found in her interview. Mary had been abused by her father from the age of five until seventeen, and by her eldest brother from when she was ten until she was twenty-seven. When she was a teenager, this brother had often abused her along with a girlfriend from the neighborhood. This girl later became his lover, developed an alcohol dependency and left the small town in late adolescence.

From the age of twelve, Mary had used drugs. She paid with money she received from two older men in her town who abused her during a span of several years. After a suicide attempt, Mary was in a psychiatric ward where she was in a near-psychotic state for a time and had problems talking properly. She made the same pronunciation errors as she did at five years old, the year of the onset of abuse, when she was trained by a speech therapist. During late adolescence, after her father's death, her brother had left the country and worked abroad for several years. Mary again attended a school, had met her future husband, had gotten married and become pregnant. After the first child was born, Mary and her husband started to build a house. At that time, her brother returned. In the guise of offering help, he gained access to Mary when she took her turn working on the house-to-be while her husband was at work. Unaware of the abusive relationship, Mary's husband felt grateful to his brother-in-law, who, whenever he could, forced Mary to have sexual intercourse with him. When pregnant the next time, Mary was not sure who was the father of this child. Her state became more and more disturbed, and the next delivery resulted in yet another admission to the psychiatric ward.

The house was finished the same year, the family moved in, partly joined by Mary's brother in his capacity as a helper. Shortly before Christmas, a couple moved into one of the houses in the neighborhood. It was Mary's girlfriend from childhood who had left earlier, married, and now returned, most probably unaware of Mary living close by as well as the brother, her molester. When Mary got to know this, she felt trapped. She fell sick and grew worse, day by day. On New Year's Eve, she was admitted to the hospital emergency room, suspected of having meningitis. The record note after five days of observation is as follows:

> She still had very high fever, less stiffness in her neck and back, now mentally adequate, red injected eyes, no movement to the sides. Still light-sensitive. No neurological defect. Seems somehow uncritical as to present situation, wants to leave and work. She does not remember that she fainted when on the toilet several times during these days. Some of this may be an effect of high fever, Hibanil and a meningeal irritation. Stop drug administration due to this. Her blood shows lymphocytosis above 50%, white blood cells down to 2.4, now rising to 3.7. No second spinal tap is performed since the blood cultures are negative and her neck is less stiff. Still no antibiotics.

Two days later, a note in her medical record indicates a referral to a consultant psychiatrist, as Mary has expressed suicidal fantasies and nightly panic attacks with the feeling of being suffocated. When dismissed from the medical ward, she enters the psychiatric ward where she stays for six months. From the time at the medical ward Mary recalls the following:

> **M:** I was supervised in the ward. They couldn't make a proper diagnosis. But there I saw my father again, so I must have been on the edge. Aside from this, they told me that I'd been out of bed during the night and had fainted, which I couldn't remember. I'm also sure that the head midwife, who talked a lot to me after my second delivery, came to see me. They must have asked her to come, and then they decided I ought to be referred to the psychiatric ward again.

> * Her voice turns soft and imprecise again. She tells that the conversations with the midwife had made her start to talk about her father. The midwife managed to mediate to Mary that she thought all this that Mary didn't dare to talk about made her so disturbed and exhausted. But Mary still did not dare to tell any details, or even about the other men or about her brother.

An opening is made. Someone has listened intensely and suggested an interrelatedness between Mary's powerlessness, panic and disturbances, and the great and shameful secrets about which she has not been able to speak. From then on, after one person has seen and understood, Mary can make progress. Since she was five, her life was influenced more and more by the sickening force of hidden abuse. Initially, this showed in a sudden retardation of speech development, making speech training necessary. Infections, injuries and self-molestations followed. Drug abuse evolved from early adolescence during a period of child-prostitution. Several suicide attempts lead to hospital admissions. Only in one period, after her father had died and her brother had left, did Mary prosper for some years. She engaged in education, met her future husband, married, and got pregnant. Her first delivery, logically reactivating what had been hidden, forced her into a psychotic state. Her brother's return, leading to doubts concerning the fatherhood of her next child, brought her again to "the edge."

And so did her final recognition that all her efforts to escape from abuse, shame, and the pain of imposed humiliation had been in vain. The most improbable of constellations had reappeared. Past was present again. Mary saw no further options. A physical and mental breakdown was the result. Her adaptability was completely exhausted

4. THEORIES ABOUT VICTIMS

In the following I shall review changing models and concepts concerning the impact of, especially, sexual boundary violation. Before doing so, a field of semantic tension has to be addressed, related to the Norwegian(N), German(G) and English terms linked to abuse, assault and insult. Both in the German and Norwegian language, the term abuse designates three kinds of violence: *Misbrauch* (G) or *misbruk* (N), indicating sexual abuse, *Mishandlung* (G) or *mishandling* (N), indicating physical abuse, and *Beleidigung* (G) or *fornærmelse* (N), indicating verbal abuse. Since in German and Norwegian texts these terms characterize different groups of acts, they do not require the addenda 'sexual,' 'physical' or 'verbal' in order to be discriminated. *Misbrauch* or *misbruk* has its etymological root in *brauchen - bruke,* to use, here as to 'misuse.' To misuse something means to unmake it, which includes two aspects: the object and the doing. *Mishandlung* or *mishandling* is derived from *handeln* or *handle*, to hand/handle, clearly indicating hands doing something, here, something which is inappropriate or inadequate. Different from *Misbrauch-misbruk* with its focus on the object violently unmade, *Mishandlung-mishandling* focuses on hands doing the violence. The languages guide the view differently, so to speak, and the impact on the imagination and on visualization of a somehow violated object or a somehow violent acting are different. The third English designation of abuse, verbal abuse, is *Beleidigung* in German. The root is *leiden* or *leid* (the Norwegian equivalent is *lide, lidelse*), suffer/suffering. The prefix *'be-'* gives the image that *Leid* has been inflicted, which may cause psychic pain or evoke deep embarrassment due to an insult to personal integrity. The prefix indicates a direction of inflicted harm, from the one who does to the one who is. The word leads the eye to the act between two persons; its focus is on an interaction. In Norwegian, on the other hand, verbal abuse is termed *fornærmelse*, which implies 'coming too close.' This word, too, focuses on interaction and relational meaning. Consequently, the English term abuse, in its modalities verbal, physical and sexual, is a more 'neutral' term than the German and Norwegian equivalents. Apart from *Misbrauch-misbruk*, sexual abuse, focusing on what has been objectified, the German and Norwegian words either denote a doer or an interaction between someone acting and someone acted upon.

This differentiation of focus, making sexual abuse the most depersonalized and least relational of abuse modalities, may influence how the phenomenon is conceptualized and reflected upon.

Likewise, the English word 'assault,' denoting a violent attack, has to be examined. The German word *Angriff* and the Norwegian word *angrep* are equivalents. In English, one may speak of a 'sexual assault,' but adding 'sexual' to 'attack' would require the word *Übergriff* in German and *overgrep* in Norwegian, literally "hold over." When addressing the action, however, assaulting someone sexually is called in German *sich an jemandem vergreifen*, and in Norwegian *å forgripe seg på noen*, literally "to grab onto someone for yourself." In Norwegian, a new term has recently been constructed or derived from the latter. Persons who have been violently assaulted, mainly if the attack was aimed at sexual violence, and who refuse to label and think of themselves as "victims of assault," use the term *overgrepne.* The corresponding role to that of the abuser, *overgriperen*, is the abused, *den overgrepne.* In German, however, there is no analogy to this construct. All assaulted and insulted people are simply victims, *Opfer* (G,pl.), *ofre* (N,pl.). The terms *Opfer* (G,sg.) and *offer* (N,sg.) denote victims of all kinds of 'violent' events, be they natural, such as an earthquake, a storm or a flood, or man-made, such as warfare, violence or despotism. So far, victim, *Opfer* and *offer* can be used in the same way also when speaking about sexually abused persons. In German and Norwegian, however, the word *Opfer-offer* also means a sacrifice, a connotation which is absent in English.

These etymologically as well as semantically defined designations, have to be explored in relation to the present project, although this is not immediately obvious in a text about abuse written in English, where no confusion of meaning has to be expected. In a German or Norwegian newspaper, one might, for example, read an article concerning people molested or killed in traffic accidents. One of the author's implicit intentions might be to draw attention to particular underlying factors behind the high number of traffic victims. One of these might be a political reluctance to limit the extensive use of private cars. The reluctance to address a factor might configure a cultural "holy cow," something representing central values in a given society. In such a text, the word *Opfer-offer* denotes both a person killed by a car in an accident, and a human being sacrificed to modern traffic. *"Opfer des Verkehrs" (G), "offer for trafikken" (N)*, could be identical with *"Opfer an den Verkehr" (G), "offer til trafikken" (N).* In using the double meaning of the word, depicting two apparently distinct roles, but semantically related and etymologically co-rooted, an associative field of the "gift to the modern god" is created. Thus, the number of traffic victims as representing an acceptance within a set of cultural values, and not an unavoidable fate or evil, is configured. As such, this opens up the possibility for a critique of the hierarchy of values, and for an identification of the holders of social power. This short semantic excursion must be kept in mind in the following discussion of sexual assault impact in a text written in English, but based on a collection of narratives in Norwegian, gathered and explored by a native German.

The discipline of victimology, gradually appearing in the seventies, was generated from knowledge in different fields such as criminology, sociology, and psychology. In 1975, the American sociologist Martin Seligman proposed a model to understand reactions observed in people who had become victims of violence.[270] He presented a theory addressing what was considered to be the main aspect of these reactions. Its core was the formulation of 'learned helplessness.' Its logic was built on the victim's experience of a violent event which s/he had been incapable of controlling, where resistance or self-defense was either impossible or seemed inadequate, and where the person had been completely subordinated to an offender's will and demands. According to this theory, the foremost characteristics of a victim, post-assault perceptive numbness and passivity, was what constituted being a victim. Seligman considered learned helplessness to be the core of a reactive depression. This view resulted in no other possible reactions to violence being taken into consideration. Therefore, they could not be seen. Apart from this, the absence of numbness and passivity was interpreted as an absence of depression. This led to the conclusion that victims without these signs were to be regarded as not harmed, not real victims, and consequently not in need of help. The theory about how a victim behaves and what kind of help is adequate did not take into consideration that not all victims react with depression. It also excluded any reflection upon the depressive state being transitional, a stage in a not as yet terminated process of coping. It prohibited a recognition that other reactions or developments might have been overlooked. For example, the researchers ignored completely the occurrence of certain violations which are so stigmatizing within every society that any victim does her or his utmost to prevent the violation from becoming known. Neither did they consider the probability of a depressive reaction being engendered by the victim's feelings of guilt, a consequence of the factual or expected societal attribution of responsibility to the victimized person, who bore the responsibility for "helplessness," it being "learned".[271]

The theory was built upon the observation of animals in experimental situations. It was derived from animals' reaction to systematically or randomly applied pain or electric shock. It was grounded in mechanical concepts from neurophysiology about stimulus-response or input-output, as formulated in behaviorist theory. In principle, any attempt to ground an understanding of human behavior in analogies of observations of animals is problematic, as this prohibits a range of recognitions. To consider animal behavior in experiments as equivalent to human behavior in life carries with it so many possibilities for error that one may wonder how these models could have gained such a considerable influence on interpretations of what humans do and why. Behavioral theory does not include the term "purposeful action", for example. But human beings are purposive and interpreting; they have purposes, feelings, and opinions linked to what they do or avoid doing. [272] Science which ignores that which above all characterizes humans when purporting to study human beings cannot be considered as representing science, but must rather be seen as an irrelevant or possibly ideological construct.

The theory of learned helplessness was reformulated by psychologist Christopher Peterson and Martin Seligman, due to shortcomings in its explanatory powers, although leaving unchanged the core of the concept.[273] Certain phenomena among victimized humans, such as self blame, low self-esteem, and guilt, could not be understood within the helplessness model which, until then, had been highly supported yet increasingly challenged. Psychologist Ronnie Janoff-Bulman and sociologist Irene Hanson Frieze proposed an opposing theory based on humans, and not on laboratory experiments with animals.

> It is proposed that victim's psychological distress is largely due to the shattering of basic assumptions held about themselves and their world. Three assumptions that change as a result of victimization are. 1) the belief in personal invulnerability; 2) the perception of the world as meaningful; 3) the view of the self as positive. Coping with victimization is presented as a process that involves *rebuilding of one's assumptive world.*[274] (italics mine)

Supported by theoretical knowledge from other scholars in a variety of fields, such as the stress-coping models of Lazarus,[275] the basic human assumption of the orderly and comprehensible world as proposed by Antonovsky,[276] and the relationship between distress and self-esteem as developed by Horowitz,[277] Janoff-Bulman and Frieze suggested the necessity of a *redefining of the event* in order to minimize the threats to one's assumptive world. This redefinition is an exercise of interpretation. It is also a making of meaning, to render tolerable that which, if not making any sense, would seem intolerable. The interpretation of what happened and what it meant, will guide the consecutive thoughts, actions and conducts. These can be conceptualized as strategies to rebuild the shattered assumption about the world as orderly and oneself as capable of living. Within such a concept, victim reactions as described in a variety of populations became comprehensible. Self-blame, for example, had been registered among victims of rape, diseases and accidents, and among survivors of group disasters. Janoff-Bulman and Frieze write:

> Interestingly, self-blame can be functional following victimization, particularly if it involves attributions to one's behavior rather than one's enduring personality characteristics. Behavioral self-blame involves attribution to a controllable and modifiable source, and thus provides the victim with a belief in future avoidability of victimization. Victims can believe that by altering their behavior, they will avoid being victimized in the future; the victim can maintain a belief in personal control over future misfortunes. Characterological attributions are associated with depression and helplessness. Behavioral self-blame, then, can help a victim reestablish not only a view of the world as orderly and comprehensible, but a view of oneself as relatively invulnerable. (Janoff-Bulman & Frieze 1983)

Attributing the origin of an assaultive event to oneself, thus placing the responsibility in oneself, will allow a victimized person to regard her/himself as capable of avoiding a repetition. This may be a greater gain than a loss, although such responsibility may create both guilt and shame. The next steps, the 'preventive measures,' must be in accordance with this construct of meaning and cause. The strategies applied may reestablish the view of oneself as capable, and they may prevent the destructive impact of having been overpowered and stricken by randomness, identical with unpredictable and inescapable forces. They may be expressed in a general or particular avoidance, in a change of routines or habits, in the self-limiting of mobility, activity, plans for the future and participation in social life, sports, education and so on. These 'reparative' strategies resulting from a self-attribution in the wake of assault, may, however, turn counterproductive concerning personal growth and development, or they may reveal their futility.

On the other hand, people who, prior to the victimization, have engaged in presumably 'safe and cautious behavior,' are, when victimized, totally confused with regard to the rightness or wrongness of what they did. This is addressed in a study among raped women.

> Women who have been raped are more likely to change their perceptions of the safety of the world in radical ways than women who have avoided being raped, although women who avoid being raped often make substantial reassessments as well. For both raped women and avoiders, one's likelihood of making a radical change in one's perceptions of dangerousness depended to a large degree on the sort of situation in which the attack took place. Women were less likely to change their perceptions of dangerousness if the attack took

> place in a situation where they believed they were in some danger before the attack. If the attack took place in circumstances which the women defined as safe, however, a much more extreme reaction was likely to occur. (Scheppele & Bart 1983)

The relevance of *the circumstances of victimization* and of *the victim's individual preconditions* within a society were accentuated ever more, emphasized by a shattered conviction of having taken all possible preventive measures and still being victimized. As to the shattered presupposition of one's own invulnerability, psychologist Linda Perloff, in a review of the literature on coping with victimization, writes:

> Specifically, I have argued that illusions of invulnerability prior to victimization may make actual misfortune all the more difficult to cope with. And, after victimization, those who perceive themselves as 'uniquely vulnerable' may show lower self-esteem, harsher self-criticism, and greater depression, and may have more difficulty reestablishing a sense of personal security, than victims who perceive themselves as 'universally vulnerable.' (Perloff 1983)

Another possibility for reestablishing self-confidence and self-esteem after victimization was proposed by the psychologists Shelley Taylor, Joanne Wood and Rosemary Lichtman, and termed 'selective evaluation.'

> It is maintained that the perception that one is a victim and the belief that others perceive one as a victim are aversive. Victims react to this aversive state by selectively evaluating themselves and their situation in ways that are self-enhancing. Five mechanisms of selective evaluation that minimize victimization are proposed and discussed: making social comparisons with less fortunate others (i.e., downward comparison); selectively focusing on attributes that make one appear advantaged; creating hypothetical, worse worlds; construing benefit from the victimizing event; and manufacturing normative standards of adjustment that make one's own adjustment appear exceptional. (Taylor, Wood & Lichtman 1983)

Searching for meaning in victimization can be demanding and absorb the victim's time and strength even decades after the victimization occurred. This was addressed by psychologists Roxane Silver, Cheryl Boon and Mary Stones in a study performed among adult women having experienced childhood sexual abuse by their fathers.

> Even though their encounters had terminated an average of 20 years previously, over 80% of the respondents still reported searching at least sometimes for some reason, meaning or way to make sense of their incest experience... Almost 80% of our sample agreed that making sense of the incest was still important to them. Although outward indicators should suggest recovery from their experience, for most women the search for understanding continued. (Silver, Boon & Stones 1983)

They underline the impact of the personal ability to attribute meaning:

> The mixture of women who could and could not make sense of their experience provided us with a unique opportunity to explore systematically the possible

> beneficial effects of finding meaning. Data analyses did confirm the previously untested suggestion that the ability to find some meaning in one's victimization facilitates effective coping. In fact, those women who were able to make some sense out of their experience reported less psychological distress**, better social adjustment**, higher levels of self-esteem**, and greater resolution of the experience** than those women who were not able to find any meaning but still were searching. (** all p-values below .005) (Silver, Boon & Stones 1983)

Also Janoff-Bulman and Frieze, although basing their work on general principles in coping with victimization, stressed the limitation of such a view for an understanding of individual expressions and strategies by concluding:

> Knowing how victims react in general does not enable us to predict how an individual victim will react to a particular type of victimization. (Janoff-Bulman & Frieze 1983)

Apart from the problem of the generalizability of theories derived from studies of selected populations, and from the limited insight into the process of interpretation and choice of strategy in reductionist study designs, they consider the research on victimization to be flawed by certain biases, which is addressed as follows:

> ...we tend to think of the prototypic victim as female. The term 'victim' seems to connote weakness and helplessness and may thus be stereotypically applied more readily to females than to males. [278] In fact, this connotation of weakness has led some researchers to prefer to use the term 'survivor' rather than 'victim' when describing men and women who have experienced traumatic negative events. (Janoff-Bulman & Frieze 1983)

Janoff-Bulman and Frieze outlined the reestablishing of shattered basic assumptions as according to the meaning attributed to the traumatic event. But they did not offer any conceptualization of how to understand the situation of persons who, despite their strategic precautions, become victimized again. Revictimization may be more likely to occur when the world once again proves lacking order, when the danger seems paramount, and when the personal ability to avoid assault proves to be illusory. Such an experience, of the futility of all doing or non-doing as preventive or protective, may lead to an impact out of proportions to the actual event. From my own general practice experience I have learned that a medically not verifiable trauma from a car-accident, for example, can generate an incapacitating, chronic and disabling pain. The impact of such a trauma cannot be read from x-rays, the insurance-report or a neurological examination of the person involved. Facts such as who was injured, how and where, may immediately become part of a trauma in the past, and interwoven with it. Such a traumatic past need not to be known by the health care worker or physician consulted after the accident. It need not even be acknowledged by the actual victim her/himself, as mentioned when exploring the phenomenon of 'forgetting' in the previous chapters.

The theories about adapting to trauma by reestablishing the basic assumptions do not cover traumata linked to or covered by social taboo. They seem unsatisfactory and incomplete as regards traumata which cannot be mediated or talked about. Thus, sexual integrity violations, especially in close relationships, and particularly early in life, hold some characteristics no other kind of trauma does. This was addressed by the American psychiatrists Patricia Rieker and Elaine Carmen (Rieker & Carmen 1986). They explored the adaptation to traumata which officially do not happen and consequently cannot have occurred. When a victim of socially silenced abuse tries to tell or otherwise communicate about the abuse without relevant others reacting adequately, the victim is not a victim but a person whose perception of reality is faulty. He or she has not been offended against, but is mistaken. Compensating for this incompatibility of realities demands adaptive strategies perhaps intended as distracters, but presenting themselves as irrational, unreasonable and insane. The strategies may aim to exclude what is unacceptable from the realm of confirmed experience. They may result in a splitting off of sensory perceptions or perceptivity. They may contribute to a confusion of tenses, a blurring of boundaries and an uncertainty about memories and in judgment. Such adaptation to a deviating experience is only 'cope-able' for a short period. In the longer run, the splitting off is destructive, and to the eyes and understanding of others, the resulting ways of being, and their bodily or mental expressions, become an indication of sickness. The authors write:

> ...we have been impressed with how often abused patients present what appear to be classic borderline states characterized by numbness, emptiness, splitting of ambivalence, abandonment depression, impulsivity, and multiple suicidal and self-mutilation episodes. Indeed, there is an intriguing overlap of symptomatology among borderline, multiple personality, and post-traumatic stress disorders. *It may be that severe abuse is an important antecedent in all three conditions...*
>
> ...The path from victim to patient involves a complex interplay among the personal characteristics of the victim, the nature of the abuse, and the life context in which the abuse occurs. Because of this complexity, it is not possible to predict the outcome of an individual case of abuse. (Rieker & Carmen 1986) (italics mine)

The authors proposed a development from an initial denial - imposed by others or by the self since what happened was too incredible, unbelievable, or unbearable to be conscious of - to the altering of affective responses and to the change of meaning of the abuse. This development or path, resulting in *an adaptation to chaos*, is consequently and by logic, *a maladaptation to normal order and meaning.*

The same year, sociologists David Finkelhor and Angela Browne developed a theoretical framework for understanding the long-term impact of sexual abuse in childhood, (Finkelhor & Browne 1986) which has already been mentioned in "Relational Strains." Because of its central place as one of the main references in the field of child sexual abuse trauma, however, it shall be considered once again in the present context.

Finkelhor and Browne linked the traumatogenic dynamics to four central elements in the children's experiences of abuse. The first element was *traumatic sexualization*, "which refers to a process in which a child's sexuality (including both sexual feelings and sexual attitudes) is shaped in a developmentally inappropriate and interpersonally dysfunctional fashion as a result of the sexual abuse." The second element was *betrayal*, "which refers to the dynamic in which children discover that someone on whom they are vitally dependent has caused them harm." The third element was *powerlessness*, or what might also be called disempowerment, the dynamic of rendering the victim powerless, refers "to the process in which the child's will, desires, and sense of efficacy are continually contravened." And the fourth element was *stigmatization*, which "refers to the negative connotations - for example, badness, shame, and guilt - that are communicated to the child about the experiences and that then become incorporated into the child's self-image."

Ronnie Janoff-Bulman elaborated the previously presented theory about basic assumptions being threatened or shattered by trauma experience in a study which, more than the previous studies had, accentuated the long-term effect of trauma adaptation.[279] Also, the profoundly individualized basis of trauma experience was explored in an interview study among sexually abused children aged ten to eighteen.[280] In these children, the process of victimization proved to be grounded in a three-layered matrix of the *sexualization* of the child, the abuser's tactics of *justification* of the abuse, and his techniques to sustain the child's *co-operation* in the ongoing abuse. The three elements were represented in a variety of modalities, guided by the significant circumstances of the child's life context.

The initial element of denial on the path from victim to patient as described by Rieker & Carmen, was addressed anew in a study about the matrix of what had been observed as the astounding proneness of victims of sexual abuse to become revictimized later in life. The authors write:

> The relationship between victimization history and denial can be conceptualized in the following way: Our findings may be taken as evidence that incestuous and repeated victimization present a greater need to defend, and that this occurs through the mechanisms of denial. In support of this interpretation are indications from clinical samples that incest victims commonly employ *avoidant defense strategies such as denial, repression, and dissociation.* Thus, denial may be viewed as a coping antidote to extensive sexual trauma, i.e., as evidence that this more primitive defense mechanism is needed to protect ego integrity in the face of massive or repeated trauma. It is also quite plausible that the denied effects of extensive sexual assault operate so as *to increase the likelihood of revictimization.* (Roth, Wayland & Woolsey 1990) (italics mine)

The victim's attribution of the origin of the victimization to either own character traits or own behavior had previously been addressed as having different impact. Janoff-Bulman and Frieze had noted that the first leads to depression as an expression of helplessness in the sense of Seligman. In case of the latter, behavior for avoidance or defense could be developed, which probably would counteract the feeling of having no options at all. These two attributional strategies, the internal (cause in self) and the external (cause outside self), explored in a study among female incest survivors, showed a significant relationship between internal attribution and low self-esteem and depression as contrasted to external attribution.[281] This study also indicated that the victims more probably attributed the cause to themselves (internal) when intercourse had been performed. The author considered the link between serious abuse and low self-esteem, depression, self-blame and self-molestation as possibly being mediated by this internal attribution.

Self-blame was studied in a population of male and female child sexual abuse victims in three age groups (children, adolescents, adults) with regard to gender difference. Blaming was divided into self-blame, molester-blame and family-blame. The authors write:

> The current findings suggest that cultural and developmental differences in the socialization of males and females may impact on the attributional process. It can be hypothesized from these results that males place particular emphasis on issues associated with maintaining interpersonal autonomy and control, with decreased molester blame associated with conditions where the male felt that he should have been able to protect himself (i.e., the perpetrator was closer to his age, he was unable to defend against the force, etc.). The role of perceived level of force appeared to be a very disruptive influence for males, and was positively correlated with both greater self and molester blame. Clinically, it can be speculated that the experience of force by male victims may be perceived as an indication of masculine inadequacy (i.e., weakness, vulnerability) which leads to simultaneous denigration of self and resentment of the perpetrator. (Hunter, Goodwin & Wilson 1992)

Relating to clinical presentation, a study exploring the relationship between symptomatology and self-blame in female child abuse victims was performed. In this study, it was not primarily the abuse category and severity which emerged as influential for health outcome, but the relationship to and the support given by the caretakers of the children. The authors write:

> Overall, our study highlights the complexity of factors which relate to child symptomatology and child self-blame in sexual abuse cases. Our findings support the importance of including not only abuse characteristic variables, but demographic (race and social class), environmental (e.g., caretaker support) and internal mediating variables (e.g., caretaker and child attributions) as potential predictors in future research. (Hazzard et al. 1995)

Departing from the conceptual framework Finkelhor and Browne had proposed in 1986, in a study among adult women, an attempt was made for the first time to support the traumatogenic model empirically, exploring the relationships between its four elements. The authors had renamed 'traumatic sexualization' 'perceived stigma,' as the former seemed difficult to elaborate in an adult sample. This study has also been mentioned in "Relational Strains," and its essential finding was:

> A path analysis indicated that the level of psychological distress presently experienced by adult women who had been sexually abused in childhood was mediated by feelings of stigma and self-blame. (Coffey et al. 1996)

A relationship between shame and stigmatization in the adaptation to sexual abuse was proposed by Feiring, Taska and Lewis, whose central message shall be quoted once more:

> ...the sexual abuse leads to shame through the mediation of cognitive attributions about the abuse and shame [which], in turn, leads to poor adjustment. Three factors, *social support, gender, and developmental period* are hypothesized to moderate the proposed stigmatization process... *Unless future research elucidates the process and circumstances whereby the experience of sexual abuse leads to poor adjustment, little progress will be made toward developing more effective treatment.* (Feiring, Taska & Lewis 1996) (italics mine)

The authors are among the very few in the field of sexual abuse research until now to stress explicitly *the importance of the context for an understanding of the impact* of abuse. Although promoting 'three factors' as the most probable moderating influences and thus not demanding a radical change of methodology, they are unambiguous about the crucial influence of abuse context. Exactly this message is central in a study of children, where the authors have tried to assess the influence of gender and developmental state on children aged five to fifteen during the first three months after disclosure of extra-familial sexual abuse. They write:

> Results revealed that sexually abused children, in comparison to non-abused children, suffered deleterious and clinically significant effects. Standard multiple regressions found that *the children's perception* of self-blame and guilt

> for the abuse and the extent of traumatization predicted their self-reported symptomatology of depression, social efficacy and general and abuse-related fears. As well, *the child's gender predicted* the level of general fearfulness. (Ligezinska et al. 1996) (italics mine)

That is the first study, to my knowledge, to explore a child's perception of sexual abuse in relation to her or his mental health (and, implicitly, the somatic health as well, as 'somatic complaints' were elicited in the questionnaires answered by the care-takers), and social functioning. It is also one of the few to elicit the influence of gender (40% of the children were boys) on presentation or development of symptoms. The authors conclude as follows:

> The most revealing findings suggest that evaluation of children's own perceptions of trauma is an important variable in predicting their immediate adjustment... Children may be affected differently because their developmental level may provide them with *a different understanding or interpretation* of sexual acts. Research conducted within this perspective would be most revealing if developmental and life transitions were assessed in a consistent manner. (italics mine) (Ligezinska et al. 1996)

Certainly, the children's "different understanding or interpretation of sexual acts" may be significant in its effect on the eventual impact the abuse has on them. However, it must be equally, or probably even more, significant and crucial in its overall effect on all other phenomena in the context of an abusive relationship.

When, for example, applying this statement to Bjarne's experience of one single anal rape, nothing in his recall attests to his perception of the abusive act as having been 'sexual.' It was a pain beyond comprehension, a danger to the core of his existence, a final testament to paramount threat, a fundamental betrayal, a total abandonment, an unmaking of the world. All his being as to personal, relational, intellectual, sexual, mental and physical aspects, was affected by this unmaking. And so was the case in all the other narratives explored in the preceding chapters. The same crucial influence of the qualities of personal, subjective experiencing, interpreting, understanding, and perceiving were obvious.

Consequently, this present study is a demonstration of the potential of a methodology which aims at an understanding from inside.

Such an approach "from inside" has been requested in a review on sexual abuse impact on children, based on 45 studies. The authors write:

> Researchers evince a great deal of concern about the effects of sexual abuse but disappointingly little concern about *why* the effects occur. Few studies are undertaken to establish or confirm any theory or explanation about what causes children to be symptomatic. Rather, most researchers simply document and count the existence of symptoms and some of their obvious correlates. This

> accounts for one of the main reasons that, in spite of numerous studies since the Browne and Finkelhor's (1986) review, there have been few theoretical advances. *Future studies need to turn to the development and confirmation of theory.* (Kendall-Tackett, Williams & Finkelhor 1993) (italics mine)

The need to develop theory by applying a methodology which allows theory-building, was formulated in a study among adults in therapy. The authors conclude:

> The findings of the present study illustrate and underscore the need for and importance of more extensive and detailed research in the area of childhood sexual molestation on the characteristics of the abuse itself. Obviously, greater knowledge about abuse characteristics is valuable in its own right, providing information about types of patterns childhood sexual abuse commonly takes. *Over and above this, however, more extensive knowledge about abuse characteristics may be essential to approaching investigation of abuse effects on a more sophisticated level.* (Gold, Hughes & Swingle 1996) (italics mine)

Lately, studies and theories about victimization grounded in psychology, have gradually been expanded upon, and overridden by, studies and theory in traumatology. This field was initially dominated by research in the biological aspects of trauma experience, as mentioned previously with regard to measures of distress. Several confluent research activities have clustered, guided primarily by the paramount interest in the complex state termed PTSD and related phenomena. Traumatology is at present informed by a cooperation of the neurosciences, psychiatry, immunology and endocrinology.[282] However, all these disciplines are anchored in the traditional naturalist paradigm dominated by a knowledge production according to reductionist methodology.

Consequently, this present study is the first which I am aware of, not only allowing to generate new hypotheses about sexual abuse impact, but also to build theories for new ways to regard and comprehend the individual process from violation to sickness in a contrast to the traditional epistemological framework of Western biomedicine.

5. GUNHILD

Among the authors mentioned in my review on theory, the topic of revictimization has been thematized by Roth, Wayland & Woolsey, who conceptualize it as linked to or engendered by the various strategies of denial in abused persons. The phenomenon of revictimization itself, and the broad variety of its possible modalities, has already been extensively addressed in the present study. Eighteen persons had been abused by several offenders, and twenty-two had experienced abuse in adulthood. Numerous studies among different populations have confirmed the high risk of becoming a victim again.[283]

Now, however, revictimization will be focused on with regard to its occurrence in the medical context.

Sexual abuse in therapeutic or other professional relationships has only recently become a topic of interest. In 1991, a Council Report of the American Medical Association addressed the topic and stated clearly that:

> (1) sexual contact and romantic relationship concurrent with physician-patient relationship is unethical; (2) sexual contact or a romantic relationship with a former patient may be unethical under certain circumstances; (3) education on the ethical issues involved in sexual misconduct should be included throughout all levels of medical training; and (4) in the case of sexual misconduct, reporting offending colleagues is especially important.[284] (Council Report 1991).

Only a few comprehensive studies of the impact of abused trust, termed 'sexploitation,' and of its occurrence in the counseling or health care professions, have been performed until now.[285] In the present study, the topic is introduced through the descriptions of Oda and Beate who experienced repeated sexual abuse by male nurses or by the school physician. Sexualization of a therapeutic relationship was described by Ruth and Berit. The latter terminated a therapy almost in panic as her male therapist became more and more personal with her. Grethe and Camilla experienced abuse when being examined medically by male gynecologists, the consequences of which have been explored in depth previously. Synnøve felt violated by means of the theoretical preoccupation in her psychologist's approach to her problem, and by his view persistently overriding hers. Ruth also addressed the male gaze and the male norm in psychiatry, making her the deviant as a consequence of an unreflected gender bias embedded in professional judgment.

Revictimization in therapeutic or counseling relationships is a gendered phenomenon. Resulting from asymmetries in the current gender roles, women are at a higher risk of lifetime sexual exploitation or violation than are men. They are more likely to define and experience themselves as in need of expert help due to an overall and culturally constituted subordination. Thus, in male dominated societies, the majority of persons explicitly or covertly consulting for sexual assault are women. Still, most consultants are men, be they lawyers, physicians, psychologists, psychiatrists or clergymen.

These societal constellations add to a fourfold asymmetry in counseling relationships, since the consultant's role ranks above that of the one consulting, and the consultant's social position is most often ranked above that of the consulting woman's. Finally, according to the gendered roles, maleness is associated with actively challenging limits and boundaries, while central signs of femaleness are linked to a self-restricting conduct, and a voluntary limiting of one's own expansion by whatever means necessary.[286]

The Norwegian psychologist, Hanne Haavind, has proposed different possibilities of revictimization in therapeutic relationships between sexually abused women and their psychotherapists. (Haavind 1994) Primarily, she addresses the most obvious form, a concrete sexual or otherwise sexualized relationship rendering the patient the object of the therapist's desire and needs. Next, she describes the woman's becoming subordinated in the extended role of a supportive, comforting and thankful patient, providing her therapist with admiration and attention. Thus, she becomes invaded in a practical manner, serving her therapist's self-image with time, empathy and adoration. Instead of being allowed to receive, she is again in the 'service position' from which she asked for help to escape.

Gunhild's interview may illustrate how the concrete and the practical form of abuse may be embedded structurally in a therapeutic relationship. She had been sexually abused from the age of six until she was twenty-three years old. Her main abuser was her stepfather, who also gave a group of male friends access to her body, since it was entirely at his disposition. At twenty-two years old, after having listened to a radio program about sexual abuse, she decided to escape. She moved far away from home and established contact with a psychotherapist. With his support, and being at a distance from the abuser, she reported her stepfather to the police. While the interrogations were taking place, she also needed medical help. This was thematized as follows:

G: I phoned him (the physician) for the first time during the evening when I had a heavy headache and asked him for medication. He said: 'I live at so and so, please come and see me.'

A: Live? Does this mean you didn't get an appointment in his office?

G: He was on all-day duty, the doctors there often are on 36-hours duty, so he was at home but on call. It was a quiet night. It resulted in us sitting and talking. We gradually got more acquainted with each other. *Then I asked my therapist to tell the physician why I was in therapy.*

A: Your therapist told your doctor about your situation. So he was almost officially informed about your condition?

G: Yes, he was. I asked him to do that. I thought they ought to cooperate. I couldn't tell him right away. He was very kind. *I wasn't accustomed to meeting kind people.* I was more or less seduced. He wasn't such a bully. I had an abuse reaction afterwards, although it wasn't really abuse, at least not physically. *But I felt dirty and disgusting afterwards.* That was during the autumn that the trial was to take place. I was balancing on a thin wire. I had to get my sleep in order to manage the days. Police interrogations took place. I traveled quite a lot. Like this, our relationship developed. In the beginning, I thought he only had me. But little by little, he had several others in addition. I found this out from other people, and sometimes I came to him when he didn't expect me.

A: In this tiny place he had found several?

G: Yes, someone warned me. But I was like this, *since someone finally had been kind to me and careful, I almost got addicted to it.* No matter what other people said. He was the rescuing angel I'd been hoping for and expecting all the time.

A: Does this mean you cared for him in a way on several levels?

G: *He was one of those who wants both mother and lover and friend and I was everything for him, in a way.*

A: The guy had many demands and you met them all. And simultaneously he was big and had prescriptions to trade. It sounds like a dangerous combination.

G: Yes. I came to know afterwards that he ... what he told me when we met again here (after I left the place) was that he couldn't bear to go on living in this little town any more. He said he preferred to have more life around him. My therapist told me secretly that the doctor had been dismissed from his job, he drank periodically. What I hadn't known was that he also abused pills - he mixed them. He was fired. Now, he's a doctor in an institution here in town.

A: You know that he's fully employed and has access to medications and to people?

G: Yes. Now there are mostly elderly people. I don't know. I have nothing to do with him any more. It ended with me quitting. *I didn't have the strength anymore.* It was after yet another abuse. Then I was twenty-three. I told him: 'Tonight, I can't handle you touching me.' But that was nothing. He'd had such a bad day in his job and felt so exhausted that he needed me and had to have intercourse. Then I exploded. *I didn't have the strength anymore.*

In Haavind's terms, Gunhild was abused "concretely" and "practically" until she was exhausted, while she was trying to escape from a relationship which had broken her and made her become more and more ill. The excerpt as such needs no further interpretation; Gunhild herself described clearly how a particular constellation caught her, although she herself had sought to open up, and had been explicit about her particular vulnerability. She was vulnerable in all respects, dependent on human support, medical help and drugs. Kindness was the most irresistible of traps she could have encountered. And, a physician's lack of professional and ethical standards was the most inescapable of dangers she could have met. Therefore, she once again became objectified, abused and exploited to the point of exhaustion.

Her sickness history was as follows: treated for urinary tract infections from early childhood, and treated frequently for somewhat serious injuries such as burns and cuts; treated for several pelvic infections, confirmed at least three times by laparoscopy; underwent cystoscopy and rectoscopy to diagnose chronic abdominal and pelvic pain; had eating disorders from childhood, anorectic periods altering with bulimic; became pregnant by her stepfather once, and was admitted for an abortion; admitted at least six times due to suicide attempts; treated in at least eight different hospitals in Norway, several of which, several times each. From the incomplete records I received it was not possible to reconstruct how many of the referrals were due to confirmed diseases. A recorded note from a gynecological ward was as follows:

> As far as I have understood she has had a psychiatric history for years, and has been treated by Dr. BB, on the background of sexual abuse by father or stepfather. Her psychiatrist is at present abroad. I have asked her before whether she wanted to talk to our psychiatrist, but till now she has not been interested, but today she comes on own initiative for a talk. *I have not touched upon these topics with her,* but have registered that *she is very strange as to behavior, and especially in relation to everything concerning gynecological examinations, which must be done under full anesthesia.* Now she has non-specific salpingitis. (italics mine)

Sexual abuse experience is known, its horror becomes obvious and is seen in every gynecological setting. Nonetheless, this is not perceived as a topic for the gynecologist to explore with his patient. It remains unthematized and left to a psychiatrist. Furthermore, although the abusive past is present in the patient's body and her reactions, *she* is called 'strange.' The impact of violence, demanding anesthesia to make examinations performable, *is attributed to the woman as 'strange behavior.'*

The wording documents the fact that knowing does not necessarily mean understanding, and that the impact of a destructive relationship is turned into an individual's failure to behave normally. Embodiment is not comprehended by the medical expert of disease of the body, although every medical approach requires that demanding medical measures be taken to overcome the embodied violation. Still, the professional response is: silence.

6. ELISABETH

The third of the possible structures for professional revictimization of victimized persons Haavind addresses is the most covert. This structure is linked to the correct application of the helping profession's theoretical models. The invalidation of women's experiences by psychiatric theory has been addressed and, partly discussed in chapter 1.5 related to the presentation of a psychiatric patient's history in Huseby's article. In the interview with Synnøve, such an invalidation of her experience by theory in her encounter with psychology has been described in "Unheard Messages." With regard to this phenomenon, Haavind writes:

> The gender-neutral pretension in modern psychology acts to subordinate the experiences of women. The gendered codes are disconnecting femininity and masculinity, and are placing masculinity in a dominant and hegemonic position. (Haavind 1994)

The topic of invalidation through theory, mirroring Synnøves experience, is also addressed in a study by Rieker & Carmen concerning a victim of father-daughter incest. The woman describes her therapy with a psychoanalyst, while she still is amnesic about the abuse. The therapy had been recommended to her because of her sexual dysfunction:

> He noticed that while I talked freely about my mother and sibs, I didn't have much to say about my father. He questioned that, session after session, and then he openly interpreted that the reason was, that I loved my father, that I longed to be with him and be like his wife, that I was envious of his penis *and* that I wished I had one. Perhaps, if he'd not forced the oedipal fantasies on me, I would have *trusted* him enough to tell him that whenever my husband and I had intercourse, my head was absolutely full of screaming rage, name-calling, etc. - that I didn't understand where all that was coming from and could he help me? (author's italics) (Rieker & Carmen 1986).

Haavind, in accordance with her professional perspective, has focused primarily on the impact of a male, theoretical hegemony on the interpretation of women's life and socially silenced abuse experiences in psychology and psychoanalysis. Its impact in psychiatry may be explored in an excerpt from Elisabeth's interview. She has been mentioned before, and the situation touched upon in the following dialogue relates to the very period during which she, diagnosed as psychotic, sees herself as having been taken from her father, as was explored in "Recognized Memories." At the end of her twenties, Elisabeth was referred to a psychiatric ward for the first time. During this stay, she had asked the doctors directly whether they thought that her twelve years of abuse experience had something to do with her present condition. The physicians had said no, they did not consider this to be the case. Gradually, her condition had improved and she had been released. Shortly after, her condition had worsened and once again she had been referred by her General Practitioner, but this time to the out-patient clinic. Once more she had mentioned the abuse, but the topic had not been explored further. Then, Elisabeth had become psychotic and had to be admitted again. The excerpt from this point is as follows:

E: The psychosis started after I'd begun to see the doctor. This was five years ago now. It was a very heavy experience. I was scared to death. I was afraid of everything, absolutely everything. And I was very word-less.

A: (almost whispering) I suggested we return to your mouth.

E: Uh-huh. (long pause) It felt very heavy to be in this ward. It was a locked ward. (her voice is almost choked by weeping)

A: Why? Why were you referred to a locked ward?

E: They thought I needed supervision. But I was no danger to myself or others. I don't know why.

A: That's tough to be on a locked ward. Was that the right place?

E: No. No. (very decisive and clear) I was forcibly committed according to §3. *They deprived me of my rights to my own life, and afterwards I reacted intensely, I felt I went five steps backwards in my development, and* ... (weeps loudly now) (she sat in front of me, bent over, her body trembling, she had covered her face with her hands and cried)

A: Say it, say it aloud ... this was abuse ...

E: Yes. That's what it was. I went backwards many steps, really, I did. Everything I'd worked through, gradually I'd regained self-confidence, everything ... everything (her voice almost bursts, strong accent on her words) broke once again. Really, I don't know why I had to be placed on a locked ward. I was scared to death, I hadn't injured myself, I hadn't tried to assault others, I hadn't attacked anybody, I was merely scared to death, that's what I was. But, of course, I had no other possibilities.

A: No, not you. But perhaps they'd had another possibility to help you than placing you on a locked ward?

E: Yes, I guess so. They might have given me medications.

The following passage has been quoted in "Recognized Experiences" and shall not be repeated. After that, the dialogue continues:

E: I've tried to get it removed, out of psychiatry. This admittance, this diagnosis. It'll follow me for ever.

A: And you want to have it annulled.

E: Yes, but I have no chance. I'm going to ask for an additional declaration. This follows me, and that's quite unreasonable.

A: You feel that the diagnosis rests on an incorrect basis?

E: Yes, I think so. I've tried to get a response to my thoughts. But I won't. It's written there. And there it will remain.

A: And it (the diagnosis) is more true than what you report?

E: Yes. (deep sigh, resigned voice) I know from inside myself, I wonder which diagnoses I've gotten through the years, because it's horrible for us who struggle with abuse experiences, that those who treat us are so keen on diagnoses. (irritated voice in the last sentence)

A: Please, would you repeat this?

E: Why do they need a diagnosis? Couldn't it be reason enough that we're sitting here with our childhood experiences? These diagnoses. They do harm, really.

A: You get a mark on you?

E: Yes. In the middle of your forehead. (loud, clear voice now) And there it will remain. And it does. And that's serious. You're harmed in addition. Yes. Then it would almost be better if you were mad, really.

A: Then you at least would match the diagnosis?

E: Yes, and I'd have been freed from this giant effort to function. That's really hard. It is. (long pause) Horrible. That's what it is.

After several attempts to tell about her abused experience but without being heard, she came into a condition in which admission to psychiatry was deemed necessary. No records are available to explore her presenting symptoms. For her, however, experience of being forcibly committed to a locked ward was insulting, especially as it came after months of struggling with deepening insight into the aftermath of sexual abuse. In addition, Elisabeth had began to see images to which she could not give meaning; she had visited her childhood home in order to find answers and had stayed for several nights in her former room where most of the abuse had occurred, crying, screaming, enraged. She had started to tell, she had spoken repeatedly, insistently, and explicitly about the abuse, and about her suspicion of a relationship between her past and her present problems.

The experts denied any relationship, and the topic was ignored. Her voice was silenced. And when she broke down because she could no longer adapt to the discrepancies between her perceptions of reality and the ignoring silence with which she was met, she was evaluated and diagnosed as psychotic. She felt a stigma was imprinted on her, visible to everyone and forever.[287] Did she feel stigmatized, as the interviewer had suggested by using the metaphor of the mark, and as Elisabeth had confirmed by expanding upon it? And how did this feeling of social stigma express itself? This was thematized in the very last passage of her interview:

E: I feel I've been working very hard on the abuse and that . . . and everything around . . .

A: But your confrontation with psychiatry still remains.

E: Yes. (breathing deeply) I don't know whether I dare take it up. I had a period when I really thought I'd go ahead with it, but I lacked strength. (pause) (weeps again) *Now I feel, well, when I meet someone from there, I feel they look upon me as an idiot.* Really. (her voice fades) (she cries loudly, cannot speak for a while) But I wonder whether I must get it out, take the confrontation.

A: The moment you said 'when I meet someone from there, they look at me like an idiot,' your whole body shrank, you lifted up your shoulders and your arms moved forward and your face turned and you became very little and almost disappeared . . . there is something left before you dare to straighten yourself up.

E: Yes. I had a very bad period after the admission, *I feel like an idiot. . .* (voice stumbling, fighting for breath, swallowing her words)

The tape-recorder is switched off.

* After a while she has calmed down. Then she tells that she has a very bad conscience because *she has let her children down* when something is going on at school or in sports. She cannot manage to take part in parents' meetings at school or other events since this involuntary admission, *because the parents of some of her children's classmates were employed on this ward and would recognize her.*

In Elisabeth's experience, the silencing force of psychiatry overpowered her. Instead of being heard, she felt humiliated by being involuntarily committed; instead of being helped, she was stigmatized by being diagnosed; instead of being understood, she was shamed by others' witnessing her regression.

Now she knows what she saw. Her 'hallucinations' were true memories, confirmed as valid by her mother, and evidence of a recalling of painful events and dramatic time. It does not help her that she knows what she knows, however. In her psychiatric record the diagnosis will remain, a documentation of madness, more true than her life and her corporeality which she needed help to understand. Once again, insulting and silencing forces have estranged her from herself and relevant others. Now she lets her children down due to the unbearable shame of feeling she is looked at as 'an idiot,' which she interprets the medical documents as saying she is. She dares not meet the eyes of others, and primarily not of those who were witness to her past as present.

Once again, shame, stigma and guilt meet in her. This time, however, the meeting of these three is forcibly imposed by the practitioners of a theory about mental disease and its presentation, and is due to a professional practice of power, as previously addressed by Saris. *Psychiatry has the power to define which topic is no topic by silencing the voices speaking it.*

7. MORBIDITY PRODUCTION

In the frame of this project, it is paramount to raise the question of whether such covert, medical revictimization of previously abused persons also can occur in *somatic medicine.* To my knowledge, this question has not previously been formulated explicitly in the research literature about the impact of silenced abuse on health.

The phenomenon of *somatization*, and the explorations of the interviews concerning the process of becoming sick will now be considered, in search for an answer. Most of our knowledge about the health impact of sexual abuse concerns mental health. The 'mental health field' comprises all kinds of sickness presentations which are, in form and content, and either by tradition or definition, systematically interpreted as being signs of mental diseases. Likewise, it includes all kinds of sickness presentations which only in content, however not in form, are defined as testifying to mental disease.

Within research of mental impact, *somatoform sickness is a consistent finding,* highly correlated to abuse experience. This is documented in many studies among adults, although problems of definition, and a confusing terminology, make measurement difficult. This is addressed by Shan A. Jumper in a 1995 review of 26 psychiatric studies. The author writes:

> For the purposes of this investigation, the term psychological symptomatology refers to those psychological difficulties experienced by individuals other than depression or impaired self-esteem, such as anxiety-related problems, personality disorders, suicidal behavior, and psychiatric illnesses, including psychotic, somatoform, and dissociative disorders. *These areas represent the majority of long-term adjustment difficulties* reported in the literature. (Jumper 1995) (italics mine)

Also in long-term studies on children, somatization has been registered, though most often termed 'somatic complaints,' as was stressed in the 1993 review by Kendall-Tackett, Williams and Finkelhor mentioned above. Lately, the number of studies about sexual abuse experience among patients in somatic health care has been increasing. Additionally, studies of non-clinical populations have shown significant differences concerning lifetime sickness among persons experiencing abuse as compared to those who do not. Consequently, researchers in the field of sexual abuse impact agree that *somatic presentations are some of the most frequent expressions of silenced sexual violation experience.* This consistent finding by sexual abuse researchers has been addressed and exemplified in most of the previous chapters in the present project.

In all the examples derived from the interviews, 'the somatic' and 'the mental,' both rooted in a corporeality informed by abuse and unmaking, *are obviously inseparable.* The traditional medical distinction as to the primary locus of sickness being the body or the mind cannot meaningfully be maintained if one is to make sense of the observations in this study. This distinction, which demands consistent criteria for what 'bodily' means and what 'mentally' means, leaves a vast number of human states of being unclassifiable. The two concepts, distinguished and separated by human intention and definition, both contribute to a problematic interface. Here lies what appears to be marginal to the field, although not representing a clear margin. By distinction, an in-between area is created, the zone of those who are marginal - the non-classifiable, the indefinable, the imprecise.

To cope with these marginals, two classes of human states of being have been engendered: the *psychosomatic* and the *somatoform.* Each of these artifacts of a dualistic order, reflecting the attempts to grasp with uniting words what has been made ungraspable through separation, leans to the opposite side. The psychosomatic is 'matter-oriented,' the somatoform is 'mind-grounded'; the demarcation line between these two is blurred. Those specializing on matter claim the visibility of structural defects as a proof of the rightness of classification; those specializing on the mind claim the probability of behavioral deviance as a proof. The sorting out of reliable proof or probability, called the diagnostical process, however, demands resources. Thus, whatever defines or limits the technical and economical possibilities of diagnosis, will shape the gray zone where resides that which is not yet classifiable. Consequently, non-medical forces steer the process of making distinguishable that which does not present itself as distinct. These create a principally unacceptable arbitrariness within the medical judgment. Such a technically or economically caused arbitrariness makes evident the ambiguity of judging in a field which is based on theoretical inconsistencies. This topic has been addressed as to its theory and practice in several of the previous chapters.

The literal fluctuation of the margins was demonstrated when exploring Line's lower back pain. Once accepted as somatic due to an anomaly verified by x-rays, it was still maintained as somatic when it reappeared despite medical intervention on its primary 'cause.' As it was considered to represent a somatic 'fact,' another, and new, anomaly had to be attributed causal function, though actually being a medical artifact of the first reparation of an anatomical anomaly. Despite a lack of verification as to causality, a completed pregnancy without aggravation of the pain in the impaired area, and a lack of any medical intervention, the disappearance of the pain was still considered proof of its somatic nature, this time explained by the non-verified second 'cause.' Then, when the pain returned yet again, no attempt to explain it was made at all. It was left to the patient, as if it were of a totally different nature and meaning than the pains before. However, this fact still did not engender questions concerning the difference - or, in case it was the same - the different reaction of the physicians. As in Veronika's history, one might suggest that Line's sickness history, as it presented in its final phase, might be interpreted as proof of the non-somatic nature of a disease. Concerning Veronika, this 'proof' was read from the disappearance of a defect; concerning Line, however, it was the reappearance of a pain which now evoked the response that medicine would refrain from intervening further.

If Line's pain were to be classified, where would it belong?

In Christine's sickness history, the movement in the gray zone was even more obvious, when comparing her first encounter with gynecological surgery with her second. In the first, all anatomical deviance was attributed causal function, due to which substantial interventions were made. In the second encounter, the remaining fallopian tube, formerly judged abnormal, was now judged to be normal, despite the same presentation of a pain which, in the first encounter, had served as the justification for the removal of the other fallopian tube. Since the first intervention had proven ineffectual, given the persistence of the pain it was intended to cure, the tube previously judged 'abnormal' was now regarded as a non-cause of the actual pain, and as such 'a normal tube.' Therefore, the pain had to be explained by something else. The explanation, implicit in the medical advice given, was that this was a pain caused by tension, making relaxation the logical treatment. Still, the pain persisted, and Christine, having repeatedly experienced its appearance as being linked to associations or memories of sexual abuse, refrained from any more consultations with medicine. Now, she was back to the situation prior to the interventions: she avoided consultations with medicine. Previously, her fear and the panic linked to genital exposure had caused her to ignore the pain of infections, and thus had prohibited her seeking appropriate consultations and proper treatment. Therefore, she had not received drugs for infections which, as was suggested, resulted in organic defects. Although these defects were not the cause of her actual pain, they 'caused' the decision to remove the organs.

If Christine' pain were to be classified, during which phase ought it to be placed where?

Susanne's pain debut during her first pregnancy was classified according to its predominant location and the concurrent situation of pelvic instability. As the pain remained after her delivery, though slightly changed, it was termed sciatica. During every pregnancy, it became 'pelvic instability' again. However, as it periodically extended into various joints, it was then termed 'arthritis.' Despite a lack of serological findings, and although Susanne continually had most of her pain in the area of her pelvis and hips, an intermittent swelling of one ankle was considered to be proof of a chronic disease, known to gradually affect and destroy the joints. The prospect for the future which this diagnosis implied, was given predominance and served to justify the medical advice to join groups for people with that disease and to behave in accordance with the expectations regarding the developmental curve of that progressive disease. This advice stood in complete contrast to Susanne's perception of being nearly pain-free after having divorced and begun living alone, and by her resistance to, or rather, strong aversion to, groups for diseased people. She still, however, accepted the diagnosis, although in doubt as to whether she really could adapt to a diagnosis while not feeling sick and while enjoying, most of all, the participation in sports with her daughters. On the background of her abuse history, reactivated in the Gestalt of her husband's physical resemblance to her abuser, a pain in the bodily area of sexual abuse and marital sexuality remained unresolved until divorce. Just as for Ruth, who had been in anxiety as long she was in sexuality, Susanne's pain was 'relieved' by divorce.

Was this two-decade long pain psychosomatic or somatoform, or were parts or phases of it somatic, and if so, which?

The literature concerning somatoform disorders is extensive. The most relevant contributions as to theory and sickness presentation of the group categorized under "Somatization Disorders", are the work of a few psychiatrists.[288]

Grounded in psychiatric theory, the group of disorders comprises seven main presentation forms, among which somatization disorder is the first mentioned. It is defined according to the DSM IV as follows:

> somatization disorder (historically referred to as hysteria or Briquet's syndrome) is a polysymptomatic disorder that begins before age 30 years, extends over a period of years, and is characterized by a combination of pain, gastrointestinal, sexual, and pseudoneurological symptoms. (DSM IV 1994)

Somatizing persons are described as follows:

> Prominent anxiety symptoms and depressed mood are very common and may be the reason for being seen in mental health settings. There may be impulsive and antisocial behavior, suicide threats and attempts, and marital discord. The lives of these individuals, particularly those with associated Personality Disorders, are often as chaotic and complicated as their medical histories. Frequent use of medications may lead to side effects and Substance-Related Disorders. *These individuals commonly undergo numerous medical examinations, diagnostic procedures, surgeries, and hospitalizations, which expose the persons to an increased risk of morbidity associated with these procedures.* Major depressive disorder, Panic Disorder, and Substance-Related Disorders are frequently associated with Somatization Disorder. Histrionic, Borderline, and antisocial Personality Disorders are the most frequently associated Personality disorders. (italics mine) (DSM IV 1994)

Concerning the gender distribution of somatizing, the following is said:

> Somatization disorder occurs only rarely in men in the United States, but the higher reported frequency in Greek and Puerto Rican men suggests that cultural factors *may* influence the sex ratio.(italics mine) (DSM IV 1994)

A critical analysis of these statements would very easily reveal their pretense of a scientific foundation and their traditional blindness as to assault embodiment under the influence of culture and gender. Such a deconstruction, however, may be done more fruitfully by exploring the problem of somatization from another perspective. The analysis will be linked to considerations about revictimization of sexually victimized persons in somatic medicine.

Somatizing patients are at above average risk for undergoing diagnostic or therapeutic interventions where findings or results do not accord with expectations, as previously quoted from DSM IV. This fact was the point of departure for a study of data from the Danish health statistics, aimed at estimating the amount of medical interventions due to somatic presentations in the general population. The author compares 'persistent somatizers' to non-somatizers with regard to utilization of health care. He writes:

> The study was carried out using the Danish national medical register to identify persons in the general population (age range 17-49 yr.) with at least 10 general admissions during an 8-yr. period. Persistent somatizers were defined as persons with more than six medically unexplained general hospital admissions in their lifetimes before 1985. Conversely, non-somatizers were patients whose admissions could mainly be ascribed to well-defined somatic disorders. (Fink 1992)

The author uses 'persistent somatizer' as a label, not as a psychiatric category. He does not reflect upon the problems of uncertainty linked to symptom presentation, to the process of diagnosing, and to the interaction between the diagnosing professional and the complaining patient. He has attached this label to persons merely according to the available data in the health register. He continues:

> The findings show that persistent somatizers had been exposed to *extensive surgery,* outnumbering the non-somatizers. Surgical operations were of several categories, with *gastrointestinal and gynecological operations being the most frequent.* (italics mine) (Fink 1992)

The two fields of surgery mentioned give immediate associations to the high occurrence of sexual abuse history among patients in secondary and tertiary care in gynecology and gastroenterology, mentioned previously in the chapters of the present study connected to somatic sickness and chronic pain. Furthermore, the author adds:

> The outcome of the surgical treatment of the persistent somatizers was, however, generally unsuccessful in that *the effect was unsatisfactory in three quarters of cases.* Similarly, *two thirds of the medical treatments were judged to be unsuccessful* in persistent somatizers. (italics mine) (Fink 1992)

Fink relates his findings to earlier studies concerning the occurrence and the description of extensive surgery in groups of patients. In psychiatric research, hysteria had been shown to be highly correlated with an above average lifetime surgery occurrence.[289] However, not all the hysterical patients were found to be somatizers, which engendered the separation between patients suffering from Briquet's syndrome,[290] which meant somatizing, and non-Briquet hysteria.[291] Fink refers to a study from 1951 which provided the basic material for the above mentioned publication by Cohen, showing an almost twofold mean of operations among hysteria patients compared to a control group. The most frequent operations in that study material were also gynecological and abdominal operations. Fink also mentions the diagnosis Munchausen Syndrome, for which surgery is reported in 61-71% of the reported case histories.

The phenomenon called Munchausen Syndrome has been mentioned in "Maladaptive Adaptations" linked to the sickness histories of Nora and Annabella. The diagnosis and the term were introduced by Asher, and later, in its presentation by-proxy, expanded on by Meadow. When reading Meadow's original publication from 1977, some of the remarks concerning the mothers allow the conclusion that these most probably had been abused as children. Additionally, Meadows writes about them:

> Some mothers who choose to stay in hospital with their child remain on the ward slightly uneasy, overtly bored, or aggressive. *These two [mothers] flourished there as if they belonged, and thrived on the attention that staff gave to them.* (italics mine) (Meadow 1977)

The description of the mothers echoes some of the wording in the interviews of Annabella and Nora. One may easily imagine these two women searching back to the only places where, according to their prior experiences, care was provided, by producing their children's symptoms and securing a stay in the ward, and by cooperating with the staff while covertly malingering the children. The Munchausen by proxy syndrome might be deconstructed from representing irrational acting and a proof of maternal psychiatric pathology, to being a logical reconstruction of abuse embodiment which has not been narrated and which therefore has remained part of a strategic, albeit destructive, repertoire, which is directed, informed and guided by silenced and unacknowledged trauma. Such a deconstruction would also render comprehensible a rather dramatic report on a mother suspected of inducing the deliveries of two of her children voluntarily so that they died shortly after, and having injured her third child repeatedly, both before attending hospitals and while staying there with him. [292] In a recent study, the family situation of children admitted to hospital with induced symptoms was studied in a total of 56 families. Of the 47 mothers in this group, 19 were interviewed as to lifetime psychiatric problems. The findings were as follows: 15 had *somatizing disorders*, partly presented to the health care system with fabricated symptoms; 5 of these had in addition *a history of seizures* without organic basis; 12 had *a history of self-harm*; 7 had *a history of substance misuse*. Somatization was linked mainly to gastrointestinal and joint symptoms, and three of the mothers had undergone surgery due to these.[293] The authors expressed their astonishment concerning the 72% occurrence of somatization among the 47 mothers, and likewise they had elicited that, among 15 who had mentioned having suffered from overall neglect during their childhood, 4 mothers had reported physical abuse, and 4 sexual abuse. The rather extensive discussion, however, offers no indication suggesting a corporeality of imposed pain of a different kind as providing the rationale for these mothers' sickness histories. The authors are more bothered by the implications of these findings concerning the accuracy of distinguishing psychiatric diagnoses, and the problems of 'lack of categories' for the variety of presentations in symptom fabrication.

In Fink's study, however, such a discussion is lacking since the author had no possibility to challenge the diagnoses he operated with, thus having to rely only on the distinction between somatic and non-somatic reasons for referral. The gender distribution in his material is not presented explicitly. It has to be searched for and is only to be found in one of the tables: the 57 non-somatizers consist of 33 females and 24 males; the 56 somatizers, however, consist of 47 females and 9 males. Five times more women than men are categorized as 'persistent somatizers,' a fact which is not commented upon at all by the author. Remarkably enough, the known asymmetric gender distribution of 'non-specific' or 'non-distinct' health problems remains unproblematized in this study on somatizing and intervention. Appendectomy was performed significantly more often on women than on men. Pelvic and abdominal surgery was performed more often among somatizers, while operations of the spine and heart/lungs were more frequent among non-somatizers.[294] Furthermore, two out of three surgical interventions among somatizers were without abnormal findings compared to one out of five among non-somatizers. Additionally, the females had significantly more operations without abnormal findings than the males. The author writes:

> Each persistent somatizer had, however, undergone a median of eight surgical procedures of which most were without abnormal findings, *which indicates that physicians tend to repeatedly and invasively evaluate all symptoms* ... The persistent somatizers did not benefit from such treatment, instituted on the basis of patient's complaints only, as the success rate of 1/4 for surgical and 1/3 for medical treatment is on a level with the placebo effect. (italics mine) (Fink 1992)

The placebo effect of surgery, especially explored with regard to appendectomy, has been reflected upon by Alan G. Johnson.[295] He points to the known connection between removal of an appendix due to pain, showing a non-pathological appendix, and still having 'cured' the pain. Within a few months after the initial operation, however, the patient may again present with abdominal pain. This time, the pain may be explained as 'adhesions' or 'irritable bowel syndrome.' The same might occur in the case of cholecystectomy due to pain and a finding of stones, interpreted as a cause of the pain. This phenomenon of attributing causality of pain to an organ which consequently has to be removed, is also mirrored in yet another piece of information in Fink's study. Of the 47 somatizing women, 15 had their uterus completely removed due to pain. Only two of these women had a satisfactory result and were relieved of their chronic pain.[296]

Fink's study, focusing on arithmetic means and presented as gender neutral, covers up the fact of the disproportional number of women among somatizers, and, consequently, the highly disproportionate number of unnecessary surgical interventions and medical treatments carried out on women. Not only are women at higher risk of being attributed a non-specific diagnosis. Women are also, as may be read from the study, at a disproportionately high risk for any kind of surgical interventions on erroneous indications, and, thus, have a completely disproportionate risk of being subjected to iatrogenic harm.

It seems that surgeons have a much lower threshold for performing surgery on women than on men, which may best be reflected in the significantly higher number of appendectomies in women among both somatizers and non-somatizers. This fact might be explored on the background of a Norwegian study about the accuracy of indications for appendectomy.[297] The overall rate of unnecessary laparotomy was 28.4%, but among women in fertile age (12 to 40 years), the rate of unnecessary surgery was 45.6%. The authors admit that such a result, confirming identical figures from several other studies, and equaling randomness, is unacceptable. They try to explain the finding as due in part to gynecological co-morbidity, which in this study accounts for 28%, though not all those cases can satisfactorily 'explain' the actual pain. A chain of arguments is linked to the surgeons' fear of causing infertility by risking perforation of an undiagnosed appendicitis. This argument, however, is not grounded in any empirical findings, as the authors also admit. One may suspect this argumentation of representing a rationalization of a medical practice wherein social gender rather than scientific evidence seems to influence professional decisions.

Fink concludes his comparison of the persistent somatizers and the non-somatizers as follows:

> The findings suggest that the costs of somatic diagnostic procedures and fruitless surgical and medical treatment attempts on persistent somatizers are enormous, and only exceeded by the risk of iatrogenic harm. This emphasizes the

> need for an early diagnosis of somatization and of treating it properly. (Fink 1992)

This statement is important enough, and it is clearly grounded in some alarming figures. In this connection, however, one might be dismayed by the author's failure to see or mention that the women bear the burden which is most problematic to calculate: the danger to their bodily health, the threat to their mental strength, and the weight of social stigma embedded in the repeated proof of non-pathology in the medical sense. Yet there is another element which has to be considered. The author does not reflect upon that the label 'persistent somatizers' not only addresses, but simultaneously judges, a group of persons. Their maintenance of presenting bodily complaints is regarded as bearing witness to mental disease. Their strategy of attending the health care system for what is non-real is judged as inappropriate. At the same time, however, the physicians whom these persons encounter practice a complementary conduct: persisting in performing invasive interventions on all kinds of bodily complaints, repeatedly and knowingly. Among the somatizers, one person had been operated twenty-four times during eight years, and five persons more than twenty times. Most operations had been unsuccessful. Nonetheless, the author does not address these remarkable parallels by labeling the doctors: 'persistent neglecters.' Nor does he identify the doctors' conduct as a main precondition for that of the patients.

Actually, Fink's study can be regarded as two valid documentations in one. The first of these is intended: making calculatable the costs generated by the somatic presentation of a mental disease. The second documentation, however, probably unintended, is implicitly provided: rendering visible *the production of somatic morbidity by medicine.*

8. GISELA

Precisely such a double reading is the point of departure for exploring the last interview which has not as yet been elucidated, and the medical records completing the abuse-sickness-history told by Gisela. Her main health problem was pelvic pain, leading to a hysterectomy as the last of several surgical interventions, which, however, did not resolve the pain. Thus, a recall of previously mentioned studies in gynecological patients with regard to the occurrence of pelvic pain and gynecological surgery is appropriate. Likewise, the previously explored interview excerpts concerning gynecological and reproductive health must be remembered. And finally, reference is made to Fink's study and the rate of hysterectomies which were unsuccessful as a surgical measure to cure pelvic pain.

Gisela had been abused by her biological father from when she was nearly three until she was five years old. From the age of seven, she was very frequently abused by a non-relative, a farmer from whom her parents used to rent a cabin for weekends and holidays. On two occasions, this man also gave three and then two of his adult friends access to Gisela. The last of these group rapes occurred on the banks of a little mountain river. The farmer silenced Gisela by pushing her head under the water, threatening to drown her if she screamed. The repeated abuse, casting a shadow over all her holidays and many weekends, terminated when Gisela was fifteen years old and her parents choose another place to spend their vacations, though still ignorant of the farmer's continuous abuse of their daughter.

Gisela had very often been sick as a child, suffering predominantly from abdominal pain and asthma before she was ten. At the age of fifteen, she had been referred to a gynecological ward due to pelvic pain, supposed to be caused by infection. Before she was eighteen, she had been admitted to the same ward three times due to pain and inflammation or infection. From that time, Gisela's pelvic pain had become constant. In addition to this pain and her asthma, a lower back pain became a problem. These three health problems remained active during her adolescence and adulthood. Gisela was pregnant three times, suffering each time from anorexia and pain termed 'pelvic instability.' She reported her pain-history as follows:

> I've had pelvic pain - I mean that I've thought of the pain as being in my pelvis and not my stomach as it was during childhood - from when I was fifteen years old, since I had these three admissions due to infections. But how they became later on, that started during my second pregnancy. I went to the doctor again and again, and every time it was an infection. And I was given drugs every time. Since then, they've decided that these (the infections) can't have occurred. My sedimentation rate was often very high. That was strange as well. They poured on the antibiotics, time and again, which I don't recall the names of... After some years, I was referred to the regional hospital. They examined me but couldn't find anything. So they decided to perform a laparoscopy. They didn't find anything wrong and couldn't understand my pain. I was sent home, but didn't improve. My doctor referred me again. And like that, I traveled back and forth quite a few times. Then a doctor at the hospital said: 'perhaps we should sterilize you, perhaps that may reduce your pain.' They thought it had something to do with my tubes. I found this difficult. Although I had three children, I was uncertain whether I wanted more. I was thirty at that time. But if it could relieve the pain, it was worth considering, and so I finally agreed. They operated on me, sterilized me and took my uterus at the same time. Psychically, it was a strain on me. And the operation didn't work as it should, the pain got worse. After the operation I felt more pain than before. But then I was called 'hysterical' and things like that, because I had to accept the pain getting worse shortly after

> such an operation. I went home and had to call the doctor several times the following week. Then he didn't want to take responsibility any more and sent me back. I was examined and the doctor stated: 'You have to be operated on, something is wrong.' But first I was given drugs, the fever decreased, so did the sedimentation rate, I was sent home with medications, but kept getting worse, was given yet another drug, was sent home again - and then operated on as an emergency. When I was waking up from the anesthesia, the doctor told me I'd had an ileus involving my appendix and one ovary ... They apologized for the problems they'd caused me by waiting. I was sent home, but I didn't recover. New examinations at the out-patient clinic, back and forth, until they considered removing everything. But first they wanted to do an ultrasound examination. That's how they found a tumor on one ovary. I was admitted again for an operation and my uterus, one ovary and both fallopian tubes were removed. They told me that I would get rid of the pain. That was that. I was thirty-two, so this was fairly early. I was sterilized and had lost everything inside. I felt only half human. But the pain remained, until two years ago when I began to tell about the abuse.

This detailed pain-sickness history was told without any interruption. It covered a time span of seventeen years, including three pregnancies and deliveries with 'their' problems, which are not mentioned here, however, as Gisela did not consider them as part of her sickness. The presentation could be compared to complete records from the hospital where all the interventions had taken place. The first laparoscopy is confirmed, as are the previous examinations and treatments performed by a local General Practitioner. The records from the next referral confirms a long history of pain and a degree of accentuation of the pain by intercourse, and that the patient has learned to avoid sexual intercourse. The examining doctor describes normal genital and pelvic conditions, especially no genital descensus, and a normal, yet very 'painful uterus,' and concludes:

> She is referred for antesuspension of her uterus since, in addition to dyspareunia, she feels pain when sitting down, and simultaneously one will be able to look into her again.

An operation is scheduled for pain during intercourse despite there being no other pelvic pathology. No other possible origin of painful sexual intercourse is probed for or even reflected upon. Since the pain has caused avoidance of coitus, an intervention seems indicated, and an operation 'correcting' the position of the uterus is deemed appropriate. In the next record note, and during preparations for the operation, mention is made that the patient has consented to a bilateral tubarectomy. No mention is made of the proposal that this might help the pain, nor is the patient's doubt linked to the decision, as mentioned by Gisela. During the operation, the pelvic organs are inspected and described as completely normal. Despite this fact, both fallopian tubes are removed. A month later, she is referred again after treatment by her General Practitioner. During the initial gynecological examination, an infection in the operation area is suspected, antibiotics are administered and the patient is dismissed. Three weeks later, an emergency operation has to be performed. The findings are described as adhesions involving the appendix, calling for an appendectomy. The tenth postoperative day, an abscess in the operation area is discovered which has to be incised under full anesthesia. This operation has not been mentioned in Gisela's report. Again, a long treatment with antibiotics is indicated. Three months later, a referral from her General Practitioner confirms constant pain and two further treatments with antibiotics. He concludes:

> The patient still has heavy pain. She is still completely incapacitated as regards work, and little by little is being severely strained psychically by her chronic problem. She is willing to be hysterectomized *if* this can enhance the situation. (italics mine)

The examining physician at the hospital cannot find any pathology in the reproductive organs or the pelvic area of the patient, apart from the fact that she complains of pain when he palpates her uterus. However, he registers that this is less so when he gets her to relax a little while examining her, a fact which is reported, yet neither problematized nor reflected upon as regards the quality or origin of the pain. He concludes that a removal of the uterus might be a premature act, and the patient is advised to await further developments. Two months later, a record note is as follows:

> The patient is still incapacitated as her pain is so severe that she hardly manages her everyday life. No irregularities of menstruation. Her pain is on her right side. Coitus is impossible, she tried three weeks ago. Exploration finding: a cystic tumor the size of a grape at left, but most pain at right, although no finding there. Her cyst may represent a pseudocyst with adhesions, but since she still has so much pain and cannot work *one is forced to do something about this.* One prepares her that one will remove both the uterus and something from the adnexae in *an operation which she will try.* (italics mine)

The lack of serious findings is reflected in a vagueness of the wording, and a rather peculiar termination of the record notes, attributing the operation to the patient's wish to 'try' it. This contrasts strongly with the imperative of acting, expressed as an almost moral claim on the physicians. The same vagueness and attribution is to be found in the operation report:

> Since she has pain all the time she wants to have something done about it. One has not found any strong indication to go ahead, since everything seems normal, but since she does not seem to get rid of her pain, she herself is ready to have her uterus removed and it may be that this may contribute to her being rid of her pain.

The imperative of acting is obviously not unambiguous in this case. The decision is constructed to be the patient's, and the action chosen by the physicians is legitimized by a double underlining of a possibility. Peroperatively, the previous suggestion of lack of pathology is confirmed, although there is a tiny irregularity around the right ovarium, which is therefore removed as well. The uterus is described as completely normal, both before removal and afterwards, when opened and inspected. Almost a month later, a record note confirms that the patient still suffers from pain in the area of intervention, but that this pain must be ascribed the operation.

Not quite a year later, the patient is referred again due to pain. A review of her history confirms *a laparoscopy, a tubal ligation, an antesuspension, a chronic salpingitis, numerous treatments with different antibiotics, an appendectomy, an abscess incision, a hysterectomy and ovarectomy.* A general and gynecological examination is performed. A *rectoscopy* is added. The record note is as follows:

> The patient has a long history with abdominal pain. Considerable *psychosocial problems* are present, something *which also her sickness history may indicate.* Still, considerable adhesions are verified which may cause her colic-like pains. Yet another operation cannot be proposed with this sickness history, but a few more examinations may be performed. such as x-rays of the intestines and possibly colonscopy. Should these tests fail to reveal special findings, *the problem should be sought solved in cooperation with a psychiatrist.* (italics mine)

And as neither colonscopy nor other diagnostical procedures can verify any pathology, the final note is:

> No cause for the patient's complaints could be found this time either, and after conference with seniors G.G. and N.N., one has decided that one has *nothing more to offer* this patient.

The term psychosocial problems has not been mentioned in any of the records before. All of a sudden, it appears as an 'explanation' of a pain which has remained the same throughout the years, and which has always been responded to with drugs and surgery. Where does this term come from? The argument following this surprising statement is tautological. How can a somatic sickness history due to pain, highly influenced by medical decisions, actions, interventions and complications, suddenly be proof of the non-somatic nature of the pain? Could this remarkable change in terminology be engendered by the fact that there were no more organs left to be removed? Thus, *one could not offer the patient yet another intervention.* This was the gynecological reading after having performed repeated '-ectomies'.

When read in a different way, as proposed in the exploration of Fink's study, one might say: *the patient had nothing more to offer to medical action.*

This situation calls for a reintroduction of reflections concerning the initially addressed semantic field of *"victim/Opfer/offer"* and *"sacrifice/Opfer/offer."* Gisela, a victim of socially silenced sexual abuse, had become revictimized by medical interventions, a professional action which contributed to a further silencing of a pain which had been forcibly inflicted upon her during childhood. Although this pain had been addressed and described repeatedly as maximally painful during sexual intercourse, sexuality had never become a topic between Gisela and her doctors. Rather the opposite was the case: as she had mentioned the pain's prohibiting of coitus, "something had to be done." This meant, however, not to talk and to listen, wonder or explore, but to invade and deprive. Sexual invasion and deprivation were the origin of the pain. Medical invasion and deprivation maintained it.

Gisela, violated during childhood and adolescence by male invasiveness, violence and disregard of her integrity, had been made more and more sick. The silencing forces in society had more and more been supported by the silencing theory and practice of medicine, regarding disembodiment as a major and appropriate mode to approach human bodies even in the face of no demonstrable pathology.[298]

9. CONCLUSIONS AND IMPLICATIONS

Since life experience embodiment is invalid in the context of medical judgment, a painful corporeality originated by imposed pain, endangered existence and forced humiliation cannot be understood in the theory and practice of biomedicine. Non-narrated, silenced violation, expressed in a wide range of indications as to the disruption of selfhood and corporeal integrity, encounters the naturalistic, objectivist and dualistic concepts of biomedicine. Particulars, mindful, lived bodies encounter general scientific theories of separated human bodies and minds. But biomedicine is ignorant as to how life is inscribed into human lived bodies, and how lived bodies are incribed in the social politics of silencing.

Violated humans are *made sick by the silence* and are sacrificed to the silence about overwhelmingly male sexual violence, which societies still resist becoming knowledgeable of and reflect upon. Both psychiatric and somatic medicine take part in the silencing, "the sickening", the sacrifice and thus, the violence. In outlining the implications of these findings, I shall argue that not only sexual violation or any other violation of personal integrity, has potentially pathogenetic impact, but also any structural humiliation of human dignity.

Consequently, I must focus on *how societal institutions themselves contribute to human suffering,* particularly those that are mandated to alleviate it. My argumentation will be exemplified with an actual case at the intersection between two highly respected societal institutions, Medicine and the Law. And it will be anchored in the work of the Hebrew University philosopher, Avishai Margalit,[299] particularly his concept of the *decent society.*

Margalit has outlined the political and societal pre-conditions for granting explicit respect for human dignity. He claims that it is not enough that a society be civilized, meaning that its members do not humiliate each other, nor that a society be just, meaning that its members have equal access to rights and support. Civility and justice are not sufficient to insure a paramount respect for human dignity. Margalit argues that we become blind to a range of significant moral and political questions when justice and/or civility alone are the main focus. In fact, he regards as futile any positively constructed argumentation for what constitutes the respectful regard for one's fellow human beings. He finds it more fruitful to construct a negative argument, based on the fact that human beings share the morally relevant characteristic of being *"something which can be humiliated."* This negative argumentation, he states, far surpasses in usefulness all of the positive ones. As we have seen in the particularly debasing twentieth century, there has been no shortage of institutionalized disrespect for all strata of populations in so many societies. According to Margalit, human beings no longer have Truth, God, Wisdom, Language, or the Laws of Nature or History in common. Paradoxically, however, they do all share the ability to be humiliated.

A decent society is reflected, according to Margalit, in the way its institutions meet the most vulnerable of its members - or its non-members. Any measures which marginalize people stigmatize them. And a stigma is the public sign of deviation from the norm, be it the norm of honor, mores, gender, race, faith or function. Regarding the concepts of honor, self-respect and self-esteem, Margalit writes: "A humiliating society is one whose institutions cause people to compromise their integrity," and, "a decent society is one whose institutions do not violate the dignity of the people in its orbit."

My study provides evidence that structural humiliation of human dignity occurs within medicine. Whenever people deviate from the norms of biomedicine, they become marginalized. This becomes the case: whenever the problems they present are not defined as medical; if their complaints cannot be objectified; if their verified symptoms do not correspond to defined diseases; if their symptoms do not meet the criteria for a somatic or a psychiatric diagnosis; if their symptoms do not respond to presumably appropriate measures; if their health does not improve as fast as expected; and, finally, whenever they return with the same presenting problem despite that, according to standard medical practice, it ought to have been solved.

These scenarios all lead to medical marginalization, regardless of their origin. It is known, however, that social stigma and shameful, silenced experiences cause health problems *but, at the same time* cannot be communicated frankly and explicitly. We must conclude, therefore, that such scenarios are likely to arise from precisely those stigmatizing and shaming experiencing. In other words: socio-culturally originating suffering and bad health are not only misunderstood in medical contexts; they will also most probably be aggravated by being responded to with "more of the same," so to speak. In fact, the public stigma of suffering from non-specific, undefined health problems is considerable. This public stigma will, in turn, be aggravated by its own consequence: undefined states of bad health interfere with a patient's accessing resources administered by social agencies.

Consequently there is a path from silenced humiliation in private to legitimatized humiliation in public. There is a link between the private experience of being made to feel worthless - through domestic abuse, subordination, exploitation, neglect or deprivation, and the public doom of being unworthy to receive help - through correct medical and legal objectification.

There is convincing evidence that certain strata of the population in each society suffer from an above average incidence of medically identifiable disease *and* from ill health of a medically indefinable kind. These strata are among the marginalized groups within the particular society, although the criteria for marginalization may differ considerably between societies. The socially accepted norm defines, by contrast, what shall be considered deviant. Thus, those who deviate suffer from the humiliation of not meeting the norm of respectability, and of not fulfilling the criteria for worthiness. Epidemiologists describe the facts and verify the figures, but they question its cause. The "cause", or rather the origin and pathogenetic matrix and force may, in fact, be collective humiliation through marginalization.

I shall now excemplify this connection with an actual case which has come to my knowledge in my role as an expert witness in 1998. Twenty-one young men from a Norwegian town, now between the ages of twenty and forty, have accused the same man of abusing them sexually during a total of at least fifteen years. The man has admitted to abusing them; yet he cannot recall abusing one of the twenty-one. None of the adults from his town seems ever to have harbored even the slightest suspicion about him. The man was shielded by his position as the moral leader of a parish, and by his charisma as, or so it seemed, a devoted friend to young boys. Because he was seen by the townspeople as upholding a high moral standard, he could grant himself the authority to criticize young people publicly for breaches of rectitude. Surely, no child dared oppose him, regardless of the circumstances. Behind this pretense of respectability, he seduced and abused any boy he could approach within the legitimating frames of parish meetings, sports, choir practices, or boy-scout excursions. He also took advantage of his roles as uncle, godfather, neighbor, and the father of his children's friends. He had legal access to nearly all places where children went, including their homes, and even their bedrooms. According to the abused men, he was completely insatiable and made use of every opportunity which fit into his pattern of conduct and sexual praxis.

He was sentenced for the abuse of two of the men while the abuse of the remaining nineteen was deemed beyond the statute of limitations for prosecution. The court sentenced him to eighteen months imprisonment. All men submitted claims requesting compensation as victims of violence. The civil authorities granted those of the claims that were connected to the man's conviction, and a few others based on documentation of psychiatric treatment sought during adolescence. These records could, in retrospect, be convincingly related to the abuse experience although none indicated any professional's suspicion that abuse might be the origin of the disorder. Fifteen claims, however, were not evaluated due to lack of evidence of any adverse, long-term impact. The men were humiliated since their demand was judged unjustified.

I met these men as a group, and most of them individually. Each had a sickness history that could be read as *a particular embodiment* of abuse the *general pattern* of which they all had in common. Each body bespoke the individual abuse perception. When each man put his own words to his experience, I, as a physician, could read the complaints and dysfunctions as inscriptions of integrity assault. The men's medically diagnosed diseases and their non-specific sicknesses followed the logic of assault embodiment.[300] *Due to theory and by tradition, however, medicine and the law remained ignorant of how they had embodied their violation.*

In almost all cases where sexual abuse is disclosed and the abuser is denounced to the police, medical evidence of assault is sought. In medicine, evidence of trauma is conceptualized as the visible and measurable impact of a situation defined as dangerous, and categorized according to objectifiable criteria. In the Norwegian case, any search for documentation of physical or mental harm would have been in vain. None of the abused men had been physically assaulted. None had ever experienced physical violence during abuse. The abuser never forced them, although he often caught them in situational traps. Or, the boys were "caught in context," so to speak, were constrained into acquiescence by relational and societal conditions. Their abuser legitimized the abuse by terming it an act of caring. He unmade their criteria for right and wrong, unmade all safe places, and occupied their minds. He made maleness an ambiguous concept, making them ashamed of their own male bodies. He alienated them from their own arousal, affecting every later sexual relation. He interfered in the education of at least two boys. He strained the adaptability of them all to such an extent that several of the boys developed chronic health problems with which they are still struggling as adults. The problems led to periods of incapacitation, frequent medical examinations, hospital referrals, treatment with drugs, physiotherapy, psychotherapy, and counseling.

From the perspective of scientific medical evidence, however, none of the above mentioned effects of sexual abuse can be proved to have been caused by abuse. In the sense of the positivist concept of causality, none of the men's sickness histories would meet the criteria of satisfactory probability. And, in the sense of the philosophy of objectivist science, no scientific proof is possible. Scientists can falsify a zero-hypothesis, or in an opposite reading, they can estimate a degree of probability. Proof is impossible. Uncertainty diminished is the best proper science could offer. However unavoidable uncertainty may be in correct scientific reasoning, it accrues another meaning when introduced into a legal context. Whenever medical uncertainty is a central premise of a legal procedure, a categorical mistake takes place. *A scientific category, probability, translates to a moral category, doubt.*

Given the importance attributed to medical expertise in prosecutions of sexual abuse, this categorical mistake is crucial. Given the impact of doubt on the practice of law in the West, this categorical mistake becomes the deciding factor, strongly influencing the outcome of the case. Given the inadequacy of the current biomedical knowledge of the impact of silenced abuse, this categorical mistake is nothing less than tragic. Consequently, a revolution in knowledge production in medicine and law is required if societal decency is to be the aim. And, the need is urgent.

10. EPILOGUE

In my gradual approach to human suffering grounded in violations by significant others, I encountered powerful forces which structure human perception of reality. The most powerful of these seemed to be polarized along a continuum of human communicative interaction, the essence of which was the *unmaking and making of meaning.* I became aware of the overt destructiveness of the hidden, wordless, unspeakable, untold. This awareness allowed me to uncover the core of the destructive power, *silencing*, however practiced and by whomever.. On the other hand, I was shown how constructive the told, the spoken, and the narrated can be. This rendered evident the core of the constructive power, *speaking.*

The *destructiveness* of the political or social silencing of voices has been attested to quite recently by observers of the impact of ethnic conflict in Guatemala. As the politically imposed wall of silence around an oppressed segment of the population begins to crack due to external pressure, the impact of decades of not being able to speak about death and destruction is becoming visible.

Comparable in effect, the strategic, multiple rapes of women, and some men, in a recent multi-ethnic war in Europe, have been met with social silencing; the impact of this suppression has been attested to by all the involved helping agencies as being not only destructive to health, but also to the practice of justice on behalf of those violated and silenced.

The *constructiveness* of political or social speaking has been explored, regarding varying mortality rates among males in Eastern Europe during the last two decades, where periods of political openness for the people's voices coincided with reduced mortality.

Comparable in effect, the mortality differences between the mothers fathers of Latin American rebels who "disappeared" during a time of political despotism, demonstrated the creative force of speaking. While the mothers, even under threats to their own life, voiced their demand to know about their adult children's fate, in public weekly demonstrations, the silent fathers died.

However, speaking out about violation can only wield its constructive power in the presence of a crucial precondition: *trust.* When humans are or feel distrusted when telling about violations, the healing and protecting potential of speaking seems to be unmade, and speaking can become destructive in a double sense: being disbelieved is silencing, and not to be believed shatters ones foundations. More destructive, perhaps, than any silence imposed by threat or force during violation, is not to be believed when voicing the violation. While the explicit aim of most societies is to create and maintain good conditions for health, a pervasive distrust of narratives about sexual violation as the normative position acts as a pathogenetic force, creating and maintaining sickness.

When entering the present project, I entered a field of that which is silenced. I felt prepared for this by virtue of my previous encounters with abused and violated persons during my professional life. However,

I was not prepared for what I gradually had to acknowledge, namely, in which ways and to which extent my own profession contributes to this silence, by silencing and promoting the chronicity of the abused.

Upon becoming aware that the very foundation for the silencing forces in the medical profession is grounded in medical theory, which means in the biomedical concepts about human bodies and human minds, I experienced a professional crisis. The awareness of representing a profession which contributes to the silencing of violence by making the violated sick was totally intolerable. I saw myself confronted with yet a deeper conflict than that which initially motivated my research. The question arose, whether learning to think differently had alienated me within, and estranged me from my profession, so that I might feel forced to leave it.

However, I realized that to leave would mean not to voice my conviction of the urgent need for radical changes in the theoretical framework of my profession; and, even more decisively, to leave would mean acting contrary regarding my hypothesis of the destructiveness of silence. So, I had better keep speaking.

Impressions

Impression 1

to be unable to speak out
to be unable to give oneself a voice
to have been deprived of the mouth
to have been made voiceless, mouthless, wordless, speechless
that which is not said, spoken of, expressed - is not
what she says is extinguished, made untrue, invalid by
how she says it
the unarticulated, non-articulatable, unspoken, unspeakable
the not listened to, unheard
she keeps her voice - takes it back to herself
she keeps her narrative - takes it back to herself
kept back, taken back, locked, unfreed
unfreed speech, unspeakable events
how many untold narratives
how many unfreed narrators

the women dare not tell what they have experienced
they dare not give it words, names, space, sound, tune, shape
it is not carried away from them - and out
they themselves take it back
what they say is called back by
how they say it
the voices decrease gradually, turn flat, tuneless, unmodulated
the breath does not carry nor lift them

toneless unhearable monotonous unheard unsaid
speechless unspeakable wordless

the words are tuned down, lowered, weakened
and even the eyes turn the same way
down, inwards
away from me
who listens and sees

I become so obsessed with these fading voices
how the women lower their voice
and have to be listened to
they will not have spoken anything - unless
someone makes all possible efforts to listen -
what they have to tell has almost to be listened
into the world
otherwise it is not
I become almost enraged
they are forsaking themselves
they are almost not real
nor really here

I become so confused
as she takes her voice inwards
not letting it out
the more she talks to me - apparently -
the more her voice stays with her
as if to say:
I want to speak
but don't look at me while I am speaking
but don't listen, I really have nothing to say
nothing real
what I say needs not to be heard
I dare not hear myself
I dare not raise my voice
I don't want that which I say to be heard

two of them talk of their ugly voices
they call their voices ugly
what is an ugly voice?

and after having listened with all of me
with all I own
and all I am able
and after having heard what I never heard before

imagined what I never had imagined
after she who said it has gone
and finally I am back in myself
after having been "in" hers
I start freezing
I shiver from sudden cold

Impression 2

if there had been a camera
if I had been able to keep what I saw
Line telling for the first time
how the first abusive act
was done to her
how difficult it visibly felt
to express
how much she struggled
trying again and again
stumbling words, stuttering voice, as if pinned to the spot
unable to move from there
"he forced me to... forced me to... forced me...
I did not want... he forced..."
silence
mouth open
breathless
then swallowing,
again and again
face as if turned inside out
by the disgust of a taste
recalled

her eyes intensely keeping mine
until I lean forward -
her breath breaks through in a sigh
it is said
and I have not withdrawn
in disgust
as she fearfully expected

leaving Line,
my head full of her delay,

her hesitation,
her eyes,
her withheld breath and fading voice
her swallowing and struggling
I return to the center
meeting the police
escorting a woman who can hardly walk
she looks as if far away, absent, closed
not really there

later I hear that she has been there before
been battered before
but returns to him
who beats her
again and again
after a while and a rest -
I am told I shall meet her tomorrow
according to the schedule
if she will be able to speak

upstairs in my room
when writing the day
with four women's lives
deformed by imposed secrecy, permanent fear, forced silence
informed by pain, shame, guilt and humiliation
I start to shiver
more and more freezing
only after a hot bath I feel warm for sleep
why is cold my reaction?

Impression 3

the next day she comes, slowly,
as if moved by a force behind her
not as if walking from a will within her
not seeing my hand
not seeing the chair
not hearing my words that she may be seated
standing unstable she needs help to sit down

again, I miss the camera

so that it could be testified to by other than my eyes only
how distorted - a body can be
when it feels like - no place to be
as there is
barely pain and pure revulsion

half the face as if dead, blurred in contours, unshaped
half the body as boneless, floating out, shapeless,
to see her makes me seek words for horror
dropping into my mother tongue
Ich denke "misgestaltet, verunstaltet, geteilt, gespaltet, doppelt"
ich sehe "formlos, deformiert, aufgelöst, verlassen"
ich fühle "halbtot, zerschlagen, gebrochen, zerstört"
ich vernehme "umarmen und trösten"
und - gleichzeitig und stärker
"zurückweichen und beschützen"

seeing a gestalt of
turning away
from one's own body
leaving it partly
to itself - or something - or someone - or nothing
the whole of the pain is too deadly to be in
the whole of revulsion too repulsing to be there
moving out, leaving, partly
in order to survive - partly - in the rest?

no face - no gaze - no voice - no words - no body
only parts of breath, uneven
only parts of words, stumbling
only parts of movement, torn

what is left of life seems drawn into one side,
the right
the left is left
it smells, she whispers, voice harsh as if unused for long
toneless, inwards, as if behind
hands, teeth and tongue
far back in her mouth which is almost closed
it stinks, she stutters, repeatedly,
with long intervals of nothing
only the right side of the face

turning slightly
shrinking slightly

and then she mentions
electricity - coming from her spine
floating her left body in strong, pulsating, painful, pushing streams
not always, but whenever she meets people
who make her fearful
who mean a threat -
her right hand shows what she cannot say
then silence until she rises - slowly - and I lead her out.

I recall what I have been told, about abuse from young age,
two men and two women, in turn, only one of them stranger
a childhood in pain, self-inflicted injuries ever since,
living in abuse - the only way to live?
outer threat, danger, harm and abuse internalized?
the evil taken inside - into mind, senses and body?
placed in the side she leaves as long as she chooses life?

Impression 4

this disturbing difference
between the voice of abuse
and the voice of life
in the same woman
the difference of everything - strength, words, precision
the non-voice of the hidden
and the strong voice of the open

when meeting them upstairs where they talk and discuss
they are different
they speak differently
competent, skillful, knowledgeable, informed
almost sophisticated on certain topics in the center of their interest

the lawsuit of Bjugn, so close - by all means
recently terminated without a sentence
but there are many children with genital marks of assault
visible - witnessed - described - measured - photographed - exposed
proven

discussed - challenged - minimized - normalized- neglected
disproved
whose competence?
are preschool girl's genitals gynecological or pediatric areas?

what caused these children the evidence of harm?
who?
not guilty - nobody - none
the women are well-informed about the language of law and the courts
they express precisely
the core of the problem:
what is a proof in the sense of the law?
they have all been confronted with it
on behalf of themselves
on behalf of others
in their imagined procedures
when preparing mentally
but not daring
when it comes to terms
as they know what is needed

good proofs are not easy to be found
says one of them
laughter with shadows of tears, bitterness, resignation, sadness
not in these days
says another
no - and not further on
they foresee what will happen
the fear of the lacking good proof
will spread

they talk about a young woman
yesterday
she filled the news
accusing a colleague
of having raped her
several knew her
she was known as tough - one of the very few women
having advanced as an officer in the military forces
would she be able to stand it through?

would somebody dare to use

the argument of ethnicity
her color
against her?

all of them know the rules for lawsuit in case of rape
as if all of them had studied both
the law
and the procedural frame
extensively
stunningly informed
using terms with an ease which reveals
training in approach
to the topic of justice - in case of abuse

" incompatible"
who said it? - the word stands in the room - silence
some rise and leave in silence, a car outside, then talk about the snow

what did they talk about in the west - likewise knowledgeable
likewise skilled in words of precision, distinction, evaluation, concern?
therapy: aroma, psychomotoric, gestalt, psychodrama, body-oriented
the stunning difference of worlds, words, voices, reflections
for the speakable - and for the silenced

Impression 5

what did I see
adding to the spoken
a dimension of pain
beyond words
a dimension of shame
beyond reason

assault in the name of help
covered by name
only - but sufficiently
forced examination to secure a proof of
the harm
of forced penetration
forced isolation to treat the intrusive pictures of
the harm

of intrusive hands

victim again
revictimized

recalling her pain and horror
the legs tied to the chair
the threat of the needle and the forced injection
Grethe's eyes widening - nothing more - but more than enough
and knowing it was true, not a nightmare, a hallucination
part of a police protocol
documented abuse covered by the word help

recalling her shame and humiliation
forced into a locked room
the horror of the pictures which meant madness
Elisabeth broke down in her chair in front of me
crying, trembling, passing out, disappearing
as she wanted not to be seen
in her shame
with a mark on her forehead
a stigmatizing print
a diagnose

so uneasy I felt afterwards - so shivering with cold
having witnessed the harm of abuse in the name of help
inscribed in women's bodies, minds and lives

Notes

1 "Life-world" ('*Lebenswelt*' in the German tradition of phenomenology since Husserl) means the world of lived experiences. That is, the world as we experience it immediately, pre-reflectively, rather than as we conceptualize it, categorize it, or reflect upon it.

2 I feel deeply indebted to phenomenologist Elizabeth A. Behnke who, skillfully and respectfully, revised the present text before it found its final form. From her profound insight into the phenomenology of the body, she offered me "terms" which allow me to address some of the aspects or dimensions of violation inscriptions, as they have presented themselves in my study. (Personal communication) Especially the notion of the "mutely testifying body" seems to me an utmost relevant "name" when thinking of the bodily expressions as valid, unspoken messages for silenced, unspeakable experiences.

3 Bourdieu 1977.

4 Foucault 1975.

5 From Norwegian and international studies, it is well known that many women, when they have the possibility to choose, prefer consulting a female General Practitioner. Thus the majority of female General Practitioners' patients are female. This lends a very special profile to their every-day work.

6 Bagley 1989; Baker & Dunkan 1985; Finkelhor 1979; Finkelhor 1986; Mullen et al. 1988; Rush 1982; Russell 1983; Russell 1986; Sorenson et al. 1987; Saetre et al. 1986; Wyatt 1985.

7 Browne & Finkelhor 1986; Conte & Schuerman 1987; Finkelhor 1986; Lindberg & Distad 1985 a; Furniss 1991; Pierce & Pierce 1985; Sedney & Brooks 1984.

8 Bagley & Ramsay 1986; Briere & Runtz 1988; Briere & Zaidi 1989; Bryer et al. 1987; Browne & Finkelhor 1986; Coons et al. 1988; Courtois 1979; Gold 1986; Herman et al. 1986; Mullen et al. 1988; Russell 1986; Sedney & Brooks 1984; Wyatt & Powell 1988.

9 Bachmann et al. 1988; Beck & van der Kolk 1987; Briere & Zaidi 1989; Bryer et al. 1987; Bulik et al. 1989; Burnam et al. 1988; Carmen et al. 1984; Coons et al. 1988; Craine et al. 1988; Ellenson 1986; Hall et al. 1989; Herman et al. 1986; Herman et al. 1989; Jacobson & Richardson 1987; Lindberg & Distad 1985 a,b; Morrison 1989; Nathanson 1989; Oppenheimer et al. 1985; Putnam et al. 1986; Ross et al. 1989; Sedney & Brooks 1984; Shapiro 1987; Shearer & Herbert 1987.

10 Bachmann et al. 1988; Briere & Runtz 1988; Bryer et al. 1987; Carmen et al. 1984; Escobar et al. 1987 a; Gross et al. 1981; Lindberg & Distad 1985 a; Morrison 1989; Sedney & Brooks 1984; Walker 1988.

11 Briere & Zaidi 1989; Bryer et al. 1987; Cunningham et al. 1988; Draijer 1989; Escobar 1987 b; Gross et al. 1981; Haber & Ross 1985; Harrop-Griffiths et al. 1988; Lindberg & Distad 1985 a; Rimsza et al. 1988.

12 Arnold, Rogers & Cook 1990.

13 Cohen et al. 1953.

14 Drossman et al. 1990; Rapkin et al. 1990; Wurtele, Kaplan & Keairnes 1990.

15 Wakley 1991.

16 Reiter et al. 1991; Felitti 1991.

17 Albach & Everaerd 1992; Bushnell, Wells & Oakley-Browne 1992; Chu & Dill 1990; Demitrack et al. 1990; Ensink 1992; Famularo, Kinscherff & Fenton 1991; Sandberg & Lynn 1992; Shearer et al. 1990; Strick & Wilcoxon 1991; van der Kolk, Perry & Herman 1991; Walker et al. 1992; Winfield, George & Swartz 1990.

18 Scott 1992.

19 Springs & Friedrich 1992.

20 Walch & Broadhead 1992.

21 Eisenberg, Owens & Dewey 1987; Howe, Herzberger & Tennen 1988.

22 Frenken & van Stolk 1990.

23 Sugg & Iniu 1992.

24 Hamberger, Saunders & Hovey 1992,

25 Koss, Woodruff & Koss 1990; Koss 1990; Koss, Koss & Woodruff 1991. Stark, Flitcraft & Frazier 1979; Stark & Flitcraft 1988; Flitcraft 1992.

26 AMA, Council Report 1992.

27 Beitchman et al. 1991; Beitchman et al. 1992. Briere 1992 a; Browne & Finkelhor 1986; Femina, Yeager & Levis 1990; Leventhal 1988; Wyatt & Peters 1986 a; Wyatt & Peters 1986 b; Wyatt & Powell 1988.

28 Goffman 1974.

29 Albach & Everaerd 1992; Bachman et al. 1988; Beck & van der Kolk 1987; Beitchman et al. 1991; Beitchman et al. 1992; Briere & Runtz 1988; Briere & Zaidi 1989; Bryer et al. 1987; Bulik et al. 1989; Burnam et al. 1988; Bushnell et al. 1992; Carlin & Ward 1992; Chu & Dill 1990; Coons et al. 1988; Craine et al. 1988; Ellenson 1986; Ensink 1992; Gold 1986; Golding et al. 1988; Goodwin et al. 1990; Herman et al. 1986; Herman et al. 1989; Jacobson & Richardson 1987; Kinzl & Biebl 1991; Kluft 1990; Lanktree et al. 1991; Lindberg & Distad 1985 a,b; McClelland et al. 1991; Metcalfe et al. 1990; Morrison 1989; Nathanson 1989; Newberger et al. 1990; Oppenheimer et al. 1985; Palmer et al. 1992; Putnam et al. 1986; Root 1991; Ross et al. 1989; Ross et al. 1990; Sandberg & Lynn 1992; Scott 1992; Sedney & Brooks 1984; Shapiro 1987; Shapiro et al. 1992; Shearer & Herbert 1987; Sheldrick 1991; Strick & Wilcoxon 1991; van der Kolk et al. 1991; Waller 1992; Winfield et al. 1990.

30 Arnold, Rogers & Cook 1990; Briere 1992 b; Cunningham et al. 1988; Draijer 1989; Drossman et al. 1990; Felitti 1991; Gross et al. 1981; Haber & Ross 1985; Harrop-Griffiths et al. 1988; Rapkin et al. 1990; Reiter et al. 1991; Rimsza et al. 1988; Sedney & Brooks 1984; Springs & Friedrich 1992; Waigant et al. 1990; Wurtele 1990.

31 Bagley 1989; Baker & Duncan 1985; Finkelhor 1979; Finkelhor 1986; Finkelhor et al. 1990; Furniss 1991; Mullen et al. 1988; Russell 1983; Saetre et al. 1986; Sorenson et al. 1987; Walch & Broadhead 1992.

32 Beitchman et al. 1991; Beitchman et al. 1992; Briere 1992; Burnam 1992; Felitti 1991; Finkelhor 1986; Finkelhor et al. 1990; Herman 1992; Kluft 1990; Koss 1991; Koss, Koss & Woodruff 1991; Leventhal 1988; Rush 1982; Russell 1986; Wyatt & Peters 1986 a,b; Wyatt & Powell 1988.

33 Baron 1981; Baron 1985; Grene 1976; Lock & Gordon 1988; McWhinney 1983; McWhinney 1989; Pellegrino 1979; Spitzack 1987; ten Have 1987; Thomasma 1985; Toulmin 1976; van Leeuwen 1987; Wulff, Pedersen & Rosenberg 1986.

34 Merleau-Ponty 1989.

35 Heidegger 1962.

36 A shift of epistemological positions affects the researcher's role and place. The 'detached observer,' the ideal in the natural sciences of the early decades of the twentieth century, is replaced by the 'participant observer,' who entered the field of physics before other areas of natural science, as scientists were obliged to realize that any observation influenced that which was observed. Until that time, the personal presence of the researcher and his or her world view and implicit values had been conceptualized as an epistemological question for researchers of the human sciences only, where, in most disciplines, the role of the 'self-reflective participant and interactive researcher' had been acknowledged and accepted. However, the interaction between scientists and their objects has to be acknowledged as representing a "two way affair."(Toulmin 1982.) The influence of the observer has been recognized as not being illimitably reducible.

37 van Manen 1990.

38 The work of Heidegger is an outstanding example of the creativity language affords for speaking out what had not previously been articulated about the life-world's "taken-for-granted-ness."

39 Richard M. Zaner, in his book 'The problems of embodiment. Some contributions to a phenomenology of the body', has elaborated the notion of embodiment in a comparative, critical reading of the philosophy of the body in the works of Marcel, Sartre, and Merleau-Ponty. (Zaner 1971)

40 Gadamer 1975.

41 In order once more to emphasize Mishler's characteristics of interviews:
Interviews are *speech activities*, which means that they are, "particular types of discourse regulated and guided by norms of appropriateness and relevance that are part of the speakers' shared linguistic competence as members of a community."
They are *jointly constructed*, which means that interviewer and interviewee contribute to the topics of interest, and to the course of the dialogue which will find its shape in between the partners.
Their analysis and interpretation must be based on a *theory of discourse and meaning*, indicating that, "meanings emerge, develop, are shaped by and in turn shape the discourse."
The meanings of questions and answers are *contextually grounded*, which includes the socio-cultural setting of a research interview, roles and rules for such particular situations, and the socio-cultural issues linked to the topics at stake. All quotations from Mishler 1986. These elements of a research interview as both a situation and a text can only be evaluated as to reliability and validity provided they are taped and transcribed according to explicit transcription rules. This topic will be dealt with in a separate chapter.

42 Svensson 1995.

43 Kinmoth 1995.

44 Jones 1995.

45 Garfinkel 1967.

46 Goffman 1974.

47 Freire 1970.

48 Garfinkel 1956.

49 Goffman 1961.

50 The focus on a mechanistically perceived nature lends appropriateness to action and acting. This may account for medicine having deduced its "bias towards action," action being considered superior to non-action in medical practice and research.

51 The notion of a variable only makes sense in sciences which are discreteness-biased.

52 In cases where cause is unknown, but which are verifiable through physical traces (hypertension, colitis ulcerosa), causality is still valid. The disease is termed 'primary' or 'idiopathic' (caused by itself), or 'psychosomatic' (caused by the self). Primary hypertension may be regarded as psychosomatic and different from a secondary hypertension caused by other malfunctions. Colitis ulcerosa might be regarded as caused by an unverified agent and thus be non-psychosomatic, in analogy to ulcus ventriculi. Ulcus ventriculi has changed status twice, recently, from that of a psychosomatic (stress/ lifestyle) condition, to a disease caused by an infectious agens; this causality, however, came to be considered a relative one as the presence of this agens was verified in a larger part of the healthy population as well.

53 One might object that modern medicine has begun to address subjective topics recently, such as satisfaction or quality of life. Yet, these are generally treated as if they were objectifiable, as variables, as measures of physical properties in the individuals.

54 Kleinman, Eisenberg & Good 1978; Ots 1990.

55 Young 1982.

56 Kleinman 1976.

57 The influence of communities is to be read in the EM its members adopt. Thus one might say that Western practitioner's theoretical or scientific EM is slightly different from their clinical EMs, and different from laymen's popular EM, certain family EMs, or sectarian religious EMs.

58 Hahn 1985.

59 Douglas & Calvez 1990; Ogden 1990.

60 Foucault 1977.

61 The truths of nature are formulated in abstract laws and abstract terms. This has evoked a critique concerning an abstract bias in the natural sciences, where the abstraction may represent the real to such a degree that it seems to become the real. Edmund Husserl has addressed this topic as follows: "It is through the garb of ideas that we take for *true being* what is actually a *method* - a method which is designed for the purpose of progressively improving, *in infinitum*, through 'scientific' predictions, those rough predictions which are the only ones originally possible within the sphere of what is actually experienced and experienceable in the life-world. (Husserl 1970:51-52)

62 Foucault 1975.

63 Kirmayer 1988.

64 Engel, Science 196:129-36.

65 Canguilhem 1989.

66 Lloyd 1984.

67 de Beauvoir 1952.

68 Harding 1986.

69 Fausto-Sterling has reviewed the most renowned and influential works in modern biology as to research on differences between and within gender, interwoven with race. She has made clear how the modern natural sciences have, step by step, delivered the modern arguments for the ancient construction of ranked difference in favor of white, heterosexual males above other males (of color or of homosexual preference), and of males in general above females in general. (Fausto-Sterling 1992)

70 The authors explore the impact of "geneticism" on medicine, and the influence of contemporary values on the researchers in the field of genetics. Among other aspects, they explore the ethical impact of covert values and of gender bias as implicit in research concerning the genetics of homosexuality. (Hubbard & Wald 1993)

71 The author has reviewed medical textbooks with regard to the metaphors chosen in the process of naming female body parts and functions in gynecology and obstetrics. She has made evident how the biological phenomenon of change in the female body became the cultural product of pathology. (Martin 1987)

72 The author shows how the construct of female pathology has quite recently been extended into the "diagnostic" and "therapeutic" approach to menopause. (Lindenbaum & Lock 1993)

73 The author shows how the construct of female pathology has quite recently been extended into research on menstruation as a health problem. (Nicolson 1995)

74 Walker 1995.

75 McKinnon 1995.

76 Prieur 1994.

77 This will be explored in depth in reference to Elaine Scarry's theory of the world as made and unmade.

78 Sartre 1956: 252-302.

79 Foucault 1975.

80 Merleau-Ponty 1989: 413.

81 Michael Polanyi has explored the tacit ways of knowing linked to every sensory perception and its interpretation or recognition. There must be cues and conditions "from" which our attention is directed "to" the object of interest. Regarding vision, this means, *we have to decide what to look for in order to be able to see it,* so we can activate the appropriate sense modality. Perception depends on tacit "means, cues, and conditions," according to Polanyi, which includes our bodily knowledge about how to operate the sense organs and to interpret their perceptions. There must be a "from" and a "to" which, by necessity, cannot be the same. There cannot be a coincidence of "from" and "to" in perception. (Polanyi 1969)

82 Gadamer 1975.

83 Ricoeur 1971.

84 Kvale 1983.

85 Geertz 1973.

86 Kvale 1989.

87 Giorgi 1989.

88 Oakley 1981.

89 Kvale consistently uses male gender singular for both interviewee and interviewer. In the following quotes I have allowed myself to introduce female gender singular into his text.

90 Lock & Scheper-Hughes 1990.

91 The historical narrative, the stories which are lived, are, according to MacIntyre to be distinguished from the fictional narratives, the stories which are invented. Paul Ricoeur has dwelt upon this distinction of lived stories and invented stories by reflecting upon whether there is a difference, and, of which kind. (Polkinghorne 1988)

92 Human beings are knowledgeable of the richness of some moments in a particular other person's presence, and the poverty of years of indifferent cohabitation. Subjective experience of "jumps" in time during a conversation, triggered by a word, a glance or a sensory perception, may open a Japanese fan with an implied story, painted in detail and unfolding colorfully. However, when the fan is closed and the story "folded back into it," the subjective experience of having been in another time is perfectly well compatible with having physically been present in presence.

93 Bremner et al. 1995; Janoff-Bulman 1989; Freyd 1996; Melchert & Parker 1997; Roth et al. 1990; Tromp et al. 1995; van der Kolk 1985; van der Kolk & Fisler 1995; van der Kolk & van der Hart 1989; Wolfe 1995.

94 Elisabeth Weber has reflected upon the impact of violence as inflicting a "break" in the cohesive lifeline. (Weber 1995)

95 Surely the awareness of historicality is framed by a knowledge about one's own birth and death. By being aware of life as a time-span, experiences, recalls and retrievals are elements in building this history, shaped as a biography, and in assigning it a certain meaning, which may be assisted by contemporary witnesses.

96 Concerning the plot, Ricoeur continues: "But a narrative conclusion can be neither deduced nor predicted. No story without surprises, coincidences, encounters, revelations, recognitions, etc. would hold our attention. This is why we have to follow it *to the conclusion.# Instead of being predictable, a conclusion must be acceptable.** (# author's italics)(* my italics) (Ricoeur 1978)

97 Morris 1994.

98 The concept of bodily identity has been increasingly challenged by the progress of modern medicine to shape or repair bodies. Even central organs of the individual body have proven to be interchangeable among humans, and exchangeable with those of animals. How much of a body can be changed before the "individual" is a composition of "other"? The utmost challenge to the body-based identity emerges in an experiment of thought: if it were possible to transplant brains, perceived of as the organic place of memory, who would remain identical, the donor or the receiver?

99 Baring & Cashford 1993.

100 This notion has been problematized by James A. Trostle as follows: "The very notion of compliance requires a dependent lay person and a dominant professional - someone to give advice, suggestions or orders, and someone to carry them out." (Trostle 1988)

101 The participation of psychiatrists in forced "treatment" of political dissidents in the former USSR has recently been documented, although unrecognized or even widely denied by Western psychiatric experts.

102 Here it might be important to consider that being heard and believed is probably not the only way for suffering persons to re-integrate disrupted memories. Receiving confirmation through other persons' witnessing seems to hold constructive power as well. There might even be a non-verbal way of acknowledgement by witnessing through others: a therapist's touch, bearing witness of her or his awareness of something inscribed in the flesh which is silently crying out. This aspect shall be elaborated in Hanna's, Eva Maria's, and Veronika's interviews.

103 This topic will be dealt with in the description of the practice of transcribing in section 1.4.

104 These references are in chronological order to reflect the development of the methodological debate: Rush 1982; Baker & Duncan 1985; Russell 1986; Finkelhor 1986; Saetre et al. 1986; Herman & Schatzow 1987; Wyatt & Peters 1986 b; Wyatt & Powell 1988; Leventhal 1988; Bagley 1989; Finkelhor et al. 1990; Agger 1992; Briere 1992; Herman 1993.

105 Polkinghorne 1991.

106 "The danger of the leading question is probably less pronounced in a qualitative interview than in questionnaires, where there are few ways of examining how the respondent has understood a question. The analysis of question-answer sequences in a qualitative interview will to a large degree be able to clarify the possible effects of different types of leading questions." (Kvale 1983:190)

107 Kvale 1989.

108 Frank 1985.

109 All the texts entitled 'Impressions' are based on written field notes of what I saw, heard or felt during my visits at the incest centers and during the interviews. Among other phenomena, the fading voices awoke my awareness of problems linked to speaking of silenced, unnamed, or unspoken experiences. These difficulties led me to exchange the technical equipment in order to assure adequate recording quality.

110 The problems connected with the transcription of dialect will be dealt with more extensively in the next chapter.

111 This interview led to a reinterview, and to cooperation resulting in lectures, as mentioned in chapter 1.3.

112 The interview situation and the preceding events from the night before are given form in Impression 3.

113 I did not see her any more before she left the center accompanied by her best friend whom she was to join for a couple of days until a contact with a psychiatric out-patient clinic had been established.

114 As this decision had to be made in advance, it was incorporated into the textual information given to the respondents in connection with the interview, underlining the fact that only a very limited number of people would have access to the tapes and would listen to the interviews. From certain comments of the interviewees regarding this very point, I feel reassured that this decision contributed to the establishment of trust in the confidentiality of the project procedures, although mainly technical and theoretical considerations had motivated it. As this information was given verbally before every interview started, and was a passus in the consent form, the assurance of a secure anonymity may have contributed to lowering reservations about revealing sensitive experiences.

115 Being limited to acoustical information only often felt almost painfully restrictive since the visual impressions recalled by listening seemed much more evocative of the impact of what was spoken. A sentence or some words could be powerfully underlined by the reaction of the person's body, visible only and so leaving no trace on the audio tape. The special communications called "Impressions" are an attempt to compensate for the sense of the relative inadequacy of mere acoustical documentation; they are referred to in certain connections in order to heighten the intensity of a message received during the interview but not present to the same expressive degree in the interview text.

116 An additional layer of communicative distance arises through the eventual translation of excerpts from the texts into English. While this had no influence on the conclusions drawn from the narratives, as all analyses of the texts were performed on the original transcriptions, it does impact on how clearly the findings may be communicated; the experience of the experience is rendered less acute by the text being put through the kinds of filtering which are inevitable with any translation. Thus, when certain metaphoric Norwegian speech has no appropriate equivalent English image, the literal Norwegian will be footnoted.

117 The interview with Veronika and the perspectives it opened generated several kinds of texts. Several lectures and publications were built specifically on her interview or the topics her history addressed.

118 Huseby 1994.

119 This sentence is the entrance into four themes which will be elaborated in depth later in this study. Their common background is psychiatric theory about what can not occur in social relations, and what, consequently, must be regarded as a evidence of mental disease. A message concerning a grave violation in the past, is turned into a proof of a grave mental disease in the present. The message itself, due to its very content, becomes proof of a mind outside the norm producing a narrative outside of normality. The repetition of the message does not cause, as might be possible, any doubts to arise as to the rightness of the professional interpretation. Rather the opposite seems to occur, as it is converted into a confirmation of its true nature as proof of mental disorder. This connection has already been mentioned in chapter 1.3 in connection with Saris' analytic contribution.

The next theme is the process of how the psychiatric profession interprets a message, according to a theory, as being proof of a disease which causes a repetition of the message. The response is a non-revision, interpreted as the confirmation of the disease. In this reading, a tautology appears: a theory confirms its constitutive elements by interpreting them as confirming the theory. This description demonstrates major characteristics of a culture and a culturally constituted framework. However, such a characteristic is hardly compatible with the scientific basis upon which psychiatry claims to build.

The third theme is the connection between a narrative about an individual experience, and it's, according to psychiatric theory, non-correspondence with collective reality. The maintenance of a theoretical denial causes a maintenance of a "disease." Such a reading allows an exploration into psychiatry's contribution to the creation of chronicity.

The fourth theme concerns the function of a cultural denial, integrated into a theory, and practiced in a systematic application by a profession endowed with social power. This allows reflections on the potential for the field of psychiatry to have a suppressive power through the attribution of mental disorders to the individual, while maintaining certain pathological social conditions. This reading allows reflections on whether victims of deprived social conditions risk being re-victimized by psychiatric practice due to the theory of the field.

120 The activity of recalling, telling, confirming and restoring has been mentioned in chapter 1.3 as part of identity establishment and confirmation. In this case, the impact of "the missing pieces" in the patient's biography becomes obvious. What she cannot recall, or can not remember in the right proportions, as demonstrated in the section, can be confirmed and corrected by her mother. Thus "put in place," the parts which appeared the most untrustworthy in the woman's narrative become the proofs that she told the truth. This "searching for the missing pieces" is a topic which emerges in almost all interviews and will be thematized in a broader connection later.

121 I owe my tutor, anthropologist Tordis Borchgrevink, gratitude for having made me aware of the way in which this case could almost be a template for a phenomenon which will become a central topic later. The narrative of the rape, the message, the key to understanding, had been there, told explicitly every time and recorded again and again. But those who heard it, heard something else because they listened from a position of theoretical preoccupation, creating the habit of a certain professional deafness. What was said but not really heard was even written down, a document in the center of the field of professional vision, uncovered, open for every authorized person to read - provided their eyes are not blinded by theoretical presupposition. Edgar Allan Poe's masterful tale, "The Purloined Letter," is a perfect illustration of this phenomenon. A highly compromising, and therefore extremely valuable, letter is searched for weeks in a systematic way by competent and skilled researchers, who, however, due to their theoretical framework, cannot see it lying there, open. I quote from Poe: "You will now understand what I meant in suggesting that, had the purloined letter been hidden anywhere within the limits of the prefect's examination - in other words, had the principle of its concealment been comprehended within the principles of the prefect - its discovery would have been a matter altogether beyond question." (Poe EA 1976:232) As an analogy, and applied on the present project, the tacitly mediated principles of hiding sexual assault may be completely, and, possibly not by chance, incongruent with the officially practiced principles of disclosing sexual assault. Therefore, the message of such abuse cannot be found in the field of vision of psychiatry, although it may lie, and even if it does lie, "in the middle of the table."

122 At this point in the interview, Veronika wants to emphasize what she has in mind and refers to a situation which occurred some years previous to our meeting. The story is transcribed in condensed form and marked. The excerpt from the interview text is as follows: "*Veronika explores in depth a situation connected to her former work among children where a little girl was in focus. The girl's sexualized language and conduct, her seductiveness towards other children and her restlessness had been a problem for a whole group of children and adults. Then Veronika experienced how several of the other adults refused to see; they ignored Veronika's concern about sexual abuse, which led to choices which became adverse for everyone." Veronika's explicit message in this passage of the interview was that there might be several layers of denial or resistance against seeing, registering, acknowledging and understanding.

123 This direct question represents a practical demonstration of how a previously received new perspective was reintegrated into a later interview. The sensation of hands holding one's head, had earlier been thematized by one of the four male interviewees, linked to forced fellatio, and to punishment for refusal to cooperate with the abusers. He could even recall how the hands had been placed, which was identical to the area where a particular pressure-like pain, linked to a pulsation inside his head, gave notice of forthcoming epileptic seizures. His description of the experience will be presented in more detail in "Lived Meanings."

124 In her study 'The Body In Pain: The making and unmaking of a World', Elaine Scarry (Scarry 1985) writes about the aftermath of torture. Her focus of interest is on, among other aspects, the very instrument of torture. The object which afflicted utmost pain and which therefore forever will represent the deepest humiliation and powerlessness, has been unmade as the thing it was before. An umbrella, a chair, a wire, or whatever might have been 'the thing,' could never again be an umbrella, a chair or a wire. Unmade as a thing and made into pain, it would forever mean pain. Could this equally be the case for a movement of a curtain, a touch on the back, a sound, a smell, a gesture, a word, a silence... Not the thing as such, but the meaning it had acquired in a situation full of pain, linked to events of powerlessness and humiliation? What if pain were an answer to an unintegrated past, to a there-and-then? What if it could be interpreted as a response to what had been unmade once and forever as things, events or gestures which, in the world of the here-and-now still carried their changed and "made" meaning? How could anyone other than the person in pain know what had been unmade from what it seems to be, and made into a pain? How could this kind of pain possibly be decoded from outside? Or even, how could it possibly be reliable? Could "things unmade" be the origin of later pain, expressing the past in the present, the painful experience embodied? Could the theory of "the things unmade" be a phenomenological key to the subjective world of pain? The theory of the world unmade shall be applied later to an array of descriptions from the interviews about sickness experiences. The topic of sensory perceptions, experienced as pain, or as unspecified sensations like nausea, shortness of breath, dizziness, blurred vision, and so on, will be dealt with in chapter 2. Such bodily symptoms, presented in medical encounters, most often lead to a symptom diagnosis according to ICPC (International Classification of Primary Care).

125 Here again arises the theme of "adaptation having turned maladaptive," which has been mentioned in connection with the pattern of hesitation and delay described in relation to the issues of tape transcription.

126 The theme of heavy blood loss and the non-contracting uterus reappeared in several of the interviews, as did reports of distancing, fading or leaving the body due to particular bodily sensations provoked by the hands or bodies or instruments of the helpers. It seemed as if hidden sexual assault could mean a special kind of risk during pregnancy and delivery, a very problematic kind of trap for the assaulted woman in situations of deeply conflicting interests, both needing help and not being able to accept it.

127 This analysis comprises both a demonstration of an approach to a "new theme," in the sense of "not yet heard or previously encountered," and a reintroduction of an "old theme," in the sense of "met before and explored previously." The new theme is "lived meanings," the old theme is "seizures." In the exploration of the "new," however, the "old" operates as a key, addressing a constellation of conflict which, at a meta-level, points to a central topic in this study, namely the breakdown of individual adaptability as the entrance into a sickness career.

128 Two index fingers pointing to the temples of one's own head do not make any sound on the tape recorder and so may not be transcribed. However, though not "spoken," the meaning of the gesture is a part of the sentence. Its central impact on the flow of speech demands its representation in the written text to render comprehensible that which follows. In the present interview, this gesture is a narrative, a key and an interface, which will become evident in the following exploration. On the methodological level, this gesture represents an essence of the shortcomings of taped speech transcripts, as shall be demonstrated.

129 Putnam 1989; van der Kolk & van der Hart 1989.

130 Bremner et al. 1993; Bremner et al. 1995; Briere & Conte 1993; Ford & Kidd 1998; Goenjian et al. 1994; Putnam et al. 1986; Rauch et al. 1996; Ross, Norton & Wozney 1989; Ross et al. 1989; Ross et al. 1990; van der Kolk et al. 1985; van der Kolk 1988; van der Kolk 1994; Wilson & Raphael 1993.

131 DSM IV 1994.

132 Agger 1989; Agger 1992; Albach & Everaerd 1992; Bass & Davis 1994; Basoglu 1992; Boney-McCoy & Finkelhor 1995; Briere & Runtz 1988; Chu & Dill 1990; Coons, Bowman & Milstein 1988; Demitrack et al. 1990; Elliott & Briere 1995; Ensink 1992; Goodwin, Cheeves & Connell 1990; Green 1993; Herman, Perry & van der Kolk 1989; Kluft 1990; Putnam 1986; Rodrigues et al. 1996; Rosen & Martin 1996; Ross, Norton & Wozney 1989; Ross et al. 1990; Sandberg & Lynn 1992; van der Kolk & Fisler 1995; Walker et al. 1992.

133 van der Kolk & Fisler 1995.

134 Here the validation of a difference is reached. The informational field in the room is addressed by the interviewer by pointing at the walls covered with sports trophies, and, at her suggestion, the interviewee agreeing to his excellent physical fitness. Therefore, the next statement is a logical one. There must be something in addition.

135 Here the second, the imagined presence is addressed. The very moment the interviewee mentioned that his main health problem had been epileptic seizures, a third person was present in the room by virtue of being imagined. This is an example of active introduction of knowledge from practical experience prior to the study. The young woman mentioned was not one of the interviewees. My encounters with her, however, were among the most powerful triggers for this study because meeting her made me intensely aware of the interrelatedness of sickness history and life history, and their reciprocally constitutional and maintaining influence. She gave me insight into the meaning of her disease, when it emerged, how it developed, why it was not properly classifiable, and why it seemed completely unresponsive to all relevant drugs. I feel deeply indebted to her, and am uncertain whether she is aware of the importance she has come to have for me both as a patient and as a person. Several motivations have brought me to work on this topic. She embodies all of them.

136 See my prior remarks concerning the transcription of non-verbal utterances. In this case, there occurred an almost pantomimic performance; I just watched while it took place, without commenting as it happened. The interviewee placed his hands alongside his head which resulted in his thumbs reaching behind his ears. This did not feel right so he tried to turn his hands upside down, which is impossible. But then he crossed his arms, and thus was able to put his thumbs in front of his ears and spread his fingers behind his ears. Obviously, he had a great interest in placing the hands the right way, that is, as if a second person were to be holding his head with thumbs in front of his ears, so that the maximum pressure from the thumbs would be in the region of both temples. Only in this way did it feel as it should. After the interview we returned to this problem of placing the hands the right way. Then I asked Fredric whether I might have my hands make the grip around his ears as he had both described and demonstrated. First he nodded, but then he quickly withdrew his head when my hands came closer.

137 The focus on the hands, and the recollecting of perceiving these abusive hands, seem to represent a certain compartementalization, in the sense of Janet, as previously mentioned.

138 Goodwin, Simms & Bergman 1979.

139 Mitchell & Gibson 1990.

140 "This finding seems unlikely to present merely the residual effect of abuse-caused physical injuries, since the seizure disorder was associated only with sexual abuse, rather than physical abuse. One hypothesis is that these were neurologically compromised children whose deficits may have contributed to the formation of inappropriate relationships in susceptible families. Another hypothesis is that dissociative symptoms, commonly associated with childhood trauma, *were mistaken as clinical evidence of complex partial seizure disorder.*" (italics mine) (Shearer et al. 1990)

141 Betts & Bogden 1992; Greig & Betts 1992.

142 Phonetic equivalent for a nodding of the head and a sound from the throat, indicating an agreement.

143 The term abjection or abject belongs to Julia Kristeva, the Bulgarian-French psychoanalyst and linguist, who, in her oeuvre, has developed a theory of the abject as that kind of otherness which threatens the individuation of the human being. Abjection expresses the horror of otherness. The abject is what repulses us, the inedible, that which threatens the orifices of our body, the zones of danger and the interfaces between self and other. The abject is that which shall not enter our bodies and which generates acts of protection, symbolizations of the forbidden, rules for the prohibited, and sensations of disgust - in certain connections felt as nausea. According to Kristeva, the feeling of anxiety and repulsion can be understood as a testimony to the continually on-going process in every individual to maintain the separation from others achieved in early infancy. In the terminology of Kristeva, *le corps propre* is both our own and our clean body as distinguished from not-own and unclean. Any threat to the distinction and separation between myself and other, tied to the zones of transition, generates anxiety and repulsion. (Kristeva 1982)

144 In Norwegian, the table for gynecological examinations is referred to as *den gynekologiske stolen*, literally, "the gynecological chair."

145 Once again the recorder proved insufficient for the production of a transcript of situational understanding. A very meaningful and impressive pantomime unfolded. Elisabeth contributed a great deal to my increasing awareness of the impact and the multiple levels of silencing in the life-world of sexual abuse experience. Some views of what happened during her interview have been channeled into the section, "Impressions," parts 1 and 5. Other parts will be dealt with in connection with the discussion concerning revictimization.

146 I myself felt a particular kind of discomfort while watching Elisabeth trying to open her mouth with the pressure from her own fingers, which I later tried to analyze. The first layer of it was certainly a feeling of indiscretion which is generated whenever one watches somebody else fail to do something adequately. This feeling concerns a reluctance to making others feel stupid or awkward. The next layer concerned a repulsion connected to seeing someone act improperly, a kind of shamefulness at having been involuntarily involved in other's intimate sphere. The third layer felt like irritation, a resistance to accepting what appears nonsensical. The next layer contained a transference of rage due to another person's subordination and lack of self-esteem. And the deepest layer was the shame of witnessing self-invasion, a violation of one's own boundaries and "complying" with therapeutic advise by acting self-destructive and self-humiliating. She offered me a Gestalt of a revictimization in the name of help, going beyond the revictimization which is embedded in self-assaultiveness, as Mary gave an example of (see below).

147 The content of his wordiness may also have masked a meaning potentially coded into the very fact of it: excessive verbalizing may be interpreted as a cloaked form of vomiting, of expelling words in lieu of whatever else one might want to spit out.

148 Basta & Peterson 1990; Perez & Windom 1994.

149 Both Veronika and the fourth man, John, had formulated in a similar way that they could not think whenever they were obliged by the circumstances or forced physically, to be immobile. They described their almost irresistible impulses to walk for hours and hours whenever they had "to think about something" in order to find a solution or to understand what was going on. Both expressed the anxiety-decreasing effect of walking alone, even at night, and even in completely unfamiliar surroundings, thus making credible that to be alone or outside or in darkness did not fuel their anxiety, as often may be hypothesized concerning anxious persons. Rather the opposite was the case: feeling caught, locked in a room, in crowded surroundings, "having no possibility to escape," was the most powerful trigger of the feeling of panic accompanied by all the bodily sensations they mentioned - or the "spacing out" which they both practiced.

150 A lecture about the topic of the impact of oral sexual abuse was given at the Nordic Semiotic Research Congress in Trondheim in 1994.

151 This change has been dealt with by, among others, Russell (1995) and Busfield (1996).

152 Bulic, Sullivan & Rorty 1989; Bushnell, Wells & Oakley-Browne 1992; Hall et al. 1989; Kinzl et al. 1994; McClelland, Mynors-Wallis & Treasure 1991; Oppenheimer et al. 1985; Palmer, Chaloner & Oppenheimer 1992; Pope & Hudson 1992; Root 1991; Shearer et al. 1990; Waller 1992; Waller et al. 1993; Waller & Ruddock 1993; Waller 1994; Welch & Fairburn 1996.

153 Herdt 1981.

154 AMA Council Report 1995; Armstrong 1991; Richards 1993.

155 This discussion is not meant to address any kind of sexual interaction between consenting adults.

156 Drossman et al. 1990.

157 Drossman et al. 1990; Drossman 1994; Drossman 1995; Drossman et al. 1996; Felitti 1991; Scarinci et al. 1994; Walker et al. 1993.

158 Cunningham, Pearce & Pearce 1988.

159 McClelland et a1. 1995; Shapiro 1987; van der Kolk, Perry & Herman 1991.

160 The body of research concerning this correlation among psychiatric patients was reviewed in 1992 by Beitchman et al.

161 Riggs, Alario & McHorney 1990.

162 Peters & Range 1995.

163 van Egmont et al. 1993.

164 All Norwegian women attending primary health care services during pregnancy have the right to free prenatal consultations and a consultation six weeks after delivery, aimed at eliciting questions and giving advice concerning contraception.

165 Hendricks-Matthews & Hoy 1993.

166 I recall, from my own general practice experience, meeting a young woman from a foreign country, who, invited to Norway by a young man, became acutely psychotic and imagined herself to be pregnant. She had to be escorted home for long-term treatment. Her case is reported in a lecture given in Innsbruck, Austria, and in an article published in Psychologie der Medizin.

167 Here Annika uses the Norwegian slang word 'pyton,' which is a very strong expression of contempt and disgust, derived from the word for a python snake.

168 This reminds me of a female patient I met in my earliest years as a physician. She represented a threat to her two small children, and so was admitted to a psychiatric ward. She perceived her children as misshapen and dressed them in clothes designed to "compensate," clothes which she had deformed by cutting one leg off the trousers or closing the opening for an arm. She knitted pullovers for the children in the shape she saw them, with two openings for a head and the arms in different lengths. I recall how I got the chills when I saw her doing this knitting. I had not thought of her for years, but Annika made the memory of her reemerge in my mind.

169 After Tanja has finished the first passage, the interviewer comments upon its content in a statement. The bed means "danger," the topic is "danger." Tanja confirms this short form by giving a sign of agreement. Then, the interviewer seems to jump. She asks for the uncle - due to her foreknowledge of the uncle as the abuser. This jump from the bed to the uncle is immediately understood as a question about location. The question in between, yet not asked, is: Were the bed and the uncle in the same house; did he live in her home; did the abuse occur there? Tanja answers the unasked question, sensitive to what the jump means. No, of course not, the uncle lives in another town and the bed was not her bed at home, but bed meant danger nonetheless, as a bed, as such. This non-verbalized interchange of contextually present no-words, reflected in the adequate response of the interviewee, shows the high degree of sensitivity for "meaning in the room" on the part of the interviewee. Several examples of such help from her side to grasp the cores of connections and relationships are to be found in this interview.

170 Bourdieu 1977.

171 Tanja uses the Norwegian word "kjedelig," which has to be understood in this context as an idiom expressing an array of emotional aspects comprising "boring," "shameful," and "unveiling," but also "troubling," "uncomfortable," "disturbing" and "bothering" in a situation where one has to face facts about former actions or conditions which one wishes undone or un-happened. The futility of this wish is translated into a term with rather passive or fatalistic connotations.

172 Tanja says "sammenkrøket," which means to have curled one's body in a certain way as if freezing or trying to become invisible by shrinking oneself, becoming little in order to disappear.

173 The ancient, male tradition of paying for sexual services which men receive from women, and thus the man's entitlement to her services, seems paramount here. It also seems integrated into judgments and strategies, even under circumstances in which the adult man ought to have been able to recognize that the situation did not carry the marks of an exchange. As it seems, Tanja's uncle behaved according to the tradition, "paying" for what he took, and therefore perhaps even convinced that no harm had been inflicted on the one who had been objectified through his sexual consumption. To buy oneself free from obligations and from a responsibility for having degraded, insulted and violated another person was his immediate impulse, although this hypothesis does not shed any light on the extent to which he recognized the harm he had caused.

174 Tanja addresses the very controversial topic of harm - and the proof of it - in the legal proceedings concerning sexual assaults and evidence of their impact on the victim. When reflecting upon her statement, and on the general tendency of trials of sexual assault cases to concern themselves merely or mainly with physical damage and the medical proof of physical harm, the importance of highlighting the consequences of "symbolic violations" in human relationships becomes obvious. The degree of violation Tanja had suffered had not been addressed at all. In the eyes of the law, no violence had occurred, no physical damage had been done, and no abuse could be proved. In addition, the accused man was known by everyone as being fond of this girl, and the girl had never refused to travel on visits to him and his wife. A twelve year old girl, obliged to describe what had been done to her without the skills required to spell out sexual acts correctly, could hardly be expected to voice either her ambivalence, what had been inflicted upon her, or such a complex and paradoxical phenomenon as the concept of the "kindly violation."

175 Beitchman et al. 1991, 1992.

176 The original wording in the interview text is: "jeg spyr av meg selv etterpå, jeg er ekkel .." Pia perceives herself as so disgusting that she is made nauseated by herself to such an extent that she has to vomit, because she is so abject. Here, Pia uses the Norwegian word 'ekkel' to describe her perception of herself. In the sense of Kristeva, she herself is the abject, and, by logical extension, she must try to get rid of herself in herself.

177 These studies shall be reflected upon in connection with the topic of strains between patients and therapists as regards influence from silenced abuse on physical and mental health.

178 Kiecolt-Glaser & Glaser 1992; Pennebaker 2000; Pennebaker, Kiecolt-Glaser & Glaser 1988; Pennebaker & Susman 1988.

179 Once again the brackets contain the short précis of a pantomimic demonstration of something on the body, once again combined with a struggling to make clear where the "it" which feels special is, as "it" is not so easy to reach with one's own hands. What Hanna points to is an area along the outer margin of her shoulderblades which one may reach by crossing the arms in front of the chest and placing the fingers behind the shoulders.

180 The same pantomime is repeated, again with the arms crossed before the chest and the thumbs pressing in front of both shoulders.

181 Here, Hanna made a very special movement with both hands which is not mentioned in the transcript but leads directly to the next question by the interviewer and therefore must be described. She shapes both hands as if grasping something which is cylindrical and which her hands "form" the shape of, with the thumbs spread from the fingers which are "collected" or closed, and the hands moving slightly. Standing in front of the interviewer while talking and demonstrating, her outstretched hands almost approach the interviewer with the intention of grasping around her arms.

182 Hanna's Norwegian word is "motkraft," directly translated to counter-force or a force/power exerted in the opposite direction. Although not a proper English term, the word counter-force may stand here as a more evocative or representing term than the correct term would be. The demand for the correct translation is, in this case as in some other examples, of a lower priority than that of giving a faithful rendering of the actual situation.

183 Annabella used the same term, "kjedelig," and in the same sense as discussed earlier in Tanja's interview about a confrontation with something one would wish could be undone.

184 At this moment in her recalling, Line makes a sound which in phonetic Norwegian is written 'æsj,' a sound expressing disgust, as if spitting something out, as if having come in contact with something greasy or rotten. She continues to talk in a very contemptuous way using the expression 'jeg tålte ikke trynet på ham,' which is a strong and rather vulgar way to say that you dislike the very sight of someone. These expressions stood in a striking contrast to her modest and correct use of language during the rest of the interview.

185 Line uses the Norwegian expression "at jeg går i stå" which gives the image of the activity of going without moving, "going to a standstill" as it says, literally. This idiom is precisely the semantic equivalent of what is going on, which can be read from the syntax and the text-as-picture, as was described in a short passage in chapter 1.4.

186 Here, an extended note is required to allow a discussion of theoretical considerations without interrupting the on-going interpretation: The field of research on sexual abuse has developed a classification system for abusive relationships. One of the most central distinctions with regard to abuse impact concerns the relationship between abuser and abused. Abuse within the family is acknowledged as being more destructive than abuse by persons from outside of it. According to this, Ruth's abuser is extra-familial. However, he belongs to the household and abuses her in her home. A distinction between family and hired help was defined precisely in Norwegian farmer's families in the first half of this century. The fact of having a hired help allows certain conclusions about the social position of the family. But still, employees were members of the household, living with the family. Thus, Ruth saw her abuser at all times of the day and almost everywhere. Simultaneously, a social distance between family and employees, as is implicit in a social hierarchy, totally excludes certain relationships and contacts. These crucial elements are not grasped by current classification instruments, and such an intra/extra distinction renders the dimensions of Ruth's estrangement within her own home and family invisible and incomprehensible. Finally, the absolute asymmetry between an adult and a child as to sexuality is overlain by a relative and "opposite" social asymmetry, which casts shame and social stigma on an improper proximity between members of ranked classes. Thus the setting creates several layers of boundaries which are violated in different ways, truly aggravating the abuse impact. This must be deciphered from the "inside of the experience," and it is completely non-existent within the current frame of theory and method in sexual abuse research.

187 Ruth's statement reflects both a cultural norm concerning female sexuality, and a social norm concerning a woman from a prosperous family. Unusually enough for her time, she was supported in seeking education and choosing a profession according to her preferences. Still, to give birth to children outside of a socially adequate wedlock was out of the question. Thus, the real choice was either to marry or remain childless.

188 Here it is necessary to add that Ruth had mentioned her childhood sexual abuse experience when entering the psychiatric ward. This will be returned to later.

189 Ruth's nonverbal communication in this short report made clear her deep indignation over what she understood as the therapist's covert invitation to let him approach her intimately.

190 The word uttered was "bevares," a Norwegian idiom verbally expressing the wish "may Heaven protect me from.." and indicating a blend of indignation, dismay and bewilderment when confronted with very discomfiting or rather shocking information. The translation "how could he dare" expresses the spontaneous response.

191 The content of this passage will be integrated into an exploration of the impact of the sexualization of therapeutic relationships in general, and the exploitation of sexually abused persons in particular. This topic will be part of an extended reflection on the phenomenon of revictimization in medicine and its different aspects.

192 Nora used the Norwegian idiom "jeg satte meg helt på bakbena." Literally, the Norwegian phrase gives the image of a person who resists being moved, in whatever sense, by placing the weight of the body on the heels in order to withstand steady pressure from in front. The image is that of defense by a particular kind of "strong" passivity, a resistance appearing as a physical regression. In Nora's case, the use of this idiom and this picture may be attributed a double meaning. The first is that she literally choose to use both her legs to stay where she is. For a person operated on for the third time because of incapacitating pain in one leg, prohibiting her from walking, this act of "standing on both legs" was an act of will and of desperation. The second is the resistance against being moved backwards. That direction meant "back home," where she did not want to go. In both senses, this utterance in this context holds almost iconographic qualities.

193 In the interview Nora says: "jeg sov meg gjennom weekenden." This is an idiom, and also an exact description of what she did. The reflexive form 'jeg sov meg,' in a literal translation 'I slept myself,' is grammatically incorrect. Nobody can 'sleep oneself,' but this was exactly what happened. Nora made herself sleep all weekend; she kept herself asleep until the weekend was over and she could leave again for the clinics. To sleep continuously meant absence in presence.

194 MB denotes a psychiatric institution of high professional standards and a rather non-tradition approach to therapeutic activity.

195 The original wording is as follows: "jeg fikk opparbeidet meg noen fridager, på en måte ..." Nora uses an idiom from the labor market, expressing how one works longer days in order to collect some compensation, or some form of free time. However, Nora does not spell out the nature of this "equivalent" overtime she put in for the right to one or two sex-free days a week.

196 See work by Paul Ricoeur and his exploration of narrative analysis with regard to unpredictable, but acceptable conclusions. (Ricoeur 1987:181-2)

197 The Norwegian idiom "å bære noen på gullfat" is an expression for adoration, protection, elevation and support of a person by others. In this case, concerning a child, the term on a semantic level indicates a "too much," an obvious spoiling which might become problematic for the child. This is what Nora emphasizes by also using "unnatural" in the preceding sentence.

198 This validating question was derived from other interviews, among them that of Veronika, Laura and Christine, who had expressed their ambivalence regarding examinations which were initiated to "find out the reason for the ..." and which they feared could unveil the real reason, as Veronika so explicitly recalled. This suggestion was negated immediately by Nora's precise denial and a new statement, which, however, generated yet another confusion to be solved.

199 In this section of the dialogue, Nora negates three of the interviewer's interpretations, which the interviewer then reformulates. This process may be understood as follows: the interviewer's repeated attempts to check out whether or not she has understood correctly may show how necessary such questioning is if communication is to take place. These inquiries may also demonstrate that meaning must be elicited in the field of human life-world experience, with all its ambiguities, inconsistencies and conflicting designations. This may also be understood as evidence of the interviewer's mental absence or lack of sensitivity vis à vis a woman with very low self-esteem who is trying to mediate the impression that her dignity has been shattered. For the sake of my own self-esteem, I prefer the first and second readings.

200 Once again, an experience from general practice and previous interviews is reintroduced, both for validation of information which has previously been acquired, and in order to check out the information provided in its deviation from what was expected, as this was "a difference which made a difference" in the sense of Gregory Bateson. In many of the previous interviews, the women had stressed their problems with male therapists, apart from, for example, Annabella, who was abused mostly by women. Even the men had emphasized that they had received most of what helped from women, and little help or rather negative reactions from men. Here was a woman abused by two men not only preferring male therapists, but asking explicitly for them. This demanded an interrogation.

201 The interviewer thinks mainly of the medical interventions as "the price to pay" for being calm and balanced. She knows in addition that Nora has been treated for years for a hypertension which, as a side effect of the medications, has caused a troubling diabetes mellitus II. The interviewee, however, is not concerned with these kinds of "costs." Her answer reveals yet another painful area in her life.

202 Only two of the interviewees reported having been molested by women, but only Annabella defined a woman as her main perpetrator. To date, one of the few studies exploring the impact of sexual abuse by women has been performed by Rudin, Zalewski & Bodmer-Turner in 1995.

203 Annabella was referred to a tertiary hospital in orthopedics and surgery. Due to her age, her case was given a lot of attention. Records from her stays over an eight year period were provided. In these, mention is made of psychiatric diagnoses, and records from two psychiatric institutions were provided, covering Annabella's frequent referrals to both institutions in the same time period, over the space of four years. Several of these admissions are not mentioned in Annabella's sickness-history at the beginning of the interview.

204 Here the original wording is "jeg hadde en knekk i ryggen" which may almost be called an idiom, mirroring the common suggestion that when there is a pain in your back it is due to a physical force from outside having caused a kind of reversible break in the spine. The anatomical orientation and the mechanical model implicit in this culturally established explanatory system is addressed in chapter 1.3. It is discussed as an over-all question of frame of reference for the understanding of body-function and functional deficiencies in the work of the Norwegian physiotherapist Eline Thornquist. (Thornquist 1994, 1995)

205 The syndrome was named, and consistently described for the first time as psychiatric theory, in an article entitled "Munchausen's syndrome," written by the English psychiatrist Richard Asher, and published in The Lancet in 1951;i:339-41. The sheer fact of an accumulation of lifetime surgical intervention far above average, however, has led to suggestions of possible causal connection to the previously reported excessive surgery in hysterical women, the so-called Briquet's syndrome, and somatization disorder. Later, an even more bizarre, exotic or irrational behavior was linked to the same name. The English pediatrician, Roy Meadow, described in The Lancet 1977;ii:343-5 two children with haematuri, which was inflicted upon them by their mothers, who then frequented the health care system on behalf of the children. Meadow's further research allowed a glimpse into a very particular kind of child abuse: the intentional engendering of sickness, which then becomes the reason for seeking medical intervention. Appearing as "Referred Munchausen Syndrome," it was termed "Munchausen by Proxy." Until now, a body of knowledge has been collected upon both topics, and mention has been made of a high occurrence of abuse experiences in the studied populations. This possible connection will be explored upon in a chapter concerning revictimization.

206 Arnold, Rogers & Cook 1990; Bergman, Brisman & Nordin 1992; Boisset-Pioro, Esdaile & Fitzcharles 1995; Cunningham, Pierce & Pierce 1988; Draijer 1989; Drossman et al. 1990; Drossman 1994; Drossman 1995; Drossman et al. 1995; Drossman et al. 1996; Felitti 1991; Golding et al. 1988; Golding 1994; Gross et al 1981; Haber & Ross 1985; Harrop-Griffiths et al. 1988; Kimerling & Kalhoun 1994; Kirkengen, Schei & Steine 1993; Laws 1993 a,b; Morrison 1989; Rapkin et al. 1990; Reiter et al. 1991; Rimsza, Berg & Locke 1988; Scarinci et al. 1994; Schei & Bakketeig 1989; Schei 1991; Springs & Friedrich 1992; Waigant et al. 1990; Walker et al. 1992; Walker et al. 1993; Walker et al. 1988; Wood, Wiesner & Reiter 1990; Wurtele, Kaplan & Keairnes 1990.

207 Golding 1994, 1996, 1999; Golding et al. 1998.

208 The latest studies reviewed concern low back pain (Linton 1997; Schofferman et al. 1993), generalized pain or fibromyalgia (Aaron et al. 1997; Finestone et al. 2000; Goldberg et al. 1999; McCauley et al. 1997; Taylor et al. 1995), headache (Golding 1999) and gynecological pain presentation (Golding & Taylor 1996; Golding et al. 1998).

209 The working title of the present project was: "What Did Really Happen? Can One Understand A Sickness-History Without Knowing The Life-History?"

210 In Norwegian, Oda says, "en ekkel opplevelse." The adjective, "ekkel," is most often attributed to what is inedible, but also to seeing something repulsive, smelling something rotten, and so on. Sensory perceptions which create an immediate repulsion, a refusal "to take in" what is, in the sense of Kristeva, abject, are "ekkel." Oda perceived of their touching and washing her not as bothersome and uncomfortable, but as repulsive.

211 I am indebted to Christian Krohn-Hansen for guiding my interest in Bourdieu's concept of "symbolic violence" to include the question of legitimacy in relation to violence. I have already leaned upon his study in "Confused Judgments." His Working Paper 1993.6 from the Centre for Development and the Environment at the University of Oslo, is the source of my following reflections.

212 The original wording is "han prøvde seg på meg i en bil," literally "he tried himself on me in a car," an idiom expressing a stretching of limits, a testing of boundaries, a proof of strength from the perspective of the doer, though spoken by the one who is the "object." In Norwegian, this is a common way to talk about men approaching women, making visible the implicit assumption that he has the right to try and she has the obligation to withstand. This interface of the male right to expand and female obligation to withhold, as linked to the characteristics of maleness and femaleness in Western patriarchies, has been explored by the German sociologist Frigga Haug. She makes evident the construction of male honor and female honorability in a constellation of a male subject, acquiring the marks of honor, and a female object, defined by her social relationship to a man. (Haug 1988) This topic is explored in the wider context of male domestic violence as a health risk for women and children. Linked to the notion of privacy, the male rights concerning his private property, and the socially constructed distinction between the public and the private, result in domestic violence being considered a "private event," and as such "shielded" from public insight and public reaction (Jecker 1993; Flitcraft 1992; Stark & Flitcraft 1979; Stark and Flitcraft 1988).

213 As Oda did, Synnøve uses the word "ekkel" in the sense of both a repulsive situation and an abject sensation, which will become obvious in a succeeding excerpt where she talks about her reactions to her husband's approaches.

214 The Norwegian wording is the idiom "de tok seg til rette," meaning, literally, that "they took themselves the right," implicitly: which was not given to them and which was not theirs. The idiom also expresses that the one speaking conceives of the act described as not right, but, nevertheless, something one has to accept.

215 In the late seventies, many child institutions in the USA began to focus on child abuse and on what was called, "the necessity of teaching children how to say no" to uncomfortable touch. These educational attempts to make children less prone to be offended by strangers, caretakers and "kind" people, were based on certain theoretical assumptions. First, abused children were perceived of as individually abused, as if causing their own abuse. Second, all studies on abuse focused on the victims and not on the abusive relationships. Third, the abusers were as if absent in the debate on the aspects of the problem of abuse. And fourth, implicit guiding structures in the society, putting children at risk of sexual abuse, were more or less ignored. Therefore it felt logical to educate the children to say 'no.' This expressed a complete disregard for an asymmetry of power where a child's 'no' is the lowest ranking voice, so to speak. Next, by focusing on a strategy of teaching children to resist abuse verbally, and thereby, however inadvertently, making it the children's implicit responsibility if the abuse were not stopped, the really unambiguous question of responsibility for sexual abuse of children was obscured. In connection with these educational programs, however, the educators became aware that the children were already quite capable of distinguishing between good and bad touch. And, even further, due to their demand and longing for touch, children deprived of "sufficient good touch" were at high risk for accepting bad touch; it was as if they could not allow themselves to discriminate, and thus risk having to refuse, the bad touch in their search to be touched in whatever way at all. (Anderson 1979)

216 The socio-cultural construct of "the natural" with regard to sexual norms and gender roles demands a far more thorough exploration than is possible and appropriate in this part of the text. It will be dealt with in chapter 3.

217 In 1994, the Norwegian newspapers reported on a case in the Swedish courts where a man, 64 years old, was found not guilty of having raped two eight-year old girls because they had not really resisted the penetration. This lack of resistance rendered the sexual offenses a non-rape.

218 This example shows one of the origins of the inconsistencies in sexual abuse research, about the frequency of unwanted or forced participation in activities perceived of or defined as sexual. Synnøve, when asked whether she has experienced rape, answers that she has not. She might even have answered no to the question of having experienced repeated abuse by several abusers. It is even uncertain whether she would have answered yes to the question of repeated unwanted sexual approaches, as she perceives of her mother's partner as her only abuser.

219 Several times during the interview, Annika has laughed while talking about painful or serious topics. The interviewer has registered this and reflected the observation back to her in the form of a question concerning the validation of a pattern, "a habit" or "a tactical conduct." Annika has explained that she is so often so very depressed and on the brink of bursting into tears, which would only attract unwanted attention, that she has become used to smiling and laughing, the more so the more crucial the topic she deals with feels to her.

220 Here is addressed what Annika had talked about in connection with her daughter's crying when leaving Annika's mother, and the mother's accusations against Annika for being such a bad mother that she feels pain on behalf of her grandchild.

221 Here it is necessary to add two summaries from the previous interview to the excerpt quoted above:
She tells that her mother always demanded that she be kind and nice and that she was not allowed to show anger. She has thought a lot about that since, about her mother being quite egoistical, not tolerating her daughter's anger, and, in order to keep calm herself, denying her daughter the right to express adequate and normal reactions and refusals. She tells that for a long time she had been convinced that she was the origin of her family's disruption.
Her mother got pregnant at seventeen. Therefore she was very preoccupied with preventing her daughter getting into the same situation. She forbade Annika to go out in the evenings, to stay overnight with friends, or to join a club on tours. She was obviously unable to see the danger within the house, where Annika's brother was abusing her several nights a week.

222 The topic of Annika's mutilated left arm will be explored in a later chapter.

223 A similar criticism may be directed at a recent study on coping strategies in abused children aged seven to twelve. The coping of these eighty-four children clustered around, "four strategies that were labeled avoidance coping, internalized coping, angry coping, and active/social coping. Each coping strategy was found to be associated with a unique set of abuse characteristics, abuse-related social environments, and symptoms." (Chaffin, Wherry & Dykman 1997)

224 A recent study of potential risk factors for sexual abuse in childhood among adult women focused on family conditions. Apart from physical abuse which was a risk factor for all, the death of the mother was significantly associated with abuse of children under twelve, and mental sickness of the mother with abuse of children over twelve. Significant predictors of abuse within the family were the lack of caring female adults and there being an alcoholic father. (Fleming, Mullen & Bammer 1997)

225 The original wording is: "jeg begynte nok å knyte meg både på skolen og hjemme." Runa uses an idiom depicting the state of being 'in a knot' or knotting herself tightly, a metaphor expressing unfree walking and moving, but which is also used frequently to convey the idea of a mental state of anxiety, withdrawal, regression and introversion. It is a term for the opposite of being 'free and open-minded,' or 'relaxed and comfortable.' To be in a knot is painful, leaving little space for breathing, moving, laughing, enjoying, and so on.

226 The so-called "run-away kids and teenagers," frequently mentioned in American social reports as being an extremely large group, engendering a considerable work load for the police and the social authorities, have gradually been understood as representing predominantly abused children.

227 After the present study was terminated in 1998, a few other studies have been focusing on the abuse-chronic pain correlation: Aaron et al. 1997; Felitti et al. 1998; Finestone et al. 2000; Goldberg et al. 1999; Golding 1999; Golding et al. 1998; Linton 1997; McCauley et al. 1997; Schofferman et al. 1993; Taylor et al. 1995. None of these studies, however, has applied a methodology which would allow for an exploration of the path from violation to pain.

228 "Memory Work" is a central term in the work of the German sociologist Frigga Haug. She has developed this as a method for group activities, aimed at a critical reflection upon certain phenomena by means of personal memories and biographical data. By comparing others' and own recollections from certain periods of life or time spans, patterns are rendered visible and thus open both to introspection and a meta-perspective on one's own life as embedded in history.

229 Abbott et al. 1995; American Medical Association, Council Report 1992; Amnesty International Publications 1991; Armstrong 1991; Benum & Anstorp 1993; Bergman & Brismar 1992; Briere 1992; Burge 1989; Burnam et al. 1988; Courtois & Raley 1992; Courtois 1993; Dahl 1993; Davenport, Browne & Palmer 1994; Ferris 1992; Finkelhor & Korbin 1988; Finkelhor 1994; Flitcraft 1992; Gazmararian et al. 1996; Goldberg, Pachas & Keith 1999; Golding et al. 1988; Golding 1994; Golding 1999; Groves et al. 1993; Halpérin et al. 1996; Hamberger, Saunders & Hovey 1992; Helton, McFarlane & Anderson 1987; Hydle 1989; Jecker 1993; Koss 1990; Koss, Koss & Woodruff 1991; Koss, Woodruff & Koss 1990; Lechner et al. 1993; Leventhal 1996; Lundberg et al. 1992; McCauley et al. 1997; McFarlane 1992; Mullen et al. 1993; Newberger et al. 1992; Novello 1992; Perez & Windom 1994; Richards 1993; Rosenberg, O'Carroll & Powell 1992; Rosenthal 1988; Sariola & Uutela 1994; Schechter 1992; Schei 1991; Schofferman et al. 1993; Scott 1992; Smith et al. 1995; Stark, Flitcraft & Frazier 1979; Stark & Flitcraft 1988; Steven et al. 1988; Sugg & Inui 1992; United Nations Publication 1989; Wells et al. 1995; WHO 1996; Wurr & Partridge 1996; Wyatt & Newcomb 1990.

230 The original wording is: "jeg hadde vel delvis hallusinasjoner. .... Muligens så jeg for meg ting." Elisabeth's formulation is as if she had an observer's perspective, speaking with a double reservation which might express a doubt, a kind of disbelief, an insecurity about what really had been the case. Mainly the last formulation alludes to a position as if speaking from outside. Here, Elisabeth has taken into consideration that the interviewer might think of hallucinations as unreal pictures and a sign of insanity. It seems as if she keeps a distance to herself in the state of psychosis, as if she herself were not really sure whether she has seen something.

231 At this point, Veronika "enters the room," in the sense of a reintroduction of elements from a previous interview which seem to be thematized again. Veronika heard voices she knew, saying things she had heard before during abusive acts. She was diagnosed as psychotic. Elisabeth alluded to a parallel situation, allowing two validations simultaneously.

232 The German word *Verkörperung* designates the process of something finding its expression in a "body" of a different kind. Giving body or shape to something can be mentioned or performed in a variety of relationships and conditions. A human being can *"verkörpern,"* meaning to embody a *doing* - of traditions, feelings, groups, a type of conduct, a kind of trait - thus more *living* than *being* - kind, helpful, eccentric, beautiful, practical, romantic or monstrous. The English verb 'to embody' alludes more to the *being*, and not quite so much to the *doing*.

233 Gestalt therapy is grounded in Gestalt theory. The most central of its techniques, visualization, is informed by the acknowledgment of the impact of visual perception on memory and awareness. Merleau-Ponty has explored the theory in his book, Phenomenology of Perception (Merleau-Ponty 1989:47).

234 Christine used the term once, which will be addressed in "Unheard Messages," but continues to talk about "forgetting."

235 In chronological order: Herman & Schatzow 1987; Femina, Yeager & Lewis 1990; Herman 1992; Leitenberg, Greenwald & Cado 1992; Briere & Conte 1993; Bass & Davis 1994; Feldman-Summers 1994; Elliot & Briere 1995; Kondora 1995; van der Kolk & Fisler 1995; Tromp et al. 1995; Freyd 1996; Scheflin & Brown 1996; Melchert & Parker 1997; Andrews et al. 1999; Chu et al. 1999.

236 Williams 1995.

237 The original wording is "når jeg er nær noe jeg ikke har tak på." "Å ha tak på" means to have a grip on, both in the sense of comprehension as well as control. So Beate's statement may mean either to comprehend or to cope with. But given the present context, expressing vaguely to "be near something," the connotations may coexist. She reacts when approaching or encountering something not in her grasp, i.e., something she does not comprehend and/or does not control.

238 The Norwegian term "the South" designates all countries in Southern Europe, and other places in the world where there are beaches and sunshine. Due to Norway's geographical position, "the South" is almost synonymous with holidays, relaxation, being far away, escaping everyday tasks and climatic strain.

239 Beate uses the Norwegian folk term for herpes zoster, "helvetesild," which literally creates the image of the "pain which burns like in hell," caused by a fully blown herpes exanthema.

240 Beate echoes the previously quoted and explored statements about the public opinion of the "kind" abusers being less abusive, and "kind abuse" being less destructive, although she has a son who has experienced this type of abuse. She also mentions what has been thematized before, and which might be called an "abuse prone" social setting, where abusive adult males offend against many people in a home or a family or neighborhood for years without being disclosed.

241 The original wording in this sentence is: "Han brukte kniven noen ganger så blodspruten sto." Literally, this expression means, "He occasionally used the so that a fountain of blood stood up."

242 Al-Eissa 1995.

243 This topic is elaborated in Patricia Munhall's phenomenological study about women's anger. Focusing on "the interactions of somatization, thoughts, and social processes evolving from unrecognized, unexpressed, or repressed anger," she concludes as follows: "The most critical finding was the transformation of anger as lived by women into socially acceptable pathology. Anger was left in silence, and the possibilities for its expression were found in physiological disorders, substance abuse, self-deprecation, and affiliation problems, among other conditions." (Munhall 1993)

244 Scheper-Hughes 1990.

245 Kleinman's juxtaposition of a "personal somewhere" and the "depersonalized nowhere" alludes to a book on the impact of reductionism and objectivism in philosophy, entitled "The View from Nowhere" by the American philosopher Thomas Nagel.

246 The possibility that a strong pathogenetic impact stemming from persistent violation indeed may exist seems worth-while taking into account. A recent study exploring the assosiations between childhood experiences and adult health may indicate this. (Felitti et al. 1998)

247 The original wording is: "Men hun hjalp meg ikke på gli," a Norwegian idiom from the world of skiing. Most children learn how to ski by having an adult's hand pushing from behind and help them get the skis to glide the right way. This push or pressure must be firm but supportive, not hard or abrupt, which would make the child fall. Therefore this idiom means a careful, supportive help to go further on one's own legs and under one's own control.

248 The considerable influence of personal presuppositions and life experiences on professionals' judgments in questions of childhood sexual abuse, has been explored as to their different aspects in the following studies: Eisenberg, Owens & Dewey 1987; Holmes & Offen 1996; Howe, Herzberger & Tennen 1988; Jackson & Nuttall 1993; Nuttal & Jackson 1994.

249 While the recorder is switched off, the interviewer asks the interviewee whether she wants to continue. Christine nods and says, "jeg vil si mer," which means both "I want to say more," and "I have to say more." We agree that she chooses to continue, which she does after a while, although still struggling with her breath.

250 This is the situation which has been referred to in "Maladaptive Adaptations," and which indicated a pattern of reservation, withdrawal and regression under a certain kind of pressure. Line may well have interpreted the interviewer's intensity in these speech turns as expressing a disbelief or skepticism about her statements, while the interviewer's probing was actually driven by the negligence she heard about. What she could not believe was not Line's experience, but that the doctor had not reacted. This made her ask once more, which intensified Line's habit of hesitation. What was meant as an on-the-spot validation became, in one way, counterproductive as it was not interpreted in this sense by the interviewee. On the other hand, it made the response pattern so evident that it could be addressed later in the interview as also described in "Maladaptive Adaptations."

251 Sandra Lee Bartky, in her book 'Femininity and Domination. Studies in the phenomenology of oppression', addresses the effect of such an objectifying, and consequently alienating, male gaze on women. (Bartky 1990:27) Her reflections on the politics of objectification of women as part of the oppressive structures in patriarchal societies applies to most aspects of sexual objectification.

252 Russell 1995.

253 The topic of psychiatric diagnoses as constructs underlining gender asymmetry has been further explored in "Making us crazy" by Kutchins& Kirk, 1999.

254 Here an earlier part of the interview is alluded to, a report of a female psychologist who responded to a clue about sexual abuse in childhood with a remark that this most probably did not imply anything serious.

255 This overview is in chronological order: Pierce & Pierce 1985; Finkelhor 1986; Sebold 1987; Pierce 1987; Halpern 1987; Buckner & Johnson 1987; Briere et al. 1988; Golding et al. 1988; Metcalf et al. 1990; Finkelhor et al. 1990; Felitti 1991; Furniss 1991; Lipscomb et al. 1992; Watkins & Bentovim 1992; Briere & Conte 1993; Fry 1993; Green 1993; Briere 1994; Davenport, Browne & Palmer 1994; Finkelhor 1994; Boney-McCoy & Finkelhor 1995; Elliot & Briere 1995; Elliot, Browne & Kilcoyne 1995; Halpérin et al. 1996; Haskett, Marziano & Dover 1996; Holmes & Offen 1996; Ligezinska et al. 1996; Gorey & Leslie 1997; Morrow, Yeager & Lewis 1997; Maddocks, Griffiths & Antao 1999.

256 The original wording is here "du gikk i stå, simpelthen." This expression has been commented upon earlier in the interview with Line as an idiom of walking in place or walking while standing, which in this case was a most precise description of what had happened.

257 Psychomotoric therapy is a special kind of physiotherapy developed in cooperation between the Norwegian psychiatrist, Trygve Bråtøy, and the physiotherapist, Aadel Bülow-Hansen. The core of the therapy is not to look at the human body in a fragmenting way but rather to consider the movements and patterns of muscular tonus as interactive and expressive. This kind of therapy is very much informed by the theory of the lived body as conceptualized in phenomenology. The mode of treatment is described comprehensively by the researchers and practitioners Eline Thornquist and Berit Heir Bunkan, in their book, *What is Psychomotoric Therapy?* (Thornquist & Bunkan 1991)

258 This question, and even its formulation, is a demonstration of the reintegration of insight and understanding from earlier interviews and explorations. The interviews with Pia and Ruth were the sources of it. Jessica promptly confirmed the connection. A further exploration of this particular link was unnecessary. It could be addressed directly as to its origin and consequences in Jessica's life.

259 This two-letter configuration is meant to represent a phonetic signal which the film, Rainman, made world famous. In it, Dustin Hoffman plays an autistic adult. Dustin Hoffman's character used this double sound to express a "warning" to himself; it announced or expressed a state of awareness of a danger or a trap, and preceded the man's reactions, which often seemed strange and confusing to other people. Karla's "OH-oh," in this context, a situation of impending invasion, was a precise imitation of the man's warning to himself.

260 An indirect validation of medical interventions as too similar to abuse memories is offered in the very first study on the connection between anal sexual molestation and encopresis in boys. The authors stress that diagnostic or therapeutic interventions due to encoprisis may reactivate abuse memories, which they illustrate with an impressive example. (Morrow, Yeager & Lewis 1997)

261 As described earlier, a pantomime takes place which is not described in detail in the transcript. When Hanna, far earlier in the interview, shows the two points in her lower back, she is sitting in front of the interviewer. She turns around on her chair while seated and points to her back with both hands by pressing her thumbs into her buttocks. In the present situation, she rises from her seat, stands in front of the interviewer, turns around and draws an imaginary line across her lower back where she earlier placed her thumbs. Turning around again, while standing, she continues to tell about the areas in front of and behind her shoulders, as was thematized in "Confused Judgments."

262 Here again the combination of nausea and of anxiety as expressed in breathing problems is addressed, as was thematized earlier in connection with the alienation of the mouth. Fanny has problems breathing and becomes nauseated, "when she comes close to these things." This reaction pattern has, however, not been confirmed in Fanny's case, perhaps due to the interviewer's lack of attention.

263 Here, once again, foreknowledge from general practice is introduced into an interview: the interviewer knew several abused persons who, due to doubts in the exactness of their own memory, never dared take an examination in the course of their education and consequently stagnated in their profession or career. One of these, aiming at becoming a librarian, stated: "The worst are settings where the only thing I have to rely on is my memory - then I have actually nothing to build on."

264 The breakdown of adaptability as representing the core of sickness is one of the central theories of Pierre Janet. This theory will be the background for exploring several sickness histories in chapter 3. Fanny's contribution is a valid example of the potential for abuse to induce illness and to provoke a breakdown of general adaptability, and the aggravating impact of repeated or reexperienced trauma.

265 The original wording is "da fikk jeg en lammelse." Fanny had used the medical term "hemiparesis" when filling out the sickness list. She had mentioned having been examined by a neurologist. Here, her wording, translated literally, might be, "then I got a paralysis." She refers to it more as a state, a condition, than as a process, although a process is what she describes. The interviewer repeats her word "lammelse."

266 Based on general practice experience as a female practitioner with many female patients, I was reminded by Beate of having taken part in validating for a patient that her former experiences represented improper professional conduct. A young woman in her early twenties consulted me for the first time during her first pregnancy. When scheduling a gynecological examination, she asked for sufficient time to adapt, uncertain as to her own reactions. At her next visit, she seemed very tense. Before she undressed, I told her what I would do, as usual. I also assured her that she would be able to see my face and keep eye contact with me, if she so chose. I announced every step of the procedure, and asked her to tell me or show me any pain or discomfort. The examination proceeded without encountering any problems. She then sat up and, while leaving the chair, said loudly: "yes, I knew." When back in my room and sitting face to face, she said: "I knew it, but not with certainty. Now I'm sure. Another doctor has examined me before. I didn't know how such a check-up is done. He knew it was my first exam. The nurse wasn't present the way she is here. He watched me quite a while and touched my genital area intensely, then he inserted his fingers and told me that he had to stimulate me slightly so I would lubricate and thereby have the discomfort of the check-up reduced. I could not say no, I didn't know any better, but I felt very uneasy, and then I saw his eyes become different, as if he were absent. That made me move, he reacted, and then he finished while saying that everything was okay. I would never go back."

267 This situation, as Grethe told it and as it could be visualized, gave associations to the phenomenon of revictimization in medicine as it emerged in different ways from the interviews, including Elisabeth's narrative about an involuntary admission to a psychiatric ward. Grethe's and Elisabeth's presentations gave the impulses for Impression 5.

268 A common American abbreviation for a Dilatation and Curettage procedure.

269 The diagnoses, or the clinical concept of PTSD, accounts for a rather limited group in a broad and various specter of ways for "leaving" or "absenting" oneself. Neither the milder degrees (mediated by drugs or alcohol) nor the most extreme modes of splitting off are comprised by the current definition. This may render certain kinds of self-neglect incomprehensible (f.ex. unsafe sex among young homosexual men despite of much knowledge as to HIV-risk). To cover the most extreme states of splitting, a new diagnosis has recently been introduced: DESNOS = Disorders of Extreme Stress Not Otherwise Specifed. (Ford & Kidd 1998)

270 Seligman 1975.

271 Concerning the victim's own contribution to assault concealment, see Sandra Lee Bartky's notion of shame and guilt as linked to imposed inferiority. (Bartky 1990:83-98)

272 Cassell 1992.

273 Peterson & Seligman 1983.

274 Janoff-Bulman & Frieze 1983.

275 Lazarus & Cohen 1978.

276 Antonovsky 1979.

277 Horowitz 1982.

278 This culturally constructed eye-distracter has resulted in a rather limited body of knowledge about the impact of maltreatment of men as compared to women, and especially that of sexual abuse. The first is addressed in a review of 126 articles (between 1989 and 1994) on the topic of physical abuse impact. The authors find adult males dramatically underrepresented, both as abused, as fathers of abused children, or as abusers. (Haskett et al. 1996) The latter has been addressed in a recent review. (Watkins & Bentovim 1992) The prevalence estimates of sexual abuse of boys as compared to girls has been evaluated recently on the background of a review of 16 cross-sectional surveys from North America from 1969 to 1991. (Gorey & Leslie 1997)

279 Janoff-Bulman 1989.

280 Berliner & Conte 1990.

281 Morrow 1991.

282 The most comprehensive contributions to this field are provided by: Bremner et al. 1993; Bremner et al. 1995; Felitti et al. 1998; Goenjian et al. 1994; Green 1993; Herman 1993; Kiecolt-Glaser & Glaser 1992; Kolb 1987; Kolb 1993; Pennebaker & Susman 1988; Pennebaker 2000; Pennebaker, Kiecolt-Glaser & Glaser 1988; Putnam et al. 1986; Putnam 1989; Rauch et al. 1996; Root 1991; Rowan et al. 1994; Sandberg & Lynn 1992; Shapiro et al. 1992; van der Kolk et al. 1985; van der Kolk 1988; van der Kolk & van der Hart 1989; van der Kolk 1994; van der Kolk & Fisler 1995; Waigand et al. 1990; Wind & Silvern 1992; Wilson & Raphael 1993.

283 Boney-McCoy & Finkelhor 1995; Bryer et al. 1987; Flitcraft 1992; Gibbons 1996; Gold, Hughes & Swingle 1996; Herman, Russell & Trocki 1986; Kendall-Tackett, Williams & Finkelhor 1993; Koss, Koss & Woodruff 1991; Rosen & Martin 1996; Roth, Wayland & Woolsey 1990; Russell 1982; Russell 1986; Shearer & Herbert 1987.

284 AMA Council Report 1991.

285 In chronological order: DeYoung 1981; Gabbart 1989; Rutter 1989; Kluft 1990; Frenken & van Stolk 1990; Wilbers et al. 1992; Gonsiorek 1994.

286 Caplan 1987; Haug 1988; Russett 1991.

287 With regard to the notion of stigma as interlinked with social norms, deviation, and identity, see the comprehensive work of Erving Goffman, "Stigma. Notes on the Management of Spoiled Identity," 1968.

288 Barsky & Borus 1995; Bass 1990; Escobar 1987; Escobar et al. 1987; Ford 1983; Lipowsky 1988; Quill 1985.

289 Cohen et al. 1953.

290 Laskow et al. 1986.

291 Coryell & Norten 1981.

292 Porter, Heitsch & Miller 1994.

293 Bools, Neale & Meadow 1994.

294 Linton 1997; Schofferman et al. 1993.

295 Johnson 1994.

296 Increasing awareness of the possible connection between chronic pelvic pain due to sexual molestation and the risk of hysterectomy as a means to "cure" the pain which, yet unrecognized, is a pain of past violations, is addressed in articles written by Hendricks-Matthews in 1991 and Wukasch in 1996.

297 Andersen et al. 1992.

298 Young 1989, Hirschauer 1991.

299 Margalit 1996.

300 These men have explicitly authorized me to use material from their interviews for documentation purposes in my ongoing research and lecturing. Their predominant and common motive is to contribute to the acknowledgement of sexual violation impact among physicians, lawyers, and professionals in counselling.

References

Aaron LA, Bradley LA, Alarcon GS, Triana-Alexander M, Alexander RW, Martin MY, Alberts KR. Perceived physical and emotional trauma as precipitating events in fibromyalgia. Arthritis Rheum 1997;40:453-60.

Abbott J, Johnson R, Koziol-McLain J, Lowenstein SR. Domestic violence against women. Incidence and prevalence in an emergency department population. JAMA 1995;273:1763-7.

Agger I. Sexual torture of political prisoners: an overview. J Trauma Stress 1989;2:305-18.

Agger I. Det blå værelse. Kvindeligt vitnesbyrd fra exilet. København: Hans Reitzels Forlag, 1992.

Albach F, Everaerd W. Posttraumatic stress symptoms in victims of childhood incest. 2nd Europ Conf on Traumatic Stress Psychother Psychosom 1992;57:143-51.

Al-Eissa YA. The impact of the Gulf armed conflict on the health and behaviour of Kuwaiti children. Soc Sci Med 1995;41:1033-7.

American Medical Association. Council on Ethical and Judical Affairs. Sexual misconduct in the practice of medicine.
JAMA 1991;266:2741-5.

American Medical Association. Council on Scientific Affairs. Violence against women. Relevance for general practitioners.
JAMA 1992;267:3184-8.

American Medical Association. Council on Ethical and Judical Affairs. Physicians and domestic violence: ethical conciderations.
JAMA 1992;267:3190-3.

American Medical Association. Council on Scientific Affairs. Female genital mutilation. JAMA 1995;274:1714-6.

Amnesty International. Women in the front line: human rights violations against women. London: Amnesty International Publications, 1991.

Andersen E, Søndenaa K, Søreide JA, Nysted A. Acute appendicitis. Preoperative observation and diagnostic accuracy. Tidsskr Nor Lægeforen 1992;112:630-4.

Anderson D. Touching: when is it caring and nurturing or when is it exploitive and damaging? Child Abuse Negl 1979;3:793-4.

Andrews B, Brewin CR, Ochera J, Morton J, Bekerian DA, Davies GM, Mollon P. Characteristics, context and consequences of memory recovery among adults in therapy. Br J Psychiatry 1999;175:141-6.

Antonovsky A. Health, stress, and coping. San Francisco, CA: Jossey-Bass, 1979.

Armstrong S. Female circumcision: fighting a cruel tradition. New Scientist 1991;2:42-7.

Arnold RP, Rogers D, Cook DAG. Medical problems of adults who were sexually abused in childhood. BMJ 1990;300:705-8.

Asher R. Munchausen's syndrome. Lancet 1951;i:339-41.

Bachmann GA, Moeller TP, Benett J. Childhood sexual abuse and the consequences in adult women. Obstet Gynecol 1988;71:631-42.

Bagley C. Prevalence and correlates of unwanted sexual acts in childhood in a national Canadian sample. Can J Public Health 1989;80:295-6.

Bagley C, Ramsay R. Sexual abuse in childhood: psychosocial outcomes and implications for social work practice. J Soc Work Hum Sexuality 1986;4:33-47.

Baker A, Duncan SP. Child sexual abuse: a study of prevalence in Great Britain. Child Abuse Negl 1985;9:457-67.

Baring A, Cashford J. Myth of the goddess. Evolution of an image. London: Arkana Penguin Books, 1993.

Baron RJ. An introduction to medical phenomenology: I can't hear you while I'm listening. Ann Intern Med 1985;103:606-11.

Baron JR. Bridging the clinical distance: an empathic rediscovery of the known. J Med Philos 1981;6:5-23.

Barsky AJ, Borus JF. Somatization and medicalization in the era of managed care. JAMA 1995; 274:1931-4.

Bartky SL. Femininity and domination. Studies in the phenomenology of oppression. New York: Routledge, 1990.

Bass E, Davis L. What we do know and don't know about traumatic memory. Fam Violence Sex Assault Inst 1994;10:31-5.

Basoglu M (ed). Torture and its consequences. Cambridge: Cambridge University Press, 1992.

Bass C. Somatization: physical symptoms and psychological illness. Oxford: Blackwell Scientific,1990.

Basta SM, Peterson RF. Perpetrator status and the personality characteristics of molested children. Child Abuse Negl 1990;14:555-66.

Baszanger I. Deciphering chronic pain.
Sociology Health & Illness 1992;14:181-215.

Beck JC, van der Kolk BMD. Reports of childhood incest and current behavior of chronically hospitalized psychotic women. Am J Psychiatry 1987;144:1474-6.

Beitchman JH, Zucker KJ, Hood JE, DaCosta GA, Akman D. A review of the short-term effects of child sexual abuse. Child Abuse Negl 1991;15:537-56.

Beitchman JH, Zucker KJ, Hood JE, DaCosta GA, Akman D, Cassavia E. A review of the long-term effects of child sexual abuse. Child Abuse Negl 1992;16:101-18.

Belenky MF, Clinchy BM, Goldberger NR, Tarule JM. Women's ways of knowing. The development of self, voice, and mind. New York: Basic Books, 1986.

Benum K, Anstorp T. Understanding psychiatric symptoms as coping reactions to violence and sexual abuse. Promot Ment Health 1993;3:17-26.

Bergman B, Brismar B, Nordin C. Utilisation of medical care by abused women. BMJ 1992;305:27-8.

Berliner L, Conte JR. The process of victimization: the victim's perspective. Child Abuse Negl 1990;14:29-40.

Betts T, Bogden S. Diagnosis, management, and prognosis of a group of 128 patients with non-epileptic attack disorder. Seizure 1992;1:27-32.

Boisset-Pioro MH, Esdaile JM, Fitzcharles M-A. Sexual and physical abuse in women with fibromyalgia syndrome.
Arthritis Rheum 1995;38:235-41.

Boney-McCoy S, Finkelhor D. Prior victimization: a risk factor for child sexual abuse and for PTSD-related symptomatology among sexually abused youth. Child Abuse Negl 1995;12:1401-21.

Bools C, Neale B, Meadow R. Munchausen syndrome by proxy: a study of psychopathology. Child Abuse Negl 1994;18:773-88.

Bourdieu P. Outline of a theory of practice. Cambridge: Cambridge University Press, 1977.

Bremner JD, Southwick SM, Johnson DR, Yehuda R, Charney DS. Childhood physical abuse and combat-related posttraumatic stress disorder in Vietnam veterans. Am J Psychiatry 1993;150:235-9.

Bremner JD, Krystal JH, Southwick SM, Charney DS. Functional neuroanatomical correlates of the effect of stress on memory. J Traum Stress 1995;8:527-53.

Briere J, Runtz M. Symptomatology associated with childhood sexual victimization in a nonclinical adult sample. Child Abuse Negl 1988;12:51-9.

Briere J, Evans D, Runtz M, Wall T. Symptomatology in men who were molested as children: A comparison study. Am J Orthopsychiatry 1988;58:457-61.

Briere J, Zaidi LY. Sexual abuse histories and sequelae in female psychiatric emergency room patients. Am J Psychiatry 1989;146:1602-6.

Briere J. Methodological issues in the study of sexual abuse effects. J Consult Clin Psychology 1992;60:196-203.

Briere J. Medical symptoms, health risk, and history of childhood sexual abuse. Mayo Clin Proc 1992;67:603-4.

Briere J, Conte J. Self-reported amnesia for abuse in adults molested as children. J Traum Stress 1993;6:21-31.

Briere J. Sexual fantasies, gender, and molestation history. Child Abuse Negl 1994;18: 131-7.

Browne A, Finkelhor D. Impact of child sexual abuse: a review of the research. Psych Bull 1986;99:66-77.

Bruckner DF, Johnson PE. Treatment for adult male victims of childhood sexual abuse. Soc Casework 1987:68:81-7.

Bryer JB, Nelson BA, Miller JB, Krol PA. Childhood sexual and physical abuse as factors in adult psychiatric illness. Am J Psychiatry 1987;144:1426-30.

Bulik CM, Sullivan PF, Rorty M. Childhood sexual abuse in women with bulimia. J Clin Psychiatry 1989;50:460-4.

Burge SK. Violence against women as a health care issue. Fam Med 1989;21:368-73.

Burnam MA, Stein JA, Golding JM et al. Sexual assault and mental disorders in a community population. J Consult Clin Psychol 1988;56:843-50.

Busfield J. Men, women and madness. Understanding gender and mental disorder. London: MacMillan, 1996.

Bushnell JA, Wells JE, Oakley-Browne MA. Long-term effects of intrafamilial sexual abuse in childhood. Acta Psychiatr Scand 1992;85:136-42.

Canguilhem G. The normal and the pathological. New York: Zone Books, 1989.

Caplan P (ed). The cultural construction of sexuality. London: Tavistock, 1987.

Carlin AS, Ward NG. Subtypes of psychiatric inpatient women who have been sexually abused. J Nerv Ment Dis 1992;180:392-7.

Carmen E, Rieker P, Mills T. Victims of violence and psychiatric illness. Am J Psychiatry 1984;141:378-87.

Cassell EJ. The body of the future. In: Leder D (ed). The body in medical thought and practice. Dordrecht: Kluwer Academic Publishers, 1992.
Chaffin M, Wherry JN, Dykman R. School age children's coping with sexual abuse: abuse stresses and symptoms associated with four coping strategies. Child Abuse Negl 1997;21: 227-40.
Chu JA, Dill DL. Dissociative symptoms in relation to childhood physical and sexual abuse. Am J Psychiatry 1990;147:887-92.
Chu JA, Frey LM, Ganzel BL, Matthews JA. Memories of childhood abuse: dissociation, amnesia, and corroboration. Am J Psychiatry 1999;156:749-55.
Coffey P, Leitenberg H, Henning K, Turner T, Bennett RT. Mediators of the long-term impact of child sexual abuse: perceived stigma, betrayal, powerlessness, and self blame. Child Abuse Negl 1996;20:447-55.
Cohen ME, Robins E, Purtell JJ, Altmann MW, Reid DE. Excessive surgery in hysteria. Study of surgical procedures in 50 women with hysteria and 190 controls. JAMA 1953;151:977-86.
Conte JR, Schuerman JR. Factors associated with increased impact of child sexual abuse. Child Abuse Negl 1987;11:201-11.
Coons PM, Bowman ES, Milstein V. Multiple personality disorder. A clinical investigation in 50 cases. J Nerv Ment Dis 1988;176:519-27.
Coryell W, Norten SG. Briquet's syndrome (somatization disorder) and primary depression: comparison of background and outcome. Compr Psychiatry 1981;22:249-56.
Courtois CA. The incest experience and its aftermath. Victimology 1979; 4:337-47.
Courtois CA. Adult survivors of sexual abuse. Prim Care 1993;20:433-46.
Courtois CA, Riley CC. Pregnancy and childbirth as triggers for abuse memories: implications for care. Birth 1992;19:222-3.
Craine LS, Henson CE, Colliver JA, McLean DG. Prevalence of a history of sexual abuse among female psychiatric patients in a state hospital system. Hosp Comm Psychiatry 1988;39:300-4.
Cunningham J, Pearce T, Pearce P. Childhood sexual abuse and medical complaints in adult women. J Interpers Violence 1988;3:131-44.
Dahl S. Rape - a hazard to health. Oslo: Scandinavian University Press, 1993.
Davenport C, Browne K, Palmer R. Opinions on the effects of child sexual abuse: Evidence for consensus. Child Abuse Negl 1994;18:725-38.
de Beauvoir S. The second sex. New York: Knopf, 1952.
Demitrack MA, Putnam FW, Brewerton TD, Brandt HA, Gold PW. Relation of clinical variables to dissociative phenomena in eating disorders. Am J Psychiatry 1990;147:1184-8.
DeYoung M. Case reports: the sexual exploitation of incest victims by helping professionals. Victimology 1981;6:92-101.
Dilthey W. Poetry and experience. Selected works, Vol V., Princeton, NY: Princeton University Press, 1985.
Douglas M, Calvez M. The self as a risk taker: a cultural theory of contagion in relation to AIDS. Revue-déconomie-regionale-et-urbaine 1990;38:445-64.
Draijer N. Long-term psychosomatic consequences of child sexual abuse. In: van Hall EV, Everaerds W, eds. The free woman. Women's health in the 1990s. New Yersey: Parthenon Publishing Group, 1989.
Drossman DA, Leserman J, Nachman G et al. Sexual and physical abuse in women with functional or organic gastrointestinal disorders. Ann Intern Med 1990;113:828-33.

Drossman DA. Physical and sexual abuse and gastrointestinal illness: what is the link? Am J Med 1994;97:105-7.

Drossman DA. Sexual and physical abuse and gastrointestinal illness. Scand J Gastroenterol Suppl 208, 1995;30:90-6.

Drossman DA, Talley NJ, Leserman J, Olden KW, Barreiro MA. Sexual and physical abuse and gastrointestinal illness. Ann Intern Med 1995;123:782-94.

Drossman DA, Li Z, Leserman J, Toomey TC. Health status by gastrointestinal diagnosis and abuse history. Gastroenterology 1996;110:999-1007.

DSM-IV Classification. New York: American Psychiatric Press, 1994.

Eisenberg N, Owens GR, Dewey ME. Attitudes of health professionals to child sexual abuse and incest. Child Abuse Negl 1987;11:109-16.

Ellenson GS. Disturbance of perception in adult female incest survivors. Social Casework 1986;67:149-59.

Elliott DM, Briere J. Posttraumatic stress associated with delayed recall of sexual abuse: a general population study. J Traum Stress 1995;8: 629-47.

Elliott M, Browne K, Kilcoyne J. Child sexual abuse prevention: what offenders tell us. Child Abuse Negl 1995;19:579-94.

Engel GL. The need for a new medical model: a challenge for biomedicine. Science 196:129-36.

Ensink BJ. Confusing realities. Amsterdam: Amsterdam University Press, 1992.

Escobar JI, Burnam MA, Karno M, Forsythe A, Golding JM. Somatizing in the community. Arch Gen Psychiatry 1987;44:713-8.

Escobar JI. Cross-cultural aspects of the somatization trait. Hosp Comm Psychiatry 1987;38:174-80.

Famularo R, Kinscherff R, Fenton T. Posttraumatic stress disorder among children clinically diagnosed as borderline personality disorder. J Nerv Ment Dis 1991;179: 428-31.

Fausto-Sterling A. Myths of gender: biological theories about women and men. New York: Basic Books, 1992.

Feldman-Summers S, Pope KS. The experience of "forgetting" childhood abuse: a national survey of psychologists. J Consult Clin Psychol 1994;62:636-9.

Felitti VJ. Long-term medical consequences of incest, rape and molestation. South Med J 1991;84:328-31.

Felitti VJ, Anda RF, Nordenberg D, Williamson DF, Spitz AM, Edvards V, Koss MP, Marks JS. Relationship of childhood abuse and household dysfunction to many of the leading causes of death in adults. The adverse childhood experience (AEC) study. Am J Prev Med 1998;14:245-58.

Femina DD, Yeager GA, Levis DO. Child abuse: adolescent records vs adult recalls. Child Abuse Negl 1990;14:227-31.

Feiring C, Taska L, Lewis M. A process model for understanding adaptation to sexual abuse: the role of shame in defining stigmatization. Child Abuse Negl 1996;20:767-82.

Ferris LE, Tudiver T. Family physicians' approaches to wife abuse: a study of Ontario, Canada, practices. Fam Med 1992;24:267-82.

Finestone HM, Stenn P, Davies F, Stalker C, Fry R, Koumanis J. Chronic pain and health care utilization in women with a history of childhood sexual abuse. Child Abuse Negl 2000;24:547-67.

Fink P. Surgery and medical treatment in persistent somatizing patients. J Psychosom Res 1992;36:439-447.

Finkelhor D. Sexually victimized children. New York: Free Press, 1979.

Finkelhor D. A sourcebook on child sexual abuse. London: Sage Publications, 1986.
Finkelhor D, Korbin J. Child abuse as an international issue. Child Abuse Negl 1988;12:3-23.
Finkelhor D, Hotaling G, Lewis IA, Smith C. Sexual abuse in a national survey of adult men and women: prevalence, characteristics, and risk factors. Child Abuse Negl 1990;14:19-28.
Finkelhor D. The international epidemiology of child sexual abuse. Child Abuse Negl 1994;18:409-17.
Fleming J, Mullen P, Bammer G. A study of potential risk factors for sexual abuse in childhood. Child Abuse Negl 1997;21:49-58.
Flitcraft AH. Violence, values, and gender. JAMA 1992;267:3194-5.
Ford CV. The somatizing disorders. Illness as a way of life. New York: Elsevier Biomedical, 1983.
Ford JD, Kidd P. Early childhood trauma and disorders of extreme stress as predictors of treatment outcome with chronic posttraumatic stress disorder. J Traum Stress 1998;11:743-61.
Foucault M. The birth of the clinic: the archeology of medical perception. New York: Vintage Books, 1975.
Foucault M. Discipline and punish. New York: Vintage Books, 1977.
Frank G. "Becoming the other": empathy and biographical interpretation. Biography 1985;8:189-200.
Frenken J, van Stolk B. Incest victims: inadequate help by professionals. Child Abuse Negl 1990;14:253-63.
Freyd JJ. Betrayal trauma: the logic of forgetting childhood abuse. Cambridge, Mass: Harvard University Press, 1996.
Furniss T. The multiprofessional handbook of child sexual abuse: integrated management, therapy and legal intervention.
London: Routledge, 1991.
Gabbard GO (ed). Sexual exploitation in professional relationships.
Washington, DC: American Psychiatric Press, 1989.
Gadamer HG. Truth and method. New York: Seabury Press, 1975.
Garfinkel H. Conditions of successful degradation ceremonies. Am J Soc 1956;61:420-4.
Garfinkel H. Studies in ethnomethodology. New York: Prentice Hall, 1967.
Gazmararian JA, Lazorick S, Spitz AM, Ballard TJ, Saltzman LE, Marks JS. Prevalence of violence against pregnant women.
JAMA 1996; 275:1915-20.
Geertz C. The interpretation of cultures. New York: Basic Books, 1973.
Gibbons J. Services for adults who have experienced child sexual assault: improving agency response. Soc Sci Med 1996;43:1755-63.
Giorgi A. Validity and reliability from a phenomenological perspective. In: Baker WJ, Mos LP, Rappart HV, Stam HJ (eds). Recent trends in theoretical psychology. New York: Springer, 1988.
Giorgi A. Some theoretical and practical issues regarding the psychological phenomenological method. San Francisco, CA: Saybrook Review 1989;7:71-85.
Goenjian AK, Najarian LM, Pynoos RS et al. Posttraumatic stress reactions after single and double trauma. Acta Psychiatr Scand 1994;214-21.
Goffman E. Asylums. New York: Anchor Books, 1961.
Goffman E. Stigma: notes on the management of spoiled identity. Harmondsworth: Penguin, 1968.

Goffman E. Frame analysis. An essay on the organization of experience. New York: Harper & Row, 1974.

Gold ER. Long-term effects of sexual victimization in childhood: an attributional approach. J Consult Clin Psychol 1986;54:471-5.

Gold SN, Hughes DM, Swingle JM. Characteristics of childhood sexual abuse among female survivors in therapy. Child Abuse Negl 1996; 20:323-35.

Goldberg RT, Pachas WN, Keith D. Relationship between traumatic events in childhood and chronic pain. J Disabil Rehab 1999;21:23-30.

Golding JM, Stein JA, Siegel JM, Burnam MA, Sorenson SB. Sexual assault history and use of health and mental health services. Am J Community Psychol 1988;16:625-44.

Golding JM. Sexual assault history and physical health in randomly selected Los Angeles women. Health Psychol 1994;13:130-8.

Golding JM. Sexual assault history and limitations in physical functioning in two general population samples. Res Nurs Health 1996;19:33-44.

Golding JM. Intimate partner violence as a risk factor for mental disorders: a meta-analysis. J Fam Violence 1999;14:99-132.

Golding JM. Sexual assault history and headache. Five general population studies. J Nerv Ment Dis 1999;187:624-9.

Golding JM. Sexual assault history and medical care seeking: the roles of symptom prevalence and illness behavior. Psychology Health 1999;14: 949-57.

Golding JM, Taylor DL. Sexual assault history and premenstrual distress in two general population samples. J Womens Health 1996;5:143-52.

Golding JM, Wilsnack SC, Learman LA. Prevalence of sexual assault history among women with common gynecologic symptoms. Am J Obstet Gynecol 1998;179:1013-9.

Gonsiorek JC (ed). Breach of trust. Sexual exploitation by health care professionals and clergy. Thousand Oaks, CA: Sage, 1994.

Goodwin JM, Simms M, Bergman R. Hysterical seizures: a sequelae to incest. Am J Orthopsychiatry 1979;49:698-703.

Goodwin JM, Cheeves K, Connell V. Borderline and other severe symptoms in adult survivors of incestous abuse. Psychiatric Ann 1990;20: 22-32.

Gordon DR. Tenacious assumptions in Western medicine. In: Lock M, Gordon DR (eds). Biomedicine examined. Dordrecht: Kluwer Academic Publishers, 1988.

Gorey KM, Leslie DR. The prevalence of child sexual abuse: integrative review adjustment for potential response and measurement biases. Child Abuse Negl 1997;21:391-8.

Green A. Childhood sexual abuse: immediate and long-term effects and intervention. J Am Acad Child Adolesc Psychiatr 1993;32:890-902.

Greig E, Betts T. Epileptic seizures induced by sexual abuse. Pathogenetic and pathoplastic factors. Seizure 1992;1:269-74.

Grene M. "To have a mind..." J Med Philosophy 1976;1:177-99.

Gross RJ, Doerr H, Caldirola D, Guzinski GM, Ripley HS. Borderline syndrome and incest in chronic pelvic pain patients. Int J Psychiatry Med 1981;10:79-96.

Groves BM, Zuckerman B, Maraus S, Cohen DJ. Silent victims. Children who witness violence. JAMA 1993;269:262-4.

Haavind H. Gender as phenomenon and gender as a mode of understanding. Tidsskr Nor Psykologforen 1994;31:767-83.

Haber JD, Ross C. Effects of spouse abuse and/or sexual abuse in the development and maintenance of chronic pain in women. Adv Pain Res Ther 1985;9:889-95.

Hahn RA. Culture-bound syndromes unbound. Soc Sci Med 1985;21: 165-71.
Hall RCW, Tice L, Beresford TP, Wooley B, Hall AK. Sexual abuse in patients with anorexia nervosa and bulimia. Psychosomatics 1989;30: 73-9.
Halpern J. Family therapy in father-son incest: a case study. Soc Casework 1987;68:88-93.
Hamberger LK, Saunders DG, Hovey M. Prevalence of domestic violence in community practice and rate of physician inquiry. Fam Med 1992;24:283-7.
Halpérin DS, Bouvier P, Jaffé PD et al. Prevalence of child sexual abuse among adolescents in Geneva: results of a cross sectional study. BMJ 1996;312:1326-9.
Harding S. The science question in feminism. Milton Keynes: Open University Press, 1986.
Harrop-Griffiths J, Katon W, Walker E, Holm L, Russo J, Hickok LR. The association between chronic pelvic pain, psychiatric diagnosis, and childhood sexual abuse. Obstet Gynecol 1988;71:589-94.
Haskett M, Marziano B, Dover ER. Absence of males in maltreatment research: a survey of recent literature. Child Abuse Negl 1996;20 :1175-82.
Haug F. Female sexualisation. London: Verso, 1988.
Hazzard A, Celano M, Gould J, Lawry S, Webb C. Predicting symptomatology and selfblame among child sex abuse victims. Child Abuse Negl 1995;19:707-14.
Heidegger M. Being and time. New York: Harper & Row, 1962.
Helton A, McFarlane J, Anderson E. Battered and pregnant: a prevalence study. Am J Public Health 1987;77:1337-9.
Hendricks-Matthews MK. The importance of assessing a woman's history of sexual abuse before hysterectomy. J Fam Pract 1991;32:631-2.
Hendricks-Matthews MK, Hoy D. Pseudocyesis in an adolescent incest survivor. J Fam Pract 1993;36:97-103.
Herdt GH. Guardians of the flutes. Idioms of masculinity. New York: McGraw-Hill, 1981.
Herman J, Russell D, Trocki K. Long-term effects of incestuous abuse in childhood. Am J Psychiatry 1986;143:1293-6.
Herman JL, Schatzow E. Recovery and verification of memories of childhood sexual trauma. Psychoanal Psychol 1987;4:1-14.
Herman JL, Perry JC, van der Kolk BA. Childhood trauma in borderline personality disorder. Am J Psychiatry 1989;146:490-5.
Herman JL. Trauma and recovery. New York: Basic Books, 1992.
Herman JL. Father-daughter incest. In: Wilson JP, Raphael B (eds). International handbook of traumatic stress syndromes. New York: Plenum Press, 1993.
Hewitt SK. Preverbal sexual abuse: what two children report in later years. Child Abuse Negl 1994;18:821-6.
Hirschauer S. The manufacture of bodies in surgery. Soc Stud Sci 1991; 21:279-319.
Holmes G, Offen L. Clinicians' hypotheses regarding clients' problems: are they less likely to hypothesize sexual abuse in male compared to female clients? Child Abuse Negl 1996;20: 493-501.
Horowitz MJ. Stress response syndromes and their treatment. In: Goldberger L, Breznitz S (eds). Handbook of stress. New York: Free Press, 1982.
Howe AC, Herzberger S, Tennen H. The influence of personal history of abuse and gender on clinicians' judgement of child abuse. J Fam Violence 1988;3:105-19.
Hubbard R, Wald E. Exploding the gene myth. How genetic information is produced and manipulated by scientists, physicians, employers, insurance companies, educators, and law enforcers. Boston: Beacon Press, 1993.

Hunter JA, Goodwin DW, Wilson RJ. Attributions of blame in child sexual victims: an analysis of age and gender influences. J Child Sex Abuse 1992;1:75-89.

Huseby T. Descriptive and psychodynamic re-evaluation of "therapy-resistent" psychotic patients can open new therapeutic possibilities. Nord J Psychiatry 1994;48:63-8.

Husserl E. The crisis of European sciences and transcendental phenomenology: an introduction to phenomenological philosophy. Evanston: Northwestern University Press, 1970.

Hydle I. Violence against the elderly in Western Europe - treatment and preventive measures in the health and social fields. J Elder Abuse Negl 1989;1:75-87.

Isenberg SR, Owen DE. Bodies, natural and contrieved: the work of Mary Douglas. Religious Stud Rev 1977;3:1-16.

Jackson H, Nuttall R. Clinician responses to sexual abuse allegations. Child Abuse Negl 1993;17:127-43.

Jacobson A, Richardson B. Assault experiences of 100 psychiatric inpatients: evidence of the need for routine inquiry. Am J Psychiatry 1987;144:908-13.

Janoff-Bulman R, Frieze IH. A theoretical perspective for understanding reactions to victimization. J Soc Iss 1983;39:1-17.

Janoff-Bulman R. Assumptive worlds and the stress of traumatic events: applications of the schema construct. Soc Cogn 1989;7:113-26.

Jaspers K. Gesammelte Schriften zur Psychopathologie. Berlin: Springer, 1963.

Jecker NS. Privacy beliefs and the violent family. Extending the ethical argument for physician intervention. JAMA 1993;269:776-80.

Johnsen AG. Surgery as a placebo. Lancet 1994;344:1140-2.

Jones R. Why do qualitative research? It should begin to close the gap between the science of discovery and implementation. BMJ 1995;311:2.

Jumper SA. A meta-analysis of the relationship of child sexual abuse to adult psychological adjustment. Child Abuse Negl 1995;19:715-28.

Kaplan M. A woman's view of DSM-III. Am Psychol 1983;38:786-92.

Kendall-Tackett KA, Williams LM, Finkelhor D. Impact of sexual abuse of children: a review and synthesis of recent empirical studies. Psychol Bull 1993;113:164-80.

Kiecolt-Glaser JK, Glaser R. Psychoneuroimmunology: can psychological interventions modulate immunity? J Consult Clin Psychol 1992; 60:569-75.

Kimerling R, Calhoun KS. Somatic symptoms, social support, and treatment seeking among sexual assault victims. J Cons Clin Psychol 1994; 62:333-40.

Kinmonth A-L. Understanding and meaning in research and practice. Fam Pract 1995;12:1-2.

Kinzl J, Biebl W. Sexual abuse of girls: aspects of the genesis of mental disorders and therapeutic implications. Acta Psychiatr Scand 1991;83: 427-31.

Kinzl JF, Traweger C, Guenther V, Biebl W. Family background and sexual abuse associated with eating disorders. Am J Psychiatry 1994; 151:1127-31.

Kirkengen AL, Schei B, Steine S. Indicators of childhood sexual abuse in gynaecological patients in a general practice. Scand J Prim Health Care 1993;11:276-80.

Kirmayer LJ. Mind and body as metaphors: hidden values in biomedicine. In: Lock M, Gordon DR (eds). Biomedicine examined. Dordrecht: Kluwer Academic Publishers, 1988.

Kleinman A. Concepts and a model for the comparison of medical systems as cultural systems. Soc Sci Med 1976;12:85-93.

Kleinman A, Eisenberg L, Good B. Culture, illness and care. Clinical lessons from anthropological and cross-cultural research. Ann Intern Med 1978;88:251-8.

Kleinman A. Writing at the margin. Discourse between anthropology and medicine. Berkeley: University of California Press, 1995.
Kluft RP (ed). Incest related syndroms of adult psychopathology. Washington DC: American Psychiatric Press, 1990.
Kolb LC. A neuropsychological hypothesis explaining posttraumatic stress disorders. Am J Psychiatry 1987;144:989-95.
Kolb LC. The psychobiology of PTSD: Perspectives and reflections on the past, present and future. J Trauma Stress 1993;6:293-304.
Kondora LL. Living the coming of memories: an interpretive phenomenological study of surviving childhood sexual abuse. Health Care Women Int 1995;16:21-30.
Koss MP. The women's mental health research agenda. Am Psychologist 1990;45:374-80.
Koss MP, Koss PG, Woodruff WJ. Deleterious effects of criminal victimization on women's health and medical utilization. Arch Intern Med 1991;151:342-7.
Koss MP, Woodruff WJ, Koss PG. Relation of criminal victimization to health perceptions among women medical patients. J Consult Clin Psychol 1990;58:147-52.
Kristeva J. Powers of horror: an essay on abjection. New York: Columbia University Press, 1982.
Krohn-Hansen C. The shaping of illegitimacy. On the anthropology of violent interaction. Working paper 1993.6. Centre for development and the environment, University of Oslo, 1993.
Kutchins H, Kirk SA. Making us crazy. DMS-The psychiatric bible and the creation of mental disorders. London: Constable, 1999.
Kvale S. The qualitative research interview: a phenomenological and hermeneutical mode of understanding. J Phenom Psychol 1983;14:171-96.
Kvale S. Issues of validity in qualitative research. København: Studentlitteratur, 1989.
Kvale S. To validate is to question. In: Kvale S (ed). Issues of validity in qualitative research. København: Studentlitteratur, 1989.
Kvale S. InterViews. An introduction to qualitative research interviewing. Thousand Oaks: Sage Publications, 1996.
Lanktree C, Briere J, Zaidi L. Incidence and impact of sexual abuse in a child outpatient sample: the role of direct inquiry. Child Abuse Negl 1991;15:447-53.
Laumann EO, Paik A, Rosen RC. Sexual dysfunction in the United States. Prevalence and predictors. JAMA 1999;281:537-44.
Laws A. Perspectives. Sexual abuse history and women's medical problems. J Gen Intern Med 1993;8:441-3.
Laws A. Does a history of sexual abuse in childhood play a role in women's medical problems. A review. J Women's Health 1993;1:165-72.
Lazarus R, Cohen J. Environmental stress. In: Altman I, Wohlwill J (eds). Human behavior and environment. Vol I. New York: Plenum, 1978.
Lechner ME, Vogel ME, Garcia-Shelton LM, Leichter JL, Steibel KR. Self-reported medical problems of adult female survivors of childhood sexual abuse. J Fam Pract 1993;36:633-8.
Leitenberg H, Greenwald E, Cado S. A retrospective study of long-term methods of coping with having been sexually abused during childhood. Child Abuse Negl 1992;16:399-407.
Levenkron S. Cutting. Understanding and overcoming self-mutilation. New York: W. W. Norton, 1998.
Leventhal JM. Have there been changes in the epidemiology of sexual abuse of children during the 20th century? Pediatrics 1988;82:766-73.

Leventhal JM. Twenty years later: we do know how to prevent child abuse and neglect. Child Abuse Negl 1996;20:647-53.

Ligezinska M, Firestone P, Manion IG, McIntyre J, Ensom R, Wells G. Childrens' emotional and behavioral reactions following the disclosure of extrafamilial sexual abuse: initial effects. Child Abuse Negl 1996;20:111-25.

Lindberg FH, Distad LJ. Post-traumatic stress disorders in women who experienced childhood incest. Child Abuse Negl 1985;9:329-34.

Lindberg FH, Distad LJ. Survival responses to incest: adolescents in crisis. Child Abuse Negl 1985;9:521-6.

Linton SJ. A population-based study of the relationship between sexual abuse and back pain: establishing a link. Pain 1997;73:47-53.

Liskow B, Othmer E, Penick EC, DeSouza C, Gabrielli W. Is Briquet's syndrome a heterogeneous disorder? Am J Psychiatry 1986;143:626-9.

Lipowsky ZJ. Somatization: the concept and its clinical application. Am J Psychiatry 1988;145:1358-68.

Lipscomb GH, Muram D, Speck PM, Mercer BM. Male victims of sexual assault. JAMA 1992;267:3064-6.

Lloyd G. The man of reason. "Male" and "female" in Western philosophy. London: Methuen, 1984.

Lock M. The politics of mid-life and menopause. Ideologies for the second sex in North America and Japan. In: Lindenbaum S, Lock M (eds). Knowledge, power and practice. The anthropology of medicine in everyday life. Berkeley, CA: University California Press, 1993.

Lock M, Gordon DR. Biomedicine examined. Dordrecht: Kluwer Academic Press, 1988.

Lock M, Scheper-Hughes N. A critical-interpretive approach in medical anthropology: rituals and routines of discipline and dissent. In: Medical anthropology - Contemporary theory and method. New York: Praeger, 1990.

Lundberg GD, Young RK, Flanagin A, Koop CE (eds). Violence. A compendium from JAMA, American Medical News, and the speciality journals of the American Medical Association. Chicago: American Medical Association, 1992.

MacIntyre A. After virtue. Notre Dame, Ind.: University of Notre Dame Press, 1981.

Maddocks A, Griffiths L, Antao V. Detecting child sexual abuse in general practice: a retrospective case-control study from Wales. Scand J Prim Health Care 1999;17:210-4.

Margalit A. The decent society. Cambridge, MA: Harvard University Press, 1996.

Martin E. The woman in the body. A cultural analysis of reproduction. Boston: Beacon Press, 1987.

Mays N, Pope C. Rigour and qualitative research. BMJ 1995;311:109-12.

McCauley J, Kern DE, Kolodner K, Dill L, Schroeder AF, DeChant HK, Ryden J, et al. Clinical characteristics of women with a history of childhood abuse. Unhealed wounds. JAMA 1997;277:1362-8.

McClelland J, Adams J, Douglas D, McCurry C, Storck M. Clinical characteristics related to severity of sexual abuse: a study of serious mentally ill youth. Child Abuse Negl 1995;19: 1245-54.

McClelland L, Mynors-Wallis L, Fahy T, Treasure J. Sexual abuse, disordered personality and eating disorders. Br J Psychiatry 1991;158:63-8.

McFarlane J, Parker B, Soenken K, Bullock L. Assessing for abuse during pregnancy. Severity and frequency of injuries and associated entry into prenatal care. JAMA 1992;267:3176-8.

McKibben L, De Vos E, Newberger EH. Victimization of mothers of abused children: a controlled study. Pediatrics 1989;84:531-5.

McKinnon S. American kinship / American incest : asymmetries in a scientific discourse. In : Yanagisako S, Delaney C (eds). Naturalizing power. Essays in feminist cultural analysis. New York: Routledge, 1995.

McWhinney IR. Changing models: the impact of Kuhn's theory on medicine. Fam Pract 1983;1:3-8.

McWhinney IR. An introduction to family medicine. Oxford: Oxford University Press, 1989.

Meadow R. Munchausen syndrome by proxy. The hinterland of child abuse. Lancet 1977;ii: 343-5.

Melchert TP, Parker RL. Different forms of childhood abuse and memory. Child Abuse Negl 1997;21:125-35.

Merleau-Ponty M. Phenomenology of perception. London: Routledge, 1989.

Merleau-Ponty M. The visible and the invisible. Evanston: Northwestern University Press, 1968.

Metcalfe M, Oppenheimer R, Dignon A, Palmer RL. Childhood sexual experiences reported by male psychiatric patients. Psychol Med 1990;20:925-9.

Mishler EG. Research interviewing. Context and narrative. Cambridge, MA: Harvard University Press, 1986.

Mitchell JD, Gibson HN. Letter to the editor. BMJ 1990;300:942.

Morris B. Anthropology of the self. The individual in cultural perspective. London: Pluto Press, 1994.

Morrison J. Childhood sexual histories of women with somatization disorder. Am J Psychiatry 1989;146:239-41.

Morrow J, Yeager CA, Lewis DO. Encopresis and sexual abuse in a sample of boys in residential treatment. Child Abuse Negl 1997;21:11-8.

Morrow KB. Attributions of female adolescent incest victims regarding their molestation. Child Abuse Negl 1991;15:477-83.

Mullen PE, Romans-Clarkson SE, Walton VA, Herbison GP. Impact of sexual and physical abuse on women's mental health. Lancet 1988;i: 842-5.

Mullen PE, Martin JL, Anderson JC, Romans SE, Herbison GP. Childhood sexual abuse and mental health in adult life. Br J Psychiatry 1993;163:721-32.

Munhall P. Women's anger and its meanings: a phenomenological perspective. Health Care Women Internat 1993;14:481-91.

Nathanson DL. Understanding what is hidden: shame in sexual abuse. Psychiatric Clin North Am 1989;12:381-88.

Newberger EH, Barkan SE, Lieberman ES, McCormick MC, Yllo K, Gary LT, Schechter S. Abuse of pregnant women and adverse birth outcome. Current knowledge and implications for practice. JAMA 1992;267:121-3.

Nicolson P. The menstrual cycle, science and femininity: assumptions underlying menstrual cycle research. Soc Sci Med 1995;41:779-84.

Novello AC. From the Surgeon General, US Public Health Service. JAMA 1992;267:3007.

Nuttall R, Jackson H. Personal history of childhood abuse among clinicians. Child Abuse Negl 1994;18:455-72.

Oakley A. Interviewing women: a contradiction in terms. In: Roberts H (ed). Doing feminist research. London: Routledge & Keagan, 1984.

Ogden J. Psychosocial theory and the creation of the risky self. Soc Sci Med 1995;40:409-15.

Olafson E, Corwin DL, Summit RC. Modern history of child sexual abuse awareness: cycles of discovery and suppression. Child Abuse Negl 1993;17:7-24.

Oppenheimer R, Howells K, Palmer RL, Chaloner DA. Adverse sexual experiences in childhood and clinical eating disorders: A preliminary description. J Psychiatr Res 1985;19:357-61.

Ornstein PA. Children's long-term retention of salient personal experiences. J Traum Stress 1995;8:581-605.

Ots T. The angry liver, the anxious heart, and the melancholic spleen. The phenomenology of perceptions in Chinese culture. Culture Medicine Psychiatry;14:21-58.

Palmer RL, Chaloner DA, Oppenheimer R. Childhood sexual expriences with adults reported by female psychiatric patients. Br J Psychiatry 1992:160:261-5.

Paradise JE, Rose L, Sleeper LA, Nathanson M. Behavior, family function, school performance, and predictors of persistent disturbance in sexually abused children. Pediatrics 1994;93: 452-9.

Pellegrino E. Humanisme and the physician. Knoxville, Tennessee: University of Tennessee Press, 1979.

Pennebaker JW. Telling stories: the health benefits of narrative. Literature Med 2000;19:3-18.

Pennebaker JW, Susman JR. Disclosure of trauma and psychosomatic processes. Soc Sci Med 1988;26:327-32.

Pennebaker JW, Kiecolt-Glaser JK, Glaser R. Disclosure of traumas and immune function: health implications for psychotherapy. J Consult Clin Psychol 1988;56:239-45.

Perez CM, Widom CS. Childhood victimization and long-term intellectual and academic outcomes. Child Abuse Negl 1994;18:617-33.

Perloff LS. Perception of vulnerability to victimization. J Soc Issues 1983;39:41-61.

Peters DK, Range LM. Childhood sexual abuse and current suicidality in college women and men. Child Abuse Negl 1995;19:335-41.

Peterson C, Seligman MEP. Learned helplessness and victimization. J Soc Issues 1983;39:103-16.

Pierce R, Pierce LH. The sexually abused child: A comparison of male and female victims. Child Abuse Negl 1985;9:191-9.

Pierce LH. Father-son incest: using the literature to guide practice. Soc casework 1987;68: 67-74.

Poe AE. The purloined letter. In: The short fiction of Edgar Allen Poe. Translation Levine S, Levine S. Indianapolis: Bobbs-Merrill Company, 1976.

Polanyi M. Knowing and being. Chicago: University of Chicago Press, 1969.

Polkinghorne DE. Human existence and narrative. In: Polkinghorne DE. Narrative knowing and the human sciences. New York: State of New York University Press, 1988.

Polkinghorne DE. Qualitative procedures for counseling research. In: Watkins CE, Schneider LJ (eds). Research in counseling. Hillsdale, NY: Lawrence Erlbaum, 1991.

Pope HG, Hudson JI. Is childhood sexual abuse a risk factor for bulimia nervosa? Am J Psychiatry 1992;149:455-63.

Porter GE, Heitsch GM, Miller MD. Munchausen syndrome by proxy: unusual manifestations and disturbing sequelae. Child Abuse Negl 1994;18:789-94.

Prieur A. Iscenesettelser av kjønn. Transvestitter og machomenn i Mexico by. Oslo: Pax, 1994.

Putnam FW. Pierre Janet and modern views of dissociation. J Traum Stress 1989;2:413-29.

Putnam FW. Dissociation in children and adolescents. A developmental perspective. New York: The Guilford Press, 1997.

Putnam FW, Guroff JJ, Silberman EK, Barban L, Post RM. The clinical phenomenology of multiple personality disorder: review of 100 recent cases. J Clin Psychiatry 1986;47:285-93.

Quill TE. Somatization disorders. One of medicine's blind spots. JAMA 1985;254:3075-9.

Radomsky NA. The association of parental alcoholisme and rigidity with chronic illness and abuse among women. Fam Pract 1992;35:54-60.

Rapkin AJ, Kames LD, Darke LL, Stampler FM, Naliboff BD. History of physical and sexual abuse in women with chronic pelvic pain. Obstet Gynecol 1990;76:92-96.

Rauch SL, van der Kolk BA, Fisler RE, Alpert NM, Orr SP, Savage CR, Fischman AJ et al. A symptom provocation study of posttraumatic stress disorder using positron emission tomography and script-driven imagery. Arch Gen Psychiatry 1996;53:380-7.

Reiter RC, Shakerin LR, Gambone JC, Milburn AK. Correlation between sexual abuse and somatization in women with somatic and nonsomatic pelvic pain. Am J Obstet Gynecol 1991;165:104-9.

Richards T. Female genital mutilation condemned by WMA. BMJ 1993; 307 ii:957.

Rieker PP, Carmen EH. The victim-to-patient process: the disconfirmation and transformation of abuse. Am J Orthopsychiatry 1986;56:360-70.

Ricoeur P. The model of the text: meaningful action considered as a text. Soc Res 1971;38:529-62.

Ricoeur P. The narrative function. Semeia 1978;13:177-202.

Riggs S, Alario AJ, McHorney C. Health risk behaviors and attempted suicide in adolescents who report prior maltreatment. J Pediatr 1990; 116:815-21.

Rimsza ME, Berg RA, Locke C. Sexual abuse: Somatic and emotional reactions. Child Abuse Negl 1988;12:201-8.

Rodriguez N, Ryan SW, Rowan AB, Foy DW. Posttraumatic stress disorder in a clinical sample of adult survivors of childhood sexual abuse. Child Abuse Negl 1996;20:943-52.

Root MPP. Persistent, disordered eating as a gender-specific, post-traumatic stress response to sexual assault. Psychotherapy 1991;28:96-102.

Rosen LN, Martin L. Impact of childhood abuse history on psychological symptoms among male and female soldiers in the U.S. army. Child Abuse Negl 1996;20:1149-60.

Rosenberg ML, O'Carroll PW, Powell KE. Let's be clear: violence is a public health care problem. JAMA 1992; 267:3071-2.

Rosenthal JA. Patterns of reported child abuse and neglect. Child Abuse Negl 1988;12: 263-71.

Rosaldo R. Culture and truth. The remaking of social analysis. London: Routledge, 1989.

Ross CA, Heber S, Norton GR, Anderson G. Somatic symptoms in multiple personality disorder. Psychosomatics 1989;30:154-60.

Ross CA, Norton GR, Wozney K. Multiple personality disorder: an analysis of 236 cases. Can J Psychiatry 1989;34:413-8.

Ross CA, Miller SD, Reagor P, Bjornson L, Fraser GA, Anderson G. Structured interview data on 102 cases of multiple personality disorder from four centers. Am J Psychiatry 1990;147:596-601.

Roth S, Wayland K, Woolsey M. Victimization history and victim-assailant relationship as factors in recovery from sexual assault. J Traum Stress 1990;3:169-80.

Rowan AB, Foy DW, Rodriguez N, Ryan S. Post traumatic stress disorder in a clinical sample of adults sexually abused as children. Child Abuse Negl 1994;18:51-61.

Rudin MM, Zalewski C, Bodmer-Turner J. Characteristics of child sexual abuse victims according to perpetrator gender. Child Abuse Negl 1995;19:963-73.

Rush F. The best kept secret. Sexual abuse of children. New Jersey: Prentice Hall Inc., 1982.

Russell D. Women, madness and medicine. Oxford: Polity Press, 1994:

Russell DEH. Rape in marriage. New York: Macmillan Publishing Co Inc., 1982.

Russell DEH. The incidence and prevalence of intrafamilial and extrafamilial sexual abuse of female children. Child Abuse Negl 1983;7:133-46.

Russell DEH. The secret trauma. New York: Basic Books, 1986.

Russett CE. Sexual science. The Victorian construction of womanhood. Cambridge, MA: Harvard University Press, 1991.

Rutter P. Sex in the forbidden zone. When men in power - therapists, doctors, clergy, teachers, and others - betray women's trust. Los Angeles: Jeremy P. Tarcher, 1989.

Sætre M, Holter H, Jebsen E. Tvang til seksualitet. En undersøkelse av seksuelle overgrep mot barn. Oslo: Cappelen, 1986.

Salner M. Validity in human science research. In: Kvale S (ed). Issues of validity in qualitative research. København: Studentlitteratur, 1989.

Sandberg DA, Lynn SJ. Dissociative experiences, psychopathology and adjustment, and child and adolescent maltreatment in female college students. J Abnorm Psychol 1992;101:717-23.

Saris AJ. Telling stories: life histories, illness narratives, and institutional landscapes. Cult Med Psychiatry 1995;19:39-72.

Sariola H, Uutela A. The prevalence of child sexual abuse in Finland. Child Abuse Negl 1994;18:827-35.

Sartre JP. Being and nothingness: a phenomenological essay on ontology. New York: Pocket Books, 1956.

Scarinci IC, McDonald-Haile J, Bradley LA, Richter JE. Altered pain perception and psychosocial features among women with gastrointestinal disorders and history of abuse: a preliminary model. Am J Med 1994;97:108-18.

Scarry E. The body in pain. The making and unmaking of the world. Oxford: Oxford University Press, 1985.

Schechter S. Abuse of pregnant women and adverse outcome. Current knowledge and implications for practice. JAMA 1992;267:121-3.

Scheflin AW, Brown D. Repressed memory or dissociative amnesia: what the science says. J Psychiatry & Law 1996;24:143-88.

Schei B, Bakketeig LS. Gynaecological impact of sexual and physical abuse by spouse. A study of a random sample of Norwegian women. Br J Obstet Gynaecol 1989;96: 1379-83.

Schei B. Prevalence of sexual abuse history in a random sample of Norwegian women. Scand J Soc Med 1990;18:63-8.

Schei B. Physically abusive spouse - a risk factor of pelvic inflammatory disease? Scand J Prim Health Care 1991;9:41-5.

Scheper-Hughes N. The rebel body: the subversive meanings of illness. TAS Journal 1990;10: 3-10.

Scheper-Hughes N, Lock M. The mindful body: a prolegomenon to future work in medical anthropology. Med Anthropol Q 1987:20;6-39.

Scheppele KL, Bart PB. Through women's eyes: defining danger in the wake of sex assault. J Soc Issues 1983;39:63-81.

Schofferman J, Anderson D, Hines R, Smith G, Keane G. Childhood psychological trauma and chronic refractory low-back pain. Clin J Pain 1993;9:260-5.

Scott KD. Childhood sexual abuse: Impact on a community's mental health status. Child Abuse Negl 1992;16:285-95.

Sebold J. Indicators of child sexual abuse in males. Soc Casework 1987; 68:75-9.

Sedney MA, Brooks B. Factors associated with a history of childhood sexual experiences in a nonclinical female population. J Am Acad Child Psychiatry 1984;23:215-8.

Seligman MEP. Helplessness: on depression, development, and death. San Francisco, CA: Freeman, 1975.

Shapiro JP, Leifer M, Martone MW, Kassem L. Cognitive functioning and social competence as predictors of maladjustment in sexually abused girls. J Interpersonal Violence 1992;7:156-64.

Shapiro S. Self-mutilation and self-blame in incest victims. Am J Psychother 1987;16: 46-54.

Shearer SL, Herbert CA. Long-term effects of unresolved sexual trauma. Am Fam Physician 1987;36:169-75.

Shearer SL, Peters CP, Quaytman MS, Ogden RL. Frequency and correlates of childhood sexual and physical abuse histories in adult female borderline inpatients. Am J Psychiatry 1990;147:214-6.

Sheldrick C. Adult sequelae of child sexual abuse. Br J Psychiatry 1991; 158 (suppl.10): 55-62.

Silver RL, Boon C, Stones MH. Searching for meaning in misfortune: making sense of incest. J Soc Issues 1983;39:81-102.

Smith D, Pearce L, Pringle M, Caplan R. Adults with a history of child sexual abuse: evaluation of a pilot therapy service. BMJ 1995;310:1175-8.

Sorenson SB, Stein JA, Siegel JM, Golding JM, Burnam MA. The prevalence of adult sexual assault: the Los Angeles epidemiologic catchment area project. Am J Epidemiol 1987;126: 1154-64.

Spitzack C. Confession and signification: the systematic inscription of body consciousness. J Med Philos 1987:12:357-69.

Springs FE, Friedrich WN. Health risk behaviors and medical sequelae of childhood sexual abuse. Mayo Clin Proc 1992;67:527-32.

Stark E, Flitcraft A, Frazier W. Medicine and patriarchal violence: the social construction of a "private" event. Int J Health Serv 1979; 9:461-93.

Stark E, Flitcraft AH. Women and children at risk: a feminist perspective on child abuse. Int J Health Serv 1988;18:97-118.

Steven I, Castell-McGregor S, Francis J, Winefield H. Child sexual abuse. Aust Fam Physician 1988;17:427-33.

Strick FL, Wilcoxon SA. A comparison of dissociative experiences in adult female outpatients with and without histories of early incestuous abuse. Dissociation 1991;4:193-9.

Sugg NK, Inui T. Primary care physicians' response to domestic violence: opening Pandora's box. JAMA 1992;267:3157-60.

Svensson PG. Qualitative methodology in public health research. Eur J Publ Health 1995;5:71.

Taylor ML, Trotter DR, Csuka ME. The prevalence of sexual abuse in women with fibromyalgia. Arthritis Rheum 1995;38:229-34.

Taylor SE, Wood JV, Lichtman RR. It could be worse: selective evaluation as a response to victimization. J Soc Issues 1983;39:19-40.

ten Have HAM. Medicine and the Cartesian image of man. Theor Med 1987;8:235-46.

Thomasma DC. Philosophy of medicine in Europe: challenges for the future. Theor Medicine 1985;6:115-23.

Thornquist E, Bunkan BH. What is psychomotor therapy? Oslo: Norwegian University Press, 1991.

Thornquist E. Profession and life: separate worlds. Soc Sci Med 1994; 39:701-13.

Thornquist E. Musculo-sceletal suffering: diagnosis and a variant view. Soc Health Illness 1995;17:166-92.

Toombs SK. The meaning of illness. A phenomenological account of the different perspectives of physician and patient. Dordrecht: Kluwer Academic Publishers, 1992.

Toulmin S. On the nature of the physician's understanding. J Med Philos 1976;1:32-50.

Toulmin S. The construal of reality: critisism in modern and postmodern science. Crit Inquiry 1982;9:93-111.

Tromp S, Koss MP, Figueredo AJ, Tharan M. Are rape memories different? A comparison of rape, other unpleasant, and pleasant memories among employed women. J Traum Stress 1995;8:607-27.

Trostle JA. Medical compliance as an ideology. Soc Sci Med 1988;27:1299-1308.

United Nations Publication. Violence against women in the family. New York: United Nations Publications, 1989.

van der Kolk BA; Greenberg M, Boyd H, Krystal J. Inescapable shock, neurotransmitters, and addiction to trauma: Toward a psychobiology of post traumatic stress. Biol Psychiatry 1985;20:314-25.

van der Kolk BA. The trauma spectrum: the interaction of biological and social events in the genesis of the trauma response. J Trauma Stress 1988;1:273-90.

van der Kolk BA, van der Hart O. Pierre Janet and the breakdown of adaptation in psychological trauma. Am J Psychiatry 1989;146:1530-40.

van der Kolk BA, Perry JC, Herman JL. Childhood origin of self-destruc-tive behavior. Am J Psychiatry 1991;148:1665-71.

van der Kolk BA. The body keeps the score: memory and the evolving psychobiology of posttraumatic stress. Harvard Rev Psychiatry 1994; 1:253-65.

van der Kolk BA, Fisler R. Dissociation and the fragmentary nature of traumatic memories: overview and explanatory study. J Trauma Stress 1995;8:505-25.

van Egmont M, Garnefski N, Jonker D, Kerkhof A. The relationship between sexual abuse and female suicidal behavior. Crisis 1993;14:129-39.

van Leeuwen E. Body of knowledge and the ontology of the body. Theor Med 1987;8:105-15.

van Manen M. Researching lived experience. Human science for an action sensitive pedagogy. New York: State University of New York Press, 1990.

Waigandt A, Wallace DL, Phelps L, Miller DA. The impact of sexual assault on physical health status. J Trauma Stress 1990;3:93-102.

Wakley GM. Sexual abuse and the primary care doctor. London: Chapman & Hall, 1991.

Walch AG, Broadhead WE. Prevalence of lifetime sexual victimization among female patients. J Fam Pract 1992;35:511-6.

Walker A. Theory and methodology in premenstrual syndrome research. Soc Sci med 1995;41:793-800.

Walker E, Katon W, Harrop-Griffiths J, Holm L, Russo J, Hickok LR. Relationship of chronic pelvic pain to psychiatric diagnoses and childhood sexual abuse. Am J Psychiatry 1988;145: 75-80.

Walker EA, Katon WJ, Neraas K, Jemelka RP, Massoth D. Dissociation in women with chronic pelvic pain. Am J Psychiatry 1992;149: 534-7.

Walker EA, Katon WJ, Roy-Bryne PP, Jemelka RP, Russo J. Histories of sexual victimization in patients with irritable bowel syndrome or inflammatory bowel disease. Am J Psychiatry 1993; 150:1502-6.

Waller G. Sexual abuse and the severity of bulimic symptoms. Br J Psychiatry 1992;161:90-3.

Waller G, Hamilton K, Rose N, Sumra J, Baldwin G. Sexual abuse and body-image distortion in the eating disorder. Br J Clin Psychol 1993;32:350-2.

Waller G, Ruddock A. Experiences of disclosure of childhood sexual abuse and psychopathology. Child Abuse Rev 1993;2:185-95.

Waller G. Childhood sexual abuse and borderline personality disorder in the eating disorders. Child Abuse Negl 1994;18:97-101.

Watkins B, Bentovim A. The sexual abuse of male children and adolescents: a review of current research. J Child Psychol Psychiatry 1992;33;197-248.

Watzlawick P, Bavelas JB, Jackson DD. Pragmatics of human communication. New York: W.W. Norton & Comp., 1967.

Weber E. The notion of persecution in Levinass Otherwise than being or Beyond essences. In: Peperzak AT (ed). Ethics as first philosophy. New York: Routledge, 1995.

Welch SL, Fairburn CG. Childhood sexual and physical abuse as risk factors for the development of bulimia nervosa: a community-based case control study. Child Abuse Negl 1996;20:633-42.

Wells RD, McCann J, Adams J, Voris J, Ensign J. Emotional, behavioral, and physical symptoms reported by parents of sexually abused, nonabused, and allegedly abused prepubescent females. Child Abuse Negl 1995;19:155-63.

Wilbers D, Veenstra G, van der Wiel HBM, Weijman Schultz WCM. Sexual contact in the doctor-patient relationship in the Netherlands. BMJ 1992;304:1531-4.

Williams LM. Recall of childhood trauma: a prospective study of women's memories of child sexual abuse. J Consult Clin Psychol 1994;62: 1167-76.

Williams LM. Recovered memories of abuse in women with documented child sexual victimization histories. J Traum Stress 1995;8:649-73.

Wilson JP, Raphael B. International handbook of traumatic stress syndrome. New York: Plenum Press, 1933.

Wind TW, Silvern L. Type and extent of child abuse as predictors of adult functioning. J Fam Violence 1992;7:261-81.

Winfield I, George LK, Swartz M, Blazer DG. Sexual assault and psychiatric disorders among a community sample of women. Am J Psychiatry 1990;147:335-41.

Wolfe J. Trauma, traumatic memory, and research: where do we go from here? J Traum Stress 1995;8:717-26.

Wood DP, Wiesner MG, Reiter RC. Psychogenic chronic pelvic pain: diagnosis and management. Clin Obstet Gynecol 1990;33:179-95.

Wooley SC. Sexual abuse and eating disorders. The concealed debate. In: Fallon P, Katzman MA, Wooley SC (eds). Feminist perspectives on eating disorders. New York: Guilford Press, 1994.
World Health Organization. Violence against women. WHO Press Office: Fact Sheet No. 128, 1996.
Wukasch RN. The impact of a history of rape and incest on the posthysterectomy experience. Health Care Women Intern 1996;17:47-55.
Wulff H, Pedersen SA, Rosenberg R. Philosophy of medicine. An introduction. Oxford: Blackwell, 1986.
Wurr CJ, Partridge IM. The prevalence of a history of childhood sexual abuse in an acute adult inpatient population. Child Abuse Negl 1996;20:867-72.
Wurtele SK, Kaplan GM, Keairnes M. Childhood sexual abuse among chronic pain patients. Clin J Pain 1990;6:110-3.
Wyatt GE. The sexual abuse of Afro-American and white-American women in childhood. Child Abuse Negl 1985;9:507-19.
Wyatt GE, Peters SD. Issues in the definition of child abuse in prevalence research. Child Abuse Negl 1986;10:231-40.
Wyatt GE, Peters SD. Methodological considerations in research on the prevalence of child sexual abuse. Child Abuse Negl 1986;10:241-51.
Wyatt GE, Powell GJ. Lasting effects of child sexual abuse. Newbury Park, CA: Sage, 1988.
Wyatt GE, Newcomb M. Internal and external mediators of women's sexual abuse in childhood. J Consult Clin Psychol 1990;58:758-67.
Yanagisako S, Delaney C (eds). Naturalizing power. Essays in feminist cultural analysis. New York: Routledge, 1995.
Young A. The anthropologies of illness and sickness. Ann Rev Anthropol 1982;11:257-85.
Young K. Disembodiment: the phenomenology of the body in medical examinations. Semiotica 1989;73:43-66.
Zaner RM. The problem of embodiment. Some contributions to a phenomenology of the body. The Hague: Martinus Nijhoff, 1971.

Index